A. Ribeiro

Soft Plate and Impact Tectonics

Springer-Verlag Berlin Heidelberg GmbH

António Ribeiro

Soft Plate and Impact Tectonics

With a Contribution of António Mateus

With 112 Figures

Springer

Prof. António Ribeiro
Departamento de Geologia
e Laboratório de Tectonofísica e Tectónica Experimental
Faculdade de Ciências – Universidade de Lisboa
58, Ruada Escola Politécnica
1250-102 Lisboa, Portugal

Dr. António Mateus
Departamento de Geologia
e Centro de Recursos Minerais, Mineralogia e Cristalografia
Faculdade de Ciências – Universidade de Lisboa
C2, Piso 5, Campo Grande
1749-016 Lisboa, Portugal

ISBN 978-3-642-63235-8

Libary of Congress Cataloging in-Publication Data applied for
Die Deutsche Bibliothek – CIP-Einheitsaufnahme
Ribeiro, António: Soft plate and impact tectonics / António Ribeiro. - Berlin; Heidelberg; New York; Barcelona; Hong Kong; London; Milan; Paris; Singapore; Tokyo: Springer, 2002
ISBN 978-3-642-63235-8 ISBN 978-3-642-56396-6 (eBook)
DOI 10.1007/978-3-642-56396-6

http://www.springer.de

Originally published by Springer-Verlag Berlin Heidelberg New York in 2002
Softcover reprint of the hardcover 1st edition 2002

Production: PRO EDIT GmbH, 69126 Heidelberg, Germany
Cover Design: Erich Kirchner, Heidelberg, Germany
Typesetting: AM-productions, Wiesloch, Germany

Printed on acid-free paper SPIN: 110745767 32/3130So-5 4 3 2 1 0

Foreword

„Structural geology is a core component of knowledge needed for intelligent living with our planet, truly sustainable development."

It gives me great pleasure to write this brief foreword for the work by António Ribeiro of Lisboa. When we consider the foundations of our life support systems, climate, the biosphere, soil, materials, energy, waste management and today in the word of cities, structural stability, we require exact knowledge of the structure of our planet and change in these structures.

Homo sapiens has lived on our planet for a very short period. Our planet is cooling by conduction and convection and it is the structures that result from these processes that control the global climate, our resource base and even soil quality and food security, etc, etc. As the title of this book suggests, some processes are slow and some fast.

I was born on a farm in Southern New Zealand, quite close to the magnificent Southern Alps. I often think that watching that environment had a profound influence on my mental development.

I have been very fortunate in my academic life to meet, to work with a host of brilliant earth scientists in New Zealand people like Benson, Coombs, Wellman introduced me to structural geology. Then to California with a host of people like Griggs, Verhoogen, marine scientists, who introduced me to the basic science and new observations of earth processes. Later I moved to England and during that period I met in the laboratory and the field a wonderful group of leading earth scientists, and got to visit structures in Europe, Africa and India with them.

It was during this period that I was introduced to Brazil and with this a large number of students from Brazil and Portugal. In this group was a brilliant structural geologist, António Ribeiro. I always remember our field trips together, from the coastal blue schist and eclogites of California to the Sierra Nevada, and to places like Yucca Mountain, Nevada. Nuclear waste management on this planet nicely illustrates the need for exact knowledge of structural geology, past, present and future.

My last field trip with António and his colleagues was on the coast of Portugal, North of Lisboa. We discussed possible processes related to meteoric impacts and the possible future subduction of Portugal and its return to the mantle! Portugal is a nation well aware of seismic hazards. On every field trip with António, I was educated and noted that he was a brilliant observer of rocks and their structures.

I consider this book truly brilliant. He uses examples of key earth processes from the entire world. It is stressed that all the geospheres interact and are modified by the processes that generate what we see at the surface of our planet.

I was pleased to see the biosphere included, for as we know it goes very deep! This book is well written and one feature that is truly impressive is the huge bibliography with over 850 references to literature over the past two hundred years. I was delighted to see the works of Lyell referenced. I have a habit in that I think I have a new idea, I check with Lyell's textbook of 1830! I always appreciate Heraclitus (who wrote): more than 2500 years ago: „the earth is melted into the sea by the same reckoning whereby the sea sinks into the earth."

All scientists who are involved in earth resources, earth structures, earth dynamics, need knowledge of structural geology. For me Ribeiro's book is an ideal text. It should be used in the appropriate courses in earth and engineering sciences. Congratulations António and all your colleagues in Portugal and Europe who have contributed.

University of Western Ontario, Canada WILLIAM S. FYFE

Preface

The central objective of this book is to show that plates deform so we can propose a Soft Plate Tectonics Theory, as opposed to the standard Rigid Plate Tectonics Theory. I borrowed the adjective 'soft' from the physics of soft matter, pioneered by Pierre-Gilles de Genes. There is a thesis as the central theme, and the whole book is organised with that purpose in mind. I chose as a model for this book, Alfred Wegener's, *The Origin of Continents and Oceans*. Wegener played the same role for earth sciences that Einstein played for physics – and our view of the entire cosmos – during the 20th century. He had a central thesis – that continents drift – which was revolutionary for his time. Wegener synthesised all arguments in favour of drift, whatever their origin: on continents, on oceans, geodetic, geophysical, geological, palaeontological-biological, palaeoclimatic and mechanical. In this last aspect he was less successful because the data on oceans were almost totally missing at the time. Nevertheless, he established continental drift even if the mechanism only became available with plate tectonics. Accordingly, there is a holistic aspect to Wegener's research, even more remarkable if we think that Wegener was a meteorologist proposing a new theory for solid earth evolution. However, if nature works in parallel – and some brains do, such as Stanislaw Ulam's, with his „ten dogs theory of memory“ – then scientific argument is arranged in series and a reductionist element is inevitable, at least in the relationship between the author and the reader. We conclude that this book is about results, not methods; the reader is directed to the references for methods. Given the large range of topics discussed, we do not claim to be specialists in all of them; we hope that the most competent views, other than our own, have been selected.

We know that our approach to the central thesis of the book is not orthodox. Let me say right away that my personal motivation was curiosity, and curiosity is inherently turbulent. There are many examples in all sciences, and also in earth sciences, that show that individuals can be in the right although the Establishment does not hold the same views; however, this fact can only be recognised in the long term. Wegener had to fight strong opposition: he was an outsider; he was just wrong about the facts; he did not provide an adequate mechanism. In this case, to be an outsider was an advantage; the final outcome showed that he was right rather than wrong about facts; and the mechanism was finally found afterwards. Plate tectonics were considered to be a „Geopoem“ and later on a „Geofact“. I think it is more important to make progress in our field, even through a „drunkard's walk“, than to build „perfect crystal“ theories.

We must learn from Wegener's lesson: he was right about the fact that continents drift, even if the underlying mechanism was discovered afterwards through plate tectonics. In soft plate tectonics we conclude that plates can suffer internal deformation, an order of magnitude below deformation across plate boundaries, but not negligible. This fact has obvious consequences for plate behaviour over time and the driving mechanism for plate motion. We claim that intraplate deformation is larger in oceanic plate interiors than in stable continental interiors, a view that opposes conventional plate tectonics. We will present the evidence, direct and indirect, for these claims.

However, even if monitoring of strain by very recent but very accurate and precise geodetic spatial techniques proves or disproves soft plate tectonics, the mechanism by which the interiors of plates deform remains less certain than the fact itself. We will support the idea that the presence of widespread seafloor hydrothermal metamorphism is the main controlling factor that could explain the very slow flow of the oceanic lithosphere. This is in contrast to much less chemical reactivity on the continent lithosphere, except in the regolith at the interface with external geospheres. However, this mechanism is less certain than the facts it explains, although more probable than other, unknown mechanisms, at the present state of knowledge. Nevertheless, we must be very cautious because some 30 years ago seafloor hydrothermal metamorphism was virtually unknown! And nature never stops surprising those who study it. Our ignorance of deformation mechanisms and tectonic regime in the ocean lithosphere is still very significant and will be discussed in Section 3.12.

The reader may find it strange that so many examples of the application of the soft plate tectonics theory came from the present plate boundary zone between Eurasia and Africa in the segment that extends from the Azores triple junction to Iberia. Plate tectonics is a global theory and the soft plate version of it must be proved to be a better approximation for the whole surface of the Earth . Nevertheless, the complex dynamic evolution of both plate interiors and boundaries expresses itself in a multiple of possible sequences of tectonic regimes; therefore, there is also a component of regional tectonics in any plate tectonics theory. This aspect must be surveyed on a global scale by a team of authors with different fields of expertise, rather than by a single specialist. In the present state of the art, no such soft plate version of tectonics of the globe exists and if this book stimulates different specialists into developing the soft plate version of plate tectonics surveyed on a global scale, it will make a significant contribution to tectonics. The author assumes his limitations very clearly: the purpose of this book is not to extend soft plate tectonics to the entire globe, but rather to prove that such an extension is feasible. In fact, we do not see inherent limitations in the methodology that is proposed.

I discussed many of the ideas that I have tried to develop in this book with many colleagues. I would like to acknowledge the support that I received from many individuals. First, I think of my teachers in tectonics: André Caire and Maurice Mattauer, from France; Robert Shackleton and John Ramsay, from England; Lloyd Cluff and Bert Slemmons from the United States. My co-operation with Bill Fyfe was extremely fruitful in the evolution of my ideas on geodynamics; through him I have become aware of the role of fluids in deformation at all levels. Bill agreed to write a Foreword to this book, and I am honoured by this.

Many other colleagues, geodesists, geophysicists and geologists helped me enlarge my views on geodynamics; they come from my home country but also from abroad. To all of them my sincere thanks

I am deeply grateful to António Mateus for authorship of Section 3.12: „Time-dependent mechanical response of the oceanic lithosphere: strain partitioning and deformation mechanism", at my request. He did an excellent job, filling a gap in the original manuscript totally beyond my scope. Luís Mendes Victor, as co-ordinator of BIGSETS read parts of the manuscript concerning the geodynamics of south-west Iberia. Luís Matias, Rui Taborda and Virgílio Mendes were extremely helpful in discussing the problem of plate rigidity, especially in the Pacific. João Cabral, Pedro Terrinha, José Madeira, José Fernando Monteiro and Nuno Lourenço read the parts of the manuscript where we have been co-operating in the last years, and kindly provided original figures, as well as José Carlos Kullberg and Félix Mendes.

My editor, Dr. Wolfgang Engel, was extremely patient and committed to having this book published by Springer-Verlag. This was an opportunity to remember the research we did together in the Variscan Fold Belt in Germany and Portugal, and the enjoyable and friendly field trips. I followed his advice as closely as possible, given his competence and experience in these matters.

I am also indebted to the anonymous referee who improved the text in both science and style. I am also grateful to Cleia Ribeiro, Henrique Duarte, André Duarte and Nuno Bon de Sousa for their help in processing the figures and text from 1997 to 2001. Their youth perfectly balanced my limitations in new technologies, typical of an old man!

Their work was funded by Fundação para a Ciência e Tecnologia through LATTEX. The same institution also funded a great part of my own research and other collaborators of LATTEX. The list of references was processed by Mrs. Nídia Pereira from the Geology Department of the Faculty of Science, Lisbon University. José Croca, of the Physics Department, quantum physicist and sculptor, provided us with many quanta of aesthetics. The English revision was possible through funding from the Fundação da Faculdade de Ciências da Universidade de Lisboa and Instituto de Ciências da Terra e do Espaço (ICTE).

Finally, although I acknowledge the support of numerous colleagues whom I admire and respect, I alone am responsible for any misconceptions.

The reader may note that I write a Preface and not an Epilogue. I tried to avoid wild speculation and to remain as objective as possible, following the advice of my editor and colleague, Dr. Wolfgang Engel. However, some questions should not be eluded. Is Earth the only planet with plate tectonics, life and intelligence? And even if we prove in the future that the answer is „yes", is it the result of some unknown link between these singular attributes of our planet? In any case, earth and space sciences in particular, and science, in general, are now at the end of the beginning and not, as claimed by some, at the beginning of the end, proving that nature's imagination is greater than ours.

Lisboa, January 2002 ANTÓNIO RIBEIRO

Contents

1 Introduction

1.1 From Continental Drift to Rigid Plate Tectonics (1910–1968)

The scientific vision of the Earth as a mobile body, as opposed to a static one, was theorised mainly by Wegener at the beginning of this century (1915), although this idea had many precursors.

A period of debate followed; the opponents belonged to two categories: those who opposed the facts and those who rejected the idea because the mechanism for continental drift was unknown. Further work, showing increasing evidence for the phenomenon of continental drift and inner convection, had become a driving mechanism since the 1930s (revision in Griggs 1939; Holmes 1944). Technological progress in the scientific study of the ocean floor after World War II, led to the seafloor spreading and plate tectonics theories in the period 1962–1968. These theories imposed a dynamic approach for the evolution of the earth since that period. The classic version of plate tectonics with rigid plates could account for 95% of the features of the earth's surface and has become a paradigm since then. There is ample literature on the history of geosciences, from continental drift to plate tectonics (for reviews see Moores and Twiss 1995; Davies 1999) and we will not go into details of the general trend presented above and agreed upon by most authors. Nevertheless, we must emphasise that, from the beginning, there was a contrast between the views of Wegener and Holmes, where the seafloor was regarded as a viscous fluid and those of Hess and Wilson where the seafloor was regarded as a brittle solid (Davies 1999, p. 41)

1.2 Soft Plate Tectonics (1987–1996)

From 1987 to 1996, the postulate of plate rigidity was criticised for a number of reasons. It was realised that plate rigidity was a poor approximation for continental tectonics (Molnar 1988) but still a good one for the oceanic domain. The evidence from experimental rheology of rocks forming the lithosphere led to the establishment of rheological strength profiles (for synthesis see Ranalli 1995), with the oceanic lithosphere stronger than the continental lithosphere. However, it was gradually realised (Rutter and Brodie 1991) that we could not extrapolate directly from experiments to nature for many reasons.

Meanwhile, the discovery of seafloor hydrothermal metamorphism (for review see Fyfe et al. 1978) showed that the igneous rocks used in experiment-based rheology of the lithosphere were poor analogues for the real oceanic lithosphere. At the same time, plate boundary zones of diffuse tectonic activity were discovered in the oceanic domain, (De Mets et al. 1990), suggesting that the rheology of the continental and oceanic lithospheres did not contrast as much as previously supposed. This conviction was further strengthened by discrepancies between space geodetic measurements and plate kinematic models like NUVEL-1 (Ribeiro 1993/1994).

Finally, marine geoid satellite data, released in 1995 (Sandwell and Smith 1995; Sandwell et al. 1995), contributed strong evidence for deformation of the oceanic lithosphere. We named this version of deformable plate tectonics, as opposed to the rigid classic version, SOFT PLATE TECTONICS (Ribeiro 1996); we borrowed the notion of soft materials from physics (Gennes and Badoz 1996).

Models of viscoplastic rheology behaviour for the oceanic lithosphere were advanced (Martinod and Molnar 1995) and elastoplastic models for the continental lithosphere (Wdowinski 1998) showed that the lithosphere is a non-conservative system. To these factual objections to the classical rigid plate tectonics version we must add theoretical ones, presented by a number of authors. In fact, we can ask: is a rigid lithosphere compatible with mantle convection?

1.3 Beyond Plate Tectonics (1980–?)

The scientific exploration of oceans paved the way for the plate tectonics revolution. In a similar way, the scientific exploration of space gave us a better perspective of earth in the solar system, in our galaxy and in the entire cosmos. A global view of the earth emerged, by considering the interaction of all geospheres: lithosphere, atmosphere, hydrosphere and biosphere; the earth is a dynamic system that must be studied in its entirety. However, space science showed that this system is not isolated and suffers influences at the scale of the solar system, the galaxy and the cosmos. It becomes increasingly evident that asteroid megaimpacts could be a source of intensive forcing on the whole earth (Alvarez et al. 1980), which must be considered an open system. Some authors (Shaw 1994) suggest that plate tectonics is the last reform of classical earth sciences and a true revolution will come from integrating impact theory in earth evolution. A new catastrophism is gradually growing from the seeds of the old uniformitarianism (Ribeiro 1998)

The new views on the earth's evolution were fertilised by another conceptual revolution that touched all the branches of science, from logic and mathematics to social sciences. The classical determinism in linear systems was overthrown by new concepts of dynamic systems, such as fractals and deterministic chaos (Middleton 1991). Being a dynamic system, the earth was no exception, and earth sciences were deeply influenced by the new concepts on non-linear dynamic systems. We have to examine those paradoxes of earth sciences in the light of the dynamic systems approach. We can summarise by saying that the earth is a non-conservative open dynamic system.

Global Tectonics with Rigid Plates: Foundations and Limitations

2.1
Plate Tectonics in Oceans and Continents

The success of plate tectonics is based on two characteristics of the theory: its *generality*, because it would explain the tectonics of the entire globe, and its *simplicity*, based on the postulate of rigidity of plates: all tectonic activity should occur at plate boundaries. Scientific theories are useful because they are the foundations for the progress of knowledge. They inevitably evolve into more refined theories or into more general ones as this progress leads towards new data and new concepts. This is also the case with plate tectonics.

The main characteristic of the planet Earth is the duality in composition and behaviour of its surface layers, expressed as continents and oceans (Wegener 1915). From the beginning, the simplicity of ocean tectonics, studied mostly by geophysical methods, contrasted with the complexity of continental tectonics studied mostly by geological methods (Molnar 1988). The distinction between ocean and continental tectonics is essential. The contrast in density between oceanic crust and suboceanic mantle is negligible if compared with the contrast between continental crust and subcontinental mantle. The role of potential energy created by thickening or thinning of the lithosphere deeply influences the tectonic style of continents but has no direct effect on oceans. In addition, the continental lithosphere is preserved almost in its entirety, because it is buoyant; in contrast, the oceanic lithosphere has sources in ridges and sinks in trenches. It is also evident that plate boundaries in oceanic regions are narrow; on the other hand, active faulting and seismicity are dispersed in continental regions over domains of thousands of kilometres, sometimes as large as the smallest ocean plates. It is difficult to evoke plate rigidity in those domains. This difference in behaviour was explained, in the standard plate tectonics theory, by differences in rheology of the oceanic versus continental lithosphere; the latter is buoyant because it is thicker and lighter and was also considered to be weaker than the oceanic lithosphere. The differences in buoyancy are well established, both by direct and indirect evidence of its composition, but the differences in strength were based on inferences on profiles obtained by extrapolation of mainly experimental data (Molnar 1988; Kohlstedt et al. 1995; Ranalli 1995, 1997).

Plate tectonics are driven by thermal convection (Turcotte and Schubert 1982). In laboratory experiments convection cells form under appropriate conditions; a thin thermal boundary layer of cool fluid forms adjacent to the upper boundary of the

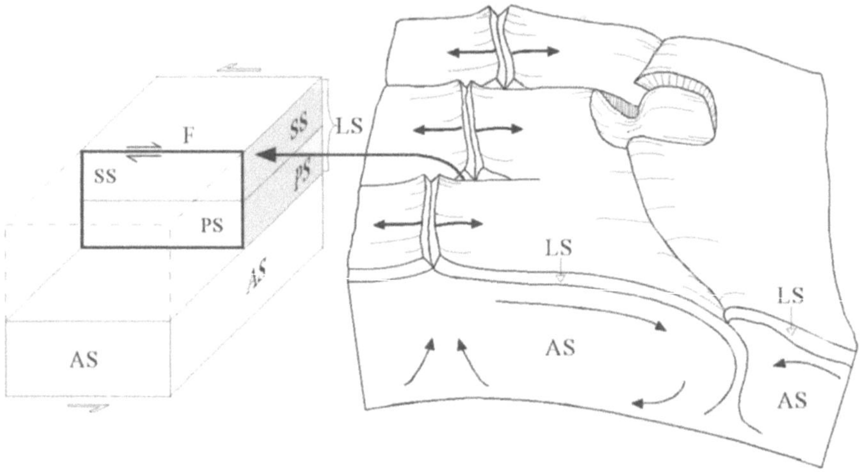

Fig. 2.1. Rheological layering of the Earth imposed by convection. The lithosphere (*LS*) is the outer thermal boundary layer above the convective asthenosphere (*AS*). The lithosphere can be subdivided in an upper schizosphere (*SS*), where slip is, at least in part, seismic and a lower plastosphere (*PS*), where slip is totally aseismic

convective fluid; this layer dissipates heat by conduction and not by convection. In the real earth the lithosphere is the thermal boundary layer; it is the outer shell of the earth, stronger and colder than the convective asthenosphere below; its thickness is 50–100 km in the oceans, depending mainly on the age of the cooling oceanic lithosphere, and 50–150 km on the continents, depending mainly on its geothermal gradient (Fig. 2.1).

The concept of lithosphere can be defined on the basis of various properties (Anderson 1989; Brown and Mussett 1993): rheology, kinematics, chemical, thermal and seismic. We are interested mainly in rheological properties of the lithosphere, so we will consider lithosphere based on rheology (Fig. 2.1).

However, even if we concentrate on rheological behaviour, we must distinguish between the *elastic lithosphere* and the plate lithosphere or *tectosphere* (Ranalli 1995; Lliboutry 1998). The *elastic lithosphere* has an average thickness of 50 km in oceans and 100 km in cratons; this is estimated on the short-term response, in the order of 10^3 years, to vertical loads, based on flexual rigidity of the elastic plate and correlation of Bouguer anomaly and topography, assuming isostatic equilibrium, as mentioned in Section 3.4. The *tectosphere* is the coherent plate of global tectonics which moves as a whole above the asthenosphere. It has a thickness of 100 km on old oceanic lithosphere and of 250 km or even more in cratons. This is estimated on the basis of seismic properties, so it is also called the *seismic lithosphere*, and is expressed by a low velocity zone (LVZ) that coincides with the top of the asthenosphere. This LVZ cannot be defined everywhere, but there is a general tendency towards a decrease in viscosity on the top of the assumed asthenosphere, even if the cause of this change in rheology behaviour is discussed (Anderson 1989) and may be of different origins at different places. The concept of tectosphere or seismic lithosphere expresses the long-term response, in the order of 1 Ma to horizontal loads related to plate movement. It assumes that the contrast with the asthenosphere in the short time scale of seismic wave propagation must be reinforced if we extend it to the long time scale of plate dynamics.

The lithosphere can be divided in two rheological layers (Scholz 1990), from top to bottom:

- The brittle schizosphere whose strength is governed by frictional strength given by Byerlee's law.
- The ductile plastosphere, whose strength is governed by flow laws of materials composing it; the flow of monomineralic rocks can be established by experimental deformation and the flow laws of rocks can be approached considering the rheology of their weakest mineral components.

It must be clearly stated that brittle deformation is characterised by loss of cohesion along discrete surfaces to form fractures and faults; ductile deformation is a permanent, coherent, solid state, in which there is no loss of cohesion on the scale of crystal grains or layers and no evidence of brittle fracturing. This is non-genetic terminology (Twiss and Moores 1992); there are several theories that can explain these well-known tectonic behaviours, which are more or less complete and substantiated. However, the two aspects must not be confused and a ductile behaviour can arise from various deformation mechanisms, brittle, plastic or a combination of both (Rutter 1986). Some authors (Kohlstedt et al. 1995) clearly differentiate the change in failure mode from localised faulting to distributed shear at brittle-ductile transition from the change in dominant deformation mechanism in the brittle-plastic transition.

There is a brittle-ductile transition separating the two rheological layers in the lithosphere, slightly below the maximum depth of hypocentres (Ranalli 1997).

Using the known data, we can establish profiles of strength expressed in differential stresses as a function of depth (Fig. 2.2).

In the schizosphere, the brittle behaviour is described by the Mohr-Coulomb law; at low loads there is a great variation among surfaces and rock types, with μ, coefficient of sliding friction ranging between 0.2 and 2. This is mainly due to the effects of surface roughness. Byerlee (1978) fit the data with

$$\tau = 0.85\,(\sigma_n - P)$$

where τ, σ_n and P are respectively the shear component of the stress tensor, the normal component of the stress tensor and the fluid pressure, expressed in megapascals (MPa).

For $\sigma_n > 200$ MPa Byerlee (1978) fit the data with

$$\tau = 50 + 0.6\,(\sigma_n - P)$$

This law controls the frictional-slip regime, where the rock that is subjected to stress is already fractured by stresses applied earlier and, consequently, has no cohesion. It is a special case of the more general Mohr-Coulomb law that controls the shear rupture regime with

$$\tau = \frac{\sigma_d}{2} = \tau_0 + \mu * \sigma_n$$

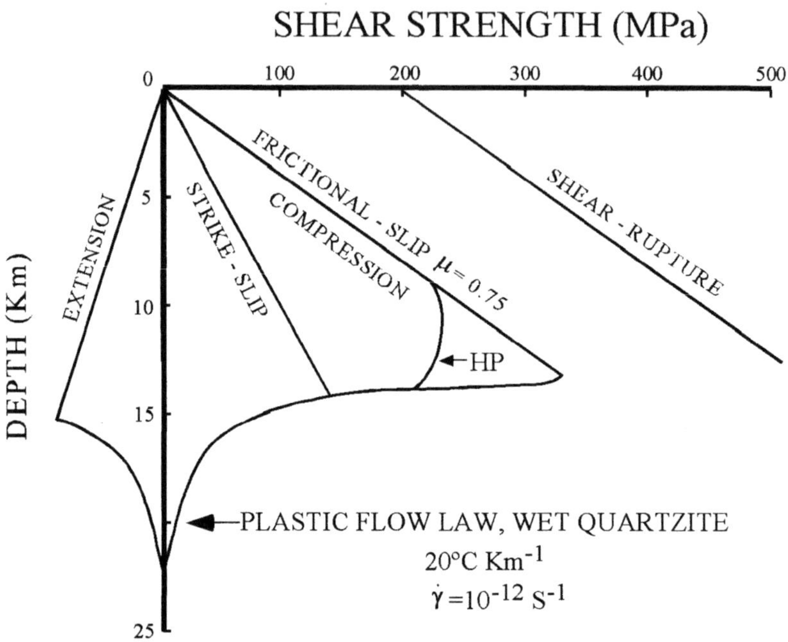

Fig. 2.2. Strength of the lithosphere. In the schizosphere optimally oriented faults are assumed, following Byerlee's law, the coefficient of sliding friction μ=0.75 and hydrostatic pore pressure. In the schizosphere, a flow law for wet quartzite, experimentally determined, is assumed for a specific strain rate and temperature gradient (after Scholz 1990). The limit for shear rupture regime is plotted, for comparison, under the same conditions (after Engelder 1993). *HP* Possible strength envelope in the higher-pressure brittle regime. (After Shimada 1993)

where τ and σ_n have the same meaning as before, σ_d is the differential stress, σ_1-σ_3, σ_1 maximum and σ_3 minimum principal stress τ_0 is the cohesion of the material and μ^* is the coefficient of internal friction. If we plot both laws in a Mohr diagram for stress it is clear that frictional slip will occur at a much lower differential stress than is required for shear rupture. If we plot both laws as a function of depth for the same effective normal stress the shear rupture regime starts ≈200 MPa above the friction slip regime (Fig. 2.2).

This friction law is, with very few exceptions, independent of lithology, sliding velocity and roughness; it is known as Byerlee's law. In this way, differential stress necessary to deform the schizosphere increases linearly with depth, and cohesion is small compared to the average level of stress.

For depths greater than 10 km the Byerlee law is invalid because shear stress necessary for failure may become depth-independent (Ord and Hobbs 1989).

In the plastosphere the steady-state creep or flow laws obey the Dorn equation (Ranalli 1995) of the form

$$\sigma = \left(\frac{\dot{\varepsilon}}{A_D}\right)^{\frac{1}{n}} \exp.\left(\frac{E}{nRT}\right)$$

where σ is differential stress and A_D is the Dorn parameter, n the stress exponent, E is the creep activation energy, R is the perfect gas constant and T is the absolute temperature. The flow laws are known for many minerals and some rocks.

As creep is a thermally activated process, the temperature distribution in the lithosphere is fundamental in controlling its rheology. We will establish the basis for heat flow in the lithosphere by considering that the main process is conduction that follows the Fourier law, where the rate of flow of heat is proportional to the temperature gradient

$$q = K\partial T/\partial x$$

where T is temperature, q is the rate of flow of heat per unit area (that is the heat flux) in the positive X direction, and K is the conductivity of the material; as heat flows from hotter to cooler domains, a negative sign results.

From the previous law we deduce that cooling time is proportional to the square of length scale. The quantity of heat is proportional to the dimension of source, L, and the rate of cooling is inversely proportional to distance so it can be demonstrated (Davies 1999) that

$$K \approx \frac{L^2}{t} \text{ or } L \approx \sqrt{Kt}$$

where t is cooling time. A more rigorous analysis shows that this equation is valid with a proportionality constant of the order of one. We conclude that the geotherms are analytical solutions of the Fourier law of heat conduction and must fit the observed heat flow (Ranalli 1995; Beekman 1994).

In the oceanic lithosphere seafloor heat flow, as well as bathymetry and lithosphere thickness, are functions of the age of the lithosphere; these features can be explained by conductive cooling of the oceanic lithosphere (Turcotte and Schubert 1982). In the first approximation geotherms are a function of the square root of the age of the lithosphere, as referred to above.

In the continental lithosphere the distribution of temperature is more complex, because its age is heterogeneous as a result of superposed tectonic cycles; it is the most recent thermal event that controls the thermal structure. The heat flow, although on average the same as in the oceanic lithosphere, is higher than in oceans, due to radioactive production.

We must conclude that the lithosphere has an elastic core; above it the stress is altered by brittle failure and below, by ductile creep (Kuznir and Park 1986; Burov and Diament 1995).

The strength envelope is dependent on stress regime; experience shows that rocks are weaker in tension than in compression (Sibson 1974; Ranalli 1995). So rocks fail first in tension, in the upper part of the lithosphere in the so-called crack-propagation regime and afterwards in the shear-rupture regime (Engelder 1993). A frictional yield criterion, valid for brittle failure on new or pre-existing fractures, can be rewritten by expressing Byerlee's law in terms of effective stress difference, lithostatic overburden pressure and pore fluid pressure (Beekmann 1994)

$$\sigma = \sigma_1 - \sigma_3 = \alpha p g z\,(1-\lambda)$$

where α is a parameter expressing the stress regime, such as

$$\alpha = \frac{R-1}{R} \text{ for normal faulting}$$

$$\alpha = R-1 \text{ for thrust faulting}$$

$$\alpha = \frac{R-1}{1+\beta(R-1)} \text{ for strike-slip faulting}$$

and where σ_1 and σ_3 are principal stresses, maximum and minimal respectively, β is rock density, ρ_w is density of water, g is gravitational acceleration, z is depth, P_H is fluid pressure,

$$\lambda = \frac{P_H}{\rho g z} = \frac{\rho_w}{\rho} \approx 0.35$$

is the mean hydrostatic pore fluid factor, variable between $\lambda=0$ for dry rock and $\lambda \approx 0.9$ in water-saturated sediments $0<\beta<1$ is a function of the magnitude of intermediate stress σ_2.

$$\sigma_2 = \sigma_3 + \beta(\sigma_1 - \sigma_3)$$

$$R = \left[\left(1+\mu^2\right)^{1/2} - \mu\right]^{-2}$$

The equations express the dependence of frictional strength on stress regime.

On the basis of the previous data, we can establish lithosphere strength profiles for various geotherms and compositions (Ranalli and Murphy 1987; Beekmann 1994) and strain rates, but large deviations are observed in the controlling parameters, such as the thermal conductivity of the rocks (Fig. 2.3).

So, in the oceanic lithosphere, the crust and uppermost mantle are brittle to a depth of 15 km if young or 50 km if old. Below the brittle-ductile transition the plastosphere is dominated by olivine flow law, a strong mineral which is the main component of peridotite (Scholz 1990).

In the continental lithosphere the upper crust also obeys Byerlee's law until the brittle ductile transition at 10 km depth if the geothermal gradient is high and the stress regime extensional, or at 20 km if it is low and the stress regime is compressional (Ranalli 1997). The lower crust is ductile and dominated by the flow laws of quartz or plagioclase, if acid or intermediate to mafic composition respectively; these minerals are notably weaker than olivine. The uppermost mantle is again strong and brittle and becomes ductile deeper down, as shown by hypocentres of earthquakes. One obtains a jam sandwich-like profile with a weak lower crust between a brittle upper crust and a strong uppermost mantle. This way, the buoyant continental crust can de-

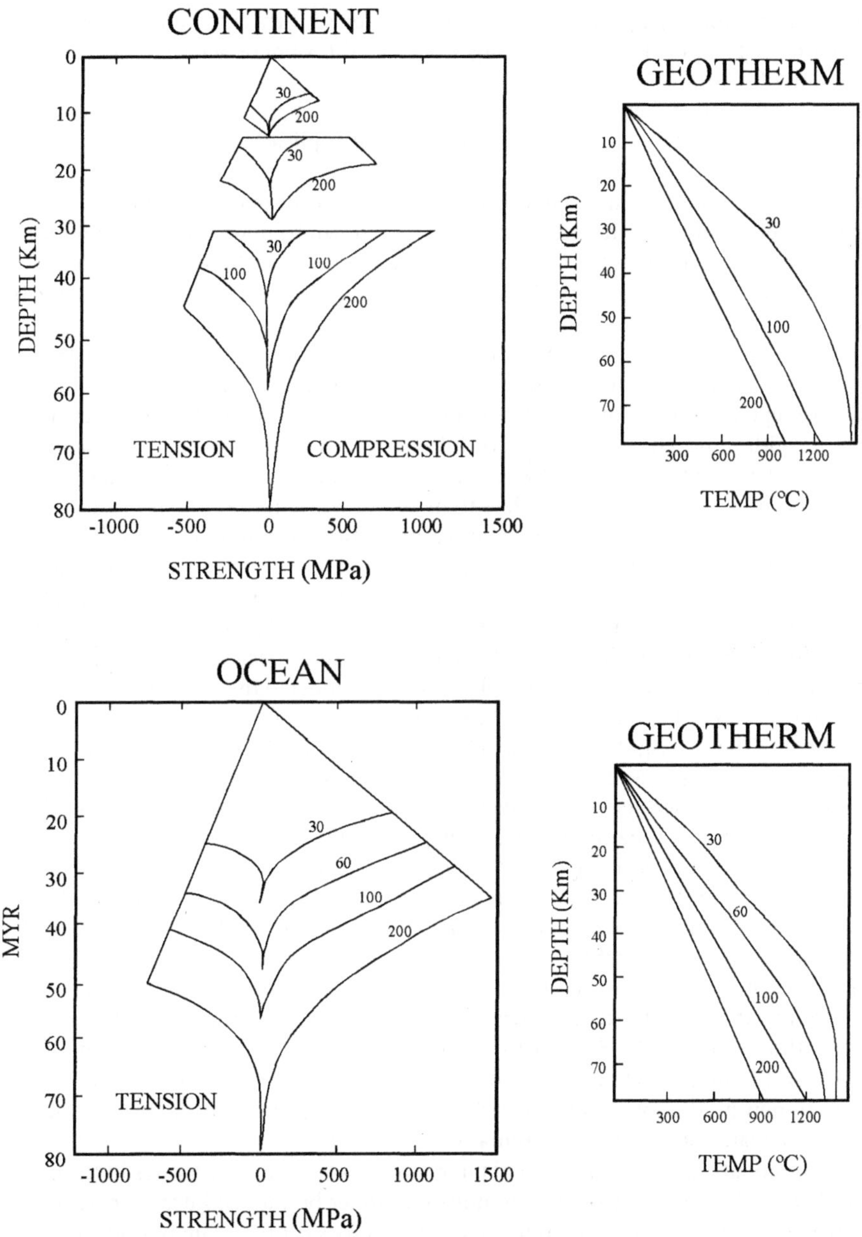

Fig. 2.3. Strength envelope for the lithosphere at several thermal ages for a ductile strain rate 10^{-15} s^{-1} for the continental lithosphere with diabase lower crust and for the oceanic lithosphere. (After Beekmann 1994)

tach from the underlying mantle to generate zones of diffuse tectonic activity and mountain ranges. In contrast, the stronger oceanic lithosphere behaves coherently as a function of depth.

These concepts of continental versus oceanic lithosphere rheology were challenged by new facts and theories after the first years of application of the plate tectonics paradigm.

One source of data consisted in the discovery of diffuse zones of seismicity and distributed deformation in the oceanic domain, as shown by anomalies in topography, gravimetry and geoid height. The most accurate and recent kinematic model, NUVEL-1 (De Mets et al. 1990), introduced plate boundary zones (PBZ) in the oceanic lithosphere. One extends between the North America and South America Plates in the central Atlantic Ocean and another one extends between the Indian and Australian Plates in the central Indian Ocean. In both cases, this was a kinematic requirement due to independent movement of the plates separated by these PBZ; these are large domains with hundreds to thousand of kilometres, as large as in the continental lithosphere. The introduction of PBZ restricted the use of the postulate of plate rigidity. Plate boundaries can be narrow, less than 20 km wide, in the case of most oceanic rifts (Gordon 1995); they are wide in convergent zones, reaching thousands of kilometres in the case of continent collision, such as India and Eurasia, or some 100 km above subduction zones. As more plate kinematic data become available, we conclude that dividing a former plate into two rigid plates with independent movement is a better model, the zone of intraplate deformation with imperfect rigidity becomes a wide PBZ. We can now apply the rigid plate tectonic model to the new plates that result from subdivision of the old plate and infer the kinematic of the PBZ from the relative movement of the two rigid plates that are put in contact by the PBZ (Gordon 1995).

However, we must ask if a more refined model, one obtained by progress on the monitoring of a denser network of space geodetic data, would lead us to a multiplication of „microplates" separated by PBZ, suggesting that: „the distinction drawn between deformation in a wide PBZ and that in plate interiors may eventually prove to be arbitrary" (Gordon 1995, p. 24, 373). In fact, the area of diffuse or wide PBZ has increased with the progress in data acquisition and is now estimated to cover 15% of the earth's surface (Gordon 1998).

The presence of PBZ showed that plate boundaries are not necessarily narrow, as the standard theory admitted. However, the question of plate rigidity remains open. Some authors (Gordon 1998) relax the condition of narrow boundaries but maintain rigidity for the subplates that result from subdivision of major plates in PBZs because it leads to new predictions contrary to the description by intraplate deformation. The concept of plate interior rigidity remains useful but uncertain. The summation of seismic moments in stable continental plate interiors leads to strain rates in the order of 10^{-20} to10^{-19} s^{-1}. Space geodetic methods put upper bounds of intraplate deformation below 2 mm $year^{-1}$ and the relative motion across diffuse plate boundaries ranges from ~2 to ~15 mm $year^{-1}$ above the previous value. The purpose of this book is to show that space geodetic methods compared with other data lead to the inference that plate interiors are not rigid and, at least for large oceanic plates, there is increasing direct evidence for it from space geodetic methods. So we relax not only the postulate of narrowing plate boundaries, but also the postulate of plate rigidity, at least for the oceanic lithosphere.

The obvious implications of the introduction of wide PBZ were that rheology of the oceanic and continental lithosphere did not contrast as much as the standard theory claimed. These findings were confirmed and extended by measurements of real displacements using satellite geodetic methods (Stein 1993). This raised serious doubts on the validity of strength profiles based only on experimental data when applied to the real lithosphere. (Ramberg 1988; Rutter and Brodie 1991; Martinod and Molnar 1995). We can summarise these objections as follows. The application of the laboratory experiment to geological deformation requires extrapolation of several orders of magnitude in strain rate from 10^{-8} (Carter and Tsenn 1987) to 10^{-13} s^{-1}; this extrapolation is based on many assumptions, leading to possible error; several deformation mechanisms involving parameters that are not quantified for the moment can be more efficient at very slow strain rates (Paterson 1987; Ord and Hobbs 1989; Gratier and Gamond 1990). Uncertainties in parameters, even when the deformation mechanism is controlled in the laboratory, suggest uncertainties in strength of at least one order of magnitude. Recent experimental evidence for granite, gabbro, dunite and eclogite (Shimada 1993) decrease deviatoric stress at the brittle-ductile transition in the high-pressure brittle regime (Fig. 2.2). It is also stressed that rheology can change during deformation, namely because anisotropy increases (Ramsay and Lisle 2000, p. 972). Finally, we do not know how to extrapolate experimental data from single-phase aggregates to the predominant polymineralic rocks that occur in nature.

The spectacular folding of the oceanic lithosphere in the central Indian Ocean stimulated comparisons between the estimation of deviatory stress in actively deforming belts in oceans and continents, without recurring to laboratory flow laws from which the strength profiles were derived. These estimations were based on surface heat flux measurements above active faults or on temperatures inferred from exhumed shear zones. (England and Molnar 1991). It was concluded that the average stress difference during compression of the oceanic lithosphere, in the range of 35–140 MPa estimated for the central Indian Ocean, was not very different from deviatory stress necessary to maintain large continental plateaux such as Tibet (Martinod and Molnar 1995). For some astute and unorthodox thinkers the very concept of lithosphere, valid for the short term, had to be revised in the long-term end of a large spectrum of rheology behaviour (Molnar et al. 1993). The rheology of the lithosphere integrated in the geologic time scale, which is much longer than the earthquake cycles, cannot be considered as elastic but rather as inelastic. The upper schizosphere deforms at least in part by a sudden slip during seismic events. Stress accumulates in the interseismic period of the cycle and drops mainly in the coseismic period. The stress necessary to cause deformation is approximately constant when averaged over long periods of time, which is a definition of plastic rheology. The plastosphere is deformed by creep and is viscous. The coupling of deformation at both levels introduces a viscous component even at a short term below the earthquake cycle, so the lithosphere should be viscoplastic rather than elastic, a topic of one section below (see Sect. 2.4).

We can summarise by saying that if our ideas on the rheology of the continental lithosphere didn't suffer radical changes in the last years, the same didn't happen with the oceanic lithosphere. We can say that the reason for this was the unpredicted discovery, in the 1970s, of hydrothermal fluid-induced seafloor metamorphism; since then the area considered to be affected by this phenomenon has increased exponentially through time. So we must refer to the effects of fluids on deformation and its implication for composition and rheology of the lithosphere. However, before this, we must consider theoretical arguments that suggest that the postulate of plate rigidity can be relaxed.

2.2 Plate Rigidity as a Postulate

In the previous section we realised right from the beginning that the theory of standard plate tectonics was a poor approximation for continents. Later, as new data accumulated, the limitations imposed by rheology of the continental lithosphere also percolated through to the oceanic lithosphere. In this section, we try to show that the postulate of plate rigidity can be dropped in favour of a more general version of the theory than the standard one.

In classic plate tectonics the rheology of the lithosphere is assumed to be rigid-elastic except in plate boundary zones. The small deformations of plate interiors should be reversible as the linear theory of elasticity implies.

Is this really necessary, or can we imagine a dynamic model of deformable moving plates? (Ribeiro 1994b)

From a strict point of view of rheology, rigidity is impossible for natural materials, because they are all liquids in the rheological sense (Fyfe et al. 1978, pp. 2–5); if we consider very large scales of time, all solids will flow (Lliboutry 1992); only on short scales is the deformation negligible for the materials we call solids in everyday life. By convention, we can put the limit between solids and liquids at a viscosity of 10^{14} Pa s for low stress (Cottrell 1964, p. 196). In fact, the deformation of an elastic solid is reversible only if it occurs at an infinitely slow rate, in order to establish a state of thermodynamic equilibrium, at any instant, on the body. However, the real movement occurs at a finite rate, so the process is, in fact, irreversible and the mechanical energy is dissipated by irreversible heat conduction between points at a different temperature and by internal friction; therefore, even solids have a very small but finite fluidity and, inversely a high viscosity (Landau and Lifshitz 1986). Soft Matter (Gennes and Badoz 1996, p. 6) is the one which „can be transformed by weak external action, much as a sculptor deforms the shape of clay by the gentle pressure of a thumb". Hence, on a geologic scale, where stress is applied during 10^6–10^2 Ma, even solids must flow and therefore must be considered soft. This is why some authors (Llyboutry 1998) restrict, solid for a crystalline state of matter and liquid for a non-crystalline condensed state of matter respectively, and elastic and viscous designate rheology models; in this sense, a slow flow of solids is inevitable.

So a strict definition of lithosphere does not not imply rigidity. In fact, „on peut considérer que la surface terrestre comme une mosa de quelques 'plaques' en déplacement relatif. Les déformations qu'elles subissent à l'échelle des temps géologiques restant du second ordre par rapport à leurs mouvements, sauf le long de leurs frontières" („we may regard the surface of the Earth as a mosaic of some plaques in relative movement. The deformations they suffer at the scale of geological time remains of the second order in relation to their movements, except along their boundaries.")(Le Pichon 1980, pp. 339–340). However, this cautious definition was rapidly forgotten. In fact, the kinematics of rigid plates were considered to have a unique property, verified in the Earth: great circles perpendicular to transform plates intersect near the Eulerian pole; and spreading rates along the same plate boundary vary as the sinus of the angular distance to the Eulerian pole. This was considered to be a test for the rigidity of plates (Le Pichon et al. 1973).

Only a very small number of authors admitted that rigidity of plates is an „artificial construct" (Van Andel 1992, p. 173). However, there is increasing evidence which we will develop below that plates can deformed internally, for a number of reasons (Ribeiro 1987, 1989).

First, plates can deform because they are large. Dimensional analysis shows that large plates must have a high, but not infinite, viscosity (Van Bemmellen 1972, pp. 249–250).

Secondly, for thermodynamic reasons, „the lithosphere deforms internally in a time-dependent manner" (Harper and Szymanski 1991, p. 12), which is a typical non-linear systems perspective. This reasoning is valid especially in the case of the oceanic lithosphere, if we apply the theory of non-equilibrium thermodynamics to rock deformation, in the same line of thought as Ramberg (1988, p. 1). „Diffusion and various kinds of chemical processes make regional metamorphic rocks yield and flow at stress much below the short-term yield stress as measured by the rock mechanicists in the laboratory". Widespread hydrothermal fluid-induced seafloor metamorphism, affecting the whole oceanic crust and partially converting the upper part of suboceanic mantle into serpentinite, must affect the rheology of the oceanic lithosphere (Fyfe et al. 1978; MacDonald and Fyfe 1985).

Thirdly, is a rigid lithosphere compatible with mantle convection? In fact, „attempts to model plate tectonics with continuum models have not successfully reproduced the observed discontinuous changes in velocity at boundaries. A temperature-dependant viscosity simply makes the cold surface boundary layer very stiff and effectively immobilises in surface" (Gable et al. 1991, p. 8391).

Or, as another author puts it (Natl Acad Sci 1992, p. 14): „The biggest problem confronting a synthesis of plate and mantle convection lies in the inability to incorporate the extremes of rheology", the study of the deformation and movement of matter. A problematic element for the mantle convection models is the quasi-rigid nature of the plates. Those models assume that temperature has a great effect on viscosity, but that would lead, he said, „to a planet completely covered with a superficial plate that has no dynamics at all", despite the convection beneath the surface, instead of the planet we actually observe. Thus, the appealing theoretical success of mantle does not begin to explain much of the data. Stevenson called it „a profound puzzle", and concluded, „We simply do not understand why earth has plate tectonics."

If we admit that the oceanic lithosphere is deformable, with a viscoelastic rheology, we can remove the difficulty of making a rigid lithosphere compatible with mantle convection. Quasi-rigid continental plates will be dragged by movement of surrounding viscoelastic oceanic plates. Are we back to a Wegenerian model of continental drift, now compatible with seafloor spreading?

If we push these ideas further should we maintain the rigid-plate kinematics model based in Eulerian poles of rotation? In this respect we should quote (Natl Acad Sci 1992, p. 14): „Yet the plate tectonics model also embodies a dilemma: not only is the cause of plate tectonics elusive, but conventional, models also do not even fully describe the phenomenon." As Stevenson said: „An interesting question is the extent to which plate tectonics is even correct. That is to say, are those plates actually rigid entities rotating on the surface of the earth about some pole?". He continued, „The answer is yes, to perhaps a 90 percent accuracy. You might say that is quite good, but that 10 percent failure is important because it leads to formation of many of the Earth's mountain ranges and other prominent geological features." Scientists modelling plate tectonics treat the plates as rigid and develop predictions about how they move over the earth's surface. The 10% failure to predict what is observed points to „deformation primarily at plate boundaries", said Stevenson, and explains why so much attention is paid to regions of the world where the plates are colliding.

One author (Keith 1993) has even proposed one model of deformable plates, the viscous flow model. It incorporates intraplate strain but considers mid-ocean ridges convergent boundaries, as an alternative view to standard plate tectonics. Many objections of the author to the standard model are valid; there is coincidence in some of his views on the rheology of the plates and our own ideas. However, we consider that there is hard evidence for seafloor spreading and for plate kinematics. Today we measure the rate of spreading in oceans by space geodesy, so we see no reason to dispute it.

2.3 Effect of Fluids on Deformation and its Implications in Composition and Rheology of Lithosphere

The role of pore pressure in faulting has been known for some decades (Hubbert and Rubey 1959) and in folding more recently (for a review see Price and Cosgrove 1990; Cosgrove 1993). However, fluids have not only a physical effect through pore pressure but also chemical effects, such as pressure solution, named by Sorby in the last century. Chemical kinetics of rock–water interactions affect the strength of rocks (for a review see Kirby 1984; Kohlstedt et al. 1995) through processes such as subcritical crack growth, hydrolytic weakening and pressure solution.

Fracture mechanics (Atkinson 1987; Scholz 1990) show that cracks can propagate rapidly above a critical stress intensity factor; but they can propagate slower below this stress intensity under long-term loading; in a brittle condition the mechanism is possibly *stress corrosion*, by which the crack tip is weakened by a chemical reaction activated by fluids.

Hydrolytic weakening (Griggs and Blacic 1964) consists in the breaking of Si–O–Si bonds when some of the O is substituted by OH because the activation barrier to dislocation motion is substantially reduced. The effect is the same as with moisture-aided crack growth, but operates mostly in the ductile regime; experiments show that creep rates increase systematically in wet samples that became weaker.

In *pressure solution* or *solution creep* mineral grains dissolve more readily at faces under high compressive stress, ions diffuse through the fluid phase on the grain boundary and they precipitate on surfaces of low compressive stress (for a review see Rutter 1983). It is the dominant deformation mechanism at low temperatures and geological strain rates.

We must notice the difference between physical and chemical effects due to the presence of fluids in rock deformation (Van der Pluijim and Marshak 1997); the physical effect of pore-fluid pressure, Pf, is to decrease confining pressure, Pc, to effective pressure, Pe, because it acts against confining pressure, such that

$$Pe=Pc-Pf$$

This decreases the elastic component and promotes fracturing, reducing ductility. However, the chemical effect reduces strength and promotes flow, which is the opposite effect.

To examine the role of fluids in the composition and rheology of the lithosphere we must keep in mind some very serious limitations in the present state of knowledge. These include theoretical problems related to the application of non-hydrostatic thermodynamics to natural processes; conceptual problems in the link between simple experiments and the behaviour of real rocks with the possible interaction of different mechanisms; heterogeneity of distribution of fluids in the schizosphere due to heterogeneities of permeability provided by fractures.

However, a general conclusion has been reinforced in the last few years; under dry conditions silicates are strong and brittle and show time-independent behaviour; but inelastic deformation in the schizosphere at low stresses will be controlled by the chemical and pore pressure effects of fluids, namely water, and environmental effects can promote time-dependent behaviour. We can also predict that the role of fluids will be more important in the oceanic than the continental lithosphere; fluid pressure is higher in oceans and more homogeneously distributed. This is confirmed by the discovery of hydrothermal seafloor metamorphism in the 1970s.

We can summarise the mechanism of hydrothermal circulation in oceanic lithosphere as follows (for review see Fowler 1990):

Hot water is less dense than cold water. If hot water penetrates the flank of a volcano near the ridge it must be replaced by cold seawater. The process is intense and drives huge convection cells going several kilometres deep. The rising water reaches near critical temperatures of 400 °C. The volume of ocean water passes through the ridge in about 10 Ma. The occurrence of hydrothermal systems was known in land volcanoes and it was predicted that they should occur at the oceanic-ridge systems. Dredged rocks showed evidence of alteration and metamorphism by reaction with hot seawater and ophiolites exposed on land confirmed these predictions. However, direct evidence was only obtained in the late 1970s.

The observed profiles for heat flow versus age of oceanic crust were very different from theoretical profiles assuming heat loss by conduction only. The measured heat flow near the ridge was much less than the theoretical prediction; and on the older crust the observed values were highly dispersed, periodically above and below theoretical values. The explanation is that the process of heat loss should be convection superposed into conduction. This would require a high thermal gradient to overcome resistance against fluid motion and permeability, providing channels in the rocks for water to move through; permeability is higher near active ridges where hot igneous rocks cause new fracturing by cooling; and also because tectonic stresses are of tensional nature. As the crust moves away from the spreading axis, the stresses became compressive, the channels are obstructed by the minerals precipitated from circulating fluids and the thickness of the less permeable sediment cover increases (Fig. 2.4).

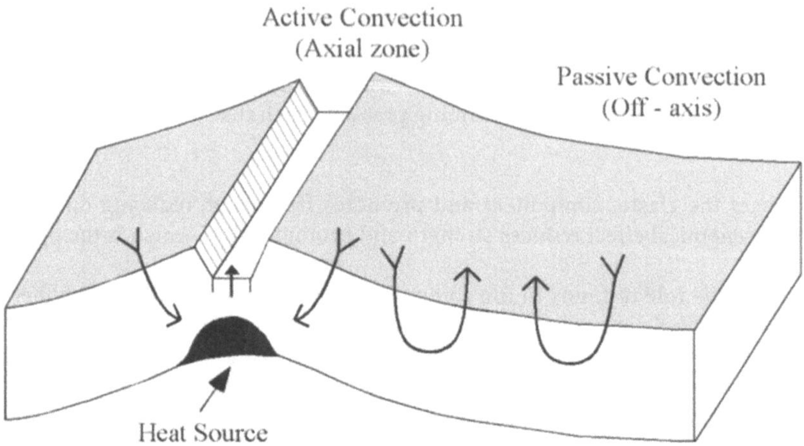

Fig. 2.4. Hydrothermal seafloor metamorphism in oceanic lithosphere. In the axial zone a heat source causes active convection; off-axis lower temperature passive convection is maintained by cooling of the oceanic lithosphere

The hydrothermal circulation changes the composition of the oceanic lithosphere because the original igneous rocks are transformed into metamorphic rocks with lower temperature mineral paragenesis; the permeability of the crust allows it to be crossed by convective fluid that converts the uppermost mantle. This should affect the rheology of the oceanic lithosphere (Fyfe et al. 1978; MacDonald and Fyfe 1985), controlling its strain regime through a variety of deformation mechanisms (see 3.12)

Hydrothermal seafloor metamorphism compels us to a revision of our ideas in the composition of oceanic lithosphere (Snow 1995). Two types must be considered (Fig. 2.5).

The *Penrose type* is based on the classical comparison between the oceanic lithosphere imaged by geophysical methods and the sequence observed by geological methods in exposed obducted ophiolites (Penrose 1972).

The *Hess type* (Hess 1962) is composed mainly of serpentinite derived from mantle peridotite altered by reaction with seawater; basalts and gabbros are minor constituents.

Until recently, it was thought that the Hess type was rare and restricted to active transforms and inactive fracture zones. However, it was found (Cannat et al. 1995) that serpentinite is widespread far from fracture zones in the ocean floor formed at slowly spreading ridges, mainly in the North Atlantic.

The mechanism underlying the differentiation between the two types of oceanic lithosphere is probably the balance between magmatic and tectonic processes in seafloor spreading. Magmatism dominates in fast spreading ridges and produces Penrose-type oceanic crust and lithosphere; tectonism dominates in slow-spreading ridges and favours the production of the Hess type. As spreading ridges are segmented, the result will be a more homogeneous Penrose-type in fast spreading oceans and an inhomogeneous lithosphere composed of rhombohedral zones of Penrose-type crust immersed in a more continuous Hess-type crust.

Fig. 2.5a–c. Schemes of slow spreading ocean crust and mantle and corresponding ophiolite sections (Snow 1995). **a** Penrose „layer-cake stratigraphy"; **b** Penrose type but with variable crustal thickness due to magmatic focusing; **c** combination of Penrose and Hess crust types due to discontinuous magmatic focusing and emplacement

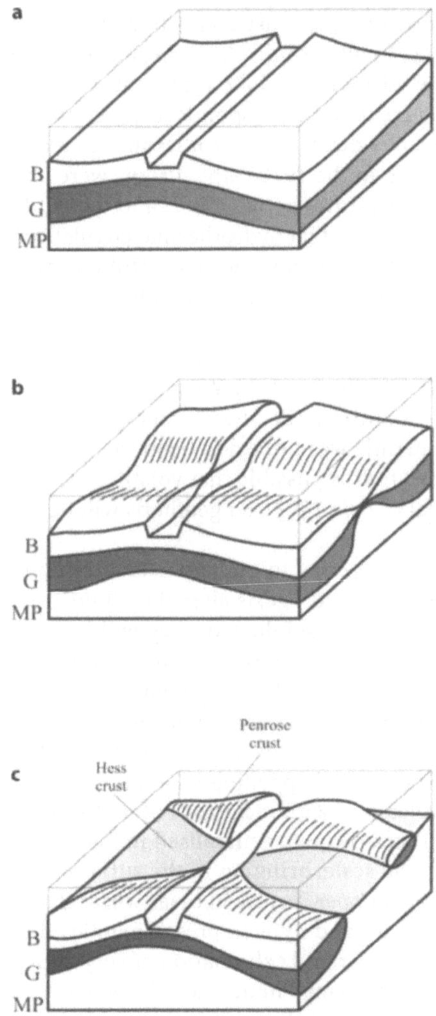

This geological model can be reconciled with indirect geophysical evidence as follows.

First, a distinction must be made between the seismologic Moho and the petrologic Moho (Clague and Straley 1977; Brocher et al. 1985). The study of ophiolites suggests, in part, that the petrologic Moho is the boundary between gabbros, the layer three of the oceanic crust and serpentinised (35%) ultramafic rocks of the upper mantle. The seismologic Moho is the boundary between serpentinised and fresh tectonised ultramafic rocks below. This is in accordance with the earlier views of Hess.

Second, it is known that magmatic chambers are continuous along fast spreading ridges, but discontinuous in slow spreading ones (reviews in Fowler 1990, p. 290)

Hence, oceanic crust seems to have more or less the same thickness and velocity structure in all oceanic plates, in spite of different magma supply rates. This suggests that the seismologic Moho represents the depth to which water circulates through the cooling crust and corresponds to the depth at which the cracks close.

Direct evidence for hydrothermal circulation in the oceanic lithosphere was discovered only in the 1970s; hot springs were observed, first of the „black smoker" type with temperatures above 350 °C and then of the „white smoker" type.

This showed that hydrothermal circulation is one of the most important earth processes. It plays a major role in controlling the chemistry of seawater, the genesis of mineral deposits and in the heat budget.

Fluids operate on a global scale (Fyfe 1988, 1992, 1994, 1997); they contribute to mixing of geospheres at crustal-mantle depth and operate in all tectonic settings and at all different phases of the Wilson cycle.

Fluids and volatile phases operate at crustal depths but also at mantle depths; the upper mantle was proved to be more hydrated than had been thought before (Carter et al. 1990). It is possible that volatiles may even promote melting at the core-mantle boundary (Fyfe 1997). So geofluids will be actively contributing to the mixing of geospheres.

We just referred to the importance of fluids at ocean ridges; but the whole ocean-floor, whatever its age, is altered by fluid circulation. In subduction zones pelagic sediments may be subducted; they will became de-watered, lubricating thrusts in the accretionary prism and above in the initial stages of subduction; when deep degassing starts, water will be injected on the upper mantle of the plate, softening it and contributing to calc-alkaline magmatism; *mantle eduction* (Barriga et al. 1992) will produce fluidised beds injected from mantle depths to the surface (Lockwood 1972); this is presently happening in some active subduction zones, producing serpentinite mud volcanoes with inclusions of high-pressure rocks (Fryer et al. 1985; Fryer 1992). This process is quite generalised in fossil subduction zones and can operate at the lithospheric scale, bringing to the surface slabs of thinned continental crust and serpentinised upper mantle in cases where a passive margin and a subduction zone collide (Ribeiro and Fyfe, in press); mantle eduction can contribute to solving the problem of every rapid exhumation of very high pressure metamorphic assemblages that characterise many Phanerozoic orogenes.

It is speculated (Fyfe 1997) that fluids will also operate in hotspots, plumes and underplating.

In continental collision zones the lower plate will be dehydrated and fluids migrate on a gigantic scale to the upper plate; they will assist in thrusting in lithosphere stacking (Fyfe and Kerrich 1985; Fyfe 1986) from the suture to the foreland, leaving markers that can be important in terms of resources such as hydrocarbon reservoir and orebodies (Oliver 1992).

In strike-slip transform boundaries, such as New-Zealand, the transpressive component can generate override processes (Allis 1981) that generate fluids which will lubricate the main transform faults.

Finally, fluids can operate mass redistributions inside the lithosphere by erosion and sedimentation; erosion rates can even influence tectonic rates and styles (Beaumont et al. 1992); so fluids will contribute to connect external and internal geodynamics in a unique global process.

We can be sure that even if the Earth becomes a dry planet – as happened to Mars? (Carr 1987; Fyfe 1997) – the dynamics of the planet will be very different. We can summarise by agreeing with Fyfe (1994) that the Earth is a water-cooled convecting body.

Fyfe (1997) asks: if a cooling Earth continues to convect, ocean volume might be expected to diminish; but it seems stable since, at least, the Phanerozoic. So could permanent small comet showers (Franck et al. 1986a,b) balance the fluid retained in successive cycles of geosphere mixing? (Fyfe 1997, pers. comm.). This is a good example of concepts on the Earth as an *open* dynamic system. A recent view of this problem (Maruyama 1999) assumes that the Earth has been leaking surface water since 750 Ma, with a decrease in sea level of 600 m. This is due to the fact that surface water consumed in subduction is only partially returned to the surface in mid-ocean ridges and hotspots; the accretion of icy meteorites does not balance this loss on the basis of the isotopic composition of meteorites and sea water. In 1 Ga the oceans will be drained and plate tectonics will cease on a dry planet.

2.4 Strain and Strain Rates in the Lithosphere: Present and Past

In the context of deformable plate tectonics strain in the lithosphere must be considered according to three aspects:

- Distribution in space: strain occurs at plate boundaries (*Interplate Strain*) or at plate interiors (*Intraplate Strain*).
- Distribution in time: in strain theory (Ramsay and Huber 1983), we distinguish between *incremental* and *finite strain*. Incremental strain can be measured by geodetic and geological methods; finite strain can be estimated by geological methods if strain markers and strain indicators are available.
- Occurrence in time: strain can occur as a discontinuous step in time, *seismic strain*, during an earthquake or as a continuous and irreversible process in time, *aseismic strain*. The seismic coupling coefficient, χ, will be defined (see also Sect. 3.5) as ,

$$\chi = \frac{M_0^s}{M_0^g},$$

where M_0^S is the seismic moment release rate and M_0^g is the moment rate estimated by geological methods from the kinematics of plates and faults. Seismic strains can be measured by seismological, geodetic and neotectonic methods; aseismic strain by geodetic and neotectonic methods. A complete spectrum of slip velocities has been given in evidence by seismology from earthquakes (sudden) to slow earthquakes (speeds of hundreds of meters per second), silent earthquakes (speeds of tens of metres per second) creep events (millimetres per year) and strain migration episodes (centimetres and millimetres per second; Yeats et al. 1997), making the distinction between seismic and aseismic strain a very subtle one.

In general, the rate at which geodynamic processes operate has a large spectrum (Lambeck 1988, pp. 1–4; Minster et al. 1990, p. 49); it varies from a length scale of

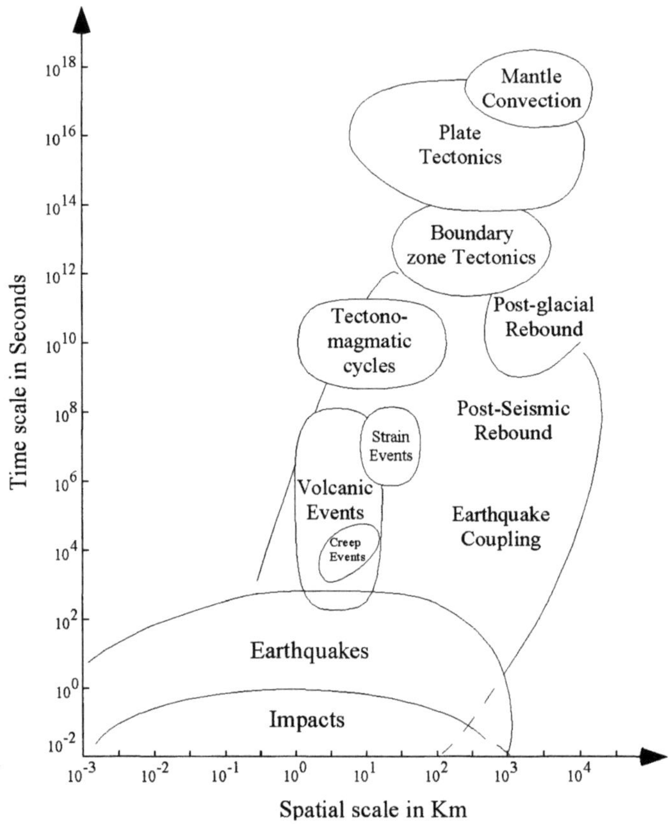

Fig. 2.6. Space-time spectrum of geodynamic processes. (Modified from Lambeck 1988; Minster et al. 1990)

10^0km and a time scale of 10^2 s to a length scale of 10^4 km and a time scale of 10^{14} s. It includes earthquake faulting, earth tides, glacial rebound, regional strain, plate tectonics and mantle convection. Different techniques must be used in different spectrum intervals. Geodetic observation can extend for a period of at most 100 years over the entire range of geodynamic processes; for the long term processes geological and geophysical methods should be used (Fig. 2.6).

Both seismic elastic strain and aseismic viscous strain can be measured by geodetic methods, depending on the particular tectonic situation of the station used for that purpose. If the station is located outside the band of elastic strain accumulation caused by an active structure, it samples the steady plate movement; the width of that band is tens of kilometres for strike-slip boundaries but hundreds of kilometres in subduction zones, as we will see in Section 3.5, which is dedicated to viscoelastic rebound.

In the last decade, spatial geodetic methods have become so accurate that they can be used to measure the instantaneous displacements between stations placed in different plates or in the same plate. These methods include very long baseline interferometry (VLBI), satellite laser ranging (SLR), global positioning system (GPS) and Doppler orbit determination and radiopositioning integrated by satellite (DORIS).

The strain measured by geodetic methods is mainly concentrated at plate boundaries, although measurable strain also occurs at diffuse zones, usually wider in continents than in oceans (Gordon and Stein 1992).

In continuum mechanics strain is a function of velocity field (Means 1990). The velocity field is known from classical and spatial geodetic methods. This way we can estimate strain inside the deformation zone along the plate boundary, using stations and baselines outside this zone. This is the theoretical basis for strain estimation from geodetic methods (Haines 1982; Turcotte and Schubert 1982; Lambeck 1988).

Given the displacement gradient tensor

$$\frac{\delta u_i}{\delta x_{,I}}$$

where u_i are displacement vectors and x_j are space co-ordinates, the infinitesimal strain tensor can be determined using the relationship

$$\mu_i = (\mu i)_o + \Omega_{ij}\delta_{xj} + \varepsilon_{ij}\delta_{xj}$$

where $(\mu_i)_o$ is a translation, Ω_{ij} is the rotation tensor and ε_{ij} is the strain tensor. Differentiating in relation to time, we obtain the strain-rate tensor, $\dot{\varepsilon}_{ij}$,

which specifies the velocity field and the vorticity tensor, $\dot{\Omega}_{ij}$

In two dimensions, such as at the surface of the Earth, the problem reduces to mapping of principal infinitesimal strains

$$\frac{\delta u_i}{\delta x_{,I}}$$

and the rotation component, ω and shear strain component, ε_{12} with

$$\omega = \frac{1}{2}\left(\frac{\delta u_2}{\delta x_1} - \frac{\delta u_1}{\delta x2}\right)$$

and

$$\varepsilon_{12} = \frac{1}{2}\left(\frac{\delta u_2}{\delta x_1} - \frac{\delta u_1}{\delta x2}\right)$$

Assuming Newtonian rheology and incompressibility the Laplace Equation satisfies compatibility conditions

$$\frac{\delta^2 \dot{\mu}_1}{\delta x_1^2} + \frac{\delta^2 \dot{\mu}_1}{\delta x_3^2} = 0$$

with velocity component $\dot{\mu}_1$

parallel to major subvertical fault. The physical significance of the Laplace equation can be synthesised as follows (Chapple 1968): a stretched membrane satisfies Laplace's equation. The displacement, *w*, obeys it in the form

$$\nabla^2 w = 0$$

where ∇ is the nabla operator and ∇^2 is the Laplacian operator. A solution to this equation is also called an harmonic function such that, in general,

$$\nabla^2 \phi = 0$$

where ϕ is some function of x and y, in two dimensions. To illustrate this concept we take a wire ring and bend it so it does not lie in a plane. Then we stretch a soap film across the wire; its displacement is smooth and approaches as closely as possible the average plane of the ring. This translates into the mathematical fact that the function ϕ, the potential, is smooth, and cannot have a maximum nor a minimum within a region. It must reach those in the boundaries of the region. In fact, in the elastic stretched membrane the elastic energy must be minimised, which is obtained by minimising the area of the surface, in accordance with a variational principle. The Laplace equation can be solved analytically or numerically (Middleton and Wilcock 1994, pp. 435–437) by using the relaxation technique.

This allows the computation of the velocity field if a set of boundary conditions are specified; the velocities in the horizontal plane are only functions of the axis parallel to the subvertical plate boundary in the reference frame chosen for the New Zealand transform boundary (Haines 1982).

Finite and incremental strain are recorded in rocks and can be estimated by geological methods (Ramsay and Huber 1983; Lisle 1994); by dating the beginning and the end of deformation episodes we can determine geological strain rates from finite strain states into the range of 10^{-13} to 10^{-15} s^{-1} (Pfiffner and Ramsay 1982); under certain favourable conditions we can even estimate strains and displacement differences in orogenic belts of any age (Ramsay 1969). However, these favourable conditions are very restrictive: strain must be removed through the deformed zone and once continuous linear features in the foreland and hinterland must be restored to their original position. The heterogeneity of deformation in time and space makes the unstraining operation complex, or even impossible if material is removed to any depth. So these palaeostrain studies allow precise and accurate estimates of strain states, and even of strain rates, but are difficult to use for rheological studies of the lithosphere because palaeostress is much more difficult to determine, as will be shown in Section 2.6 and also because the kinematics are not usually preserved in the geological record; in fact, even if we know the displacement vectors, movement paths are not recorded (Means 1976). To deal with the problem of lithosphere rheology we must follow a different approach from the pure geological one.

In a geodynamic approach we must integrate all the available information by a variety of methods. We must derive strain from kinematics as precisely as possible; for that propose we can use a kinematic model as near as possible to the present date; such a model exists and it is called NUVEL-1 (De Mets et al. 1990), based on plate kinematics of the last 3 Ma, assuming that these were minimal variations since then; hence, we have a stable kinematic framework as a foundation of the present velocity field. We will compare this theoretical work with the real velocity field measured in the last century with classical geodetic methods or in the last decade with spatial geodetic methods. The velocity field from geodetic methods and from the kinematic model must agree at a certain level of precision; if they do not agree, we must explain the origin of the anomalies between the measured real values and the theoretical ones, obtained through the kinematic model.

We must apply this strategy to plate boundaries in the first place; there are two reasons for this. One is theoretical: the kinematic model NUVEL-1 assumes plate rigidity except on plate boundaries; so it is unable to predict intraplate strains to begin with. The second reason is operational: although spatial geodetic methods are very precise the period of observation is very short, just one decade, as stated before, and no more than a century for classical methods. Intraplate strains, if they exist, are of second order to interplate strains; so they must be at least one order of magnitude below interplate strains and, in general, below the precision of the most sophisticated geodetic spatial methods in the present state of knowledge. Therefore, in the geodynamic approach, we will focus first on strain at plate boundaries, integrating geodetic, geophysical and geological information; discrepancies between different methods must be interpreted, as will be seen in Section 2.7.

Then we will move on to look at the evidence, both geophysical and geological, for intraplate strain, exploring discrepancies between geodetic methods and the kinematic model NUVEL-1. Finally, we will extend our period of observation, going back in time to look for evidence of finite strain, and assuming that displacements are known from plate kinematic direct reconstructions based on magnetic anomalies in the oceans in the last 200 Ma. Wilson cycles older than 200 Ma must be studied by other methods, as we will see in Chapter 4.

For stations outside deformed zones, when we try to compare two different scales of time, such as finite displacement accumulated in the order of 1 Ma and incremental displacement observed in the order of a decade or a century, we face a very difficult problem: can we extrapolate from 10 years to 1 Ma or 10^6 years, over five orders of magnitude? If plate movement is steady, the extrapolation is valid; if plate movement is episodic, the extrapolation is invalid. We will see in Section 3.3 that there are reasons to suppose that plate movement is steady. In any case, this problem can have a probabilistic solution; as the period of geodetic spatial measurements accumulates, the observed values must converge to the predicted values in kinematic models if plate movement is steady; the degree of convergence will give us a degree of confidence in the results. Discrepancies between observed and theoretical values must then be due to defects in plate rigidity rather than to anomalies in plate steadiness. Therefore, our basic hypothesis can be tested by observation and measurement. We tend to think that in a few years the central thesis of this book will be either refuted or confirmed.

2.5 Stress in the Lithosphere: Present and Past

In a rigid lithosphere the stress state will be indeterminate; but a stress state exists in the lithosphere, so it cannot be rigid. If we want to study the rheology of the lithosphere, we need to establish a constitutive equation relating stress and strain (Ranalli 1995). We can consider stress as a cause and strain as an effect; stress would be the independent variable or input and strain the dependent variable or output. However, this „arrow of time“ does not exist in the balance and constitutive equations (Twiss and Moores 1992, pp. 425–426) that are reversible in elastic or viscous behaviour, with no memory (Llyboutry 1992). So we can think of other ways, such as specifying the boundary conditions in terms of either strain or stress.

In fact, if we study the rheology of geological materials, we know more about strain than about stress, either in the present state or in the past. This is the difference between the study of nature and the use of experiments or simulation, where we can prescribe, from the beginning, perfectly known stress states. This is why we referred to strain first and to stress afterwards in the study of a natural object, such as the lithosphere.

The present stress state of the lithosphere can be measured by a variety of techniques, and is the object of the World Stress Map Project (Zoback 1992). However, if the „eigenvectors" of the stress tensor or principal stress directions are known for 7300 data points, only 4% of this total is available for the „eigenvalues" of the stress tensor or principal stress values or magnitudes.

Tectonic stress information is retrieved from six types of geological and geophysical data: earthquake focal mechanisms, well bore breakouts, in situ stress measurements (hydraulic fracturing and overcoring) and neotectonic data including fault-slip and volcanic alignments (Zoback 1992; Engelder 1993). However, the significance of this information has been disputed because it depends on the rheology of the stressed media that is assumed. If we assume an elastic rheology, stress and incremental strain are related by linear equations; this is the usual procedure in the standard plate tectonics theory. However, some authors (Harper and Szymanski 1991) maintain that the lithosphere is a non-equilibrium dissipative, dynamic system that deforms internally in a time-dependent manner, which is a typical non-linear systems perspective.

This perspective regards stress states as extremely variable, both in space and in time or, in other words, stress is inherently dynamic and cannot be precisely matched by any static perspective. So, according to the views that they defend, different authors put more emphasis on the uniformity of stress state in space and time or in its variability both in space and in time; the fluctuation in time would be in the order of 10–10^6 years, depending on the magnitude of stress considered. Space-time fluctuations could be self-similar (Harper and Szymanski 1991), as seismicity suggests; this aspect will be considered in Section 3.6. Even if the magnitudes of stress are generally not known, it is possible to determine the stress regimes (Anderson 1951) on the basis of stress indicators on extensional, with dip-slip normal faulting, strike-slip regime and thrust faulting regime; mixed modes can also exist with oblique-slip faulting. Constriction is rare but it exists and radial extension is even more rare. We will describe below the relationships between different stress regimes and tectonic settings.

If the rheology of the lithosphere is not elastic, we can add a doubt that some of the methods considered for stress estimation specify, in fact, an incremental strain state and not a true stress state. This is the case for earthquake focal mechanisms that have been interpreted as incremental strain indicators when they contain a significant non-dipole component and for fault slip data that can be interpreted in terms of strain inversion data rather than in terms of stress inversion data, as will be referred to in Section 2.6. We conclude that many arguments used to maintain that the lithosphere is elastic, such as its acting as a stress guide, are based on circular reasoning. The distinction between renewable and non-renewable stress (Bott and Kusznir 1984; Park 1988) is essential if we do not assume an elastic rheology from the beginning.

Renewable stresses are those that persist from the continuous application of causative forces, even if the strain energy is progressively dissipated. *Non-renewable stresses* are dissipated by release of the strain energy initially present. The relative magnitude of renewable and non-renewable stresses expresses the rheology of the lithosphere; but this is difficult to estimate even in the 4% of cases where magnitudes are known, because we must follow the fluctuations of the stress state in a period of time of the order of the seismic cycle.

If stress data have some ambiguity, as rheology indicators they are very significant because stress fields are correlated with underlying geodynamic processes. From this point of view we must consider first- and second-order patterns of stress (Zoback 1992).

In many areas tectonic stress indicators are of good quality and give congruent results by different methods. If this is the case, we can define provinces of uniform stress orientations with horizontal dimensions of 20–200 times the average thickness of the schizosphere, 20–25 km. First-order intraplate stress fields are the result of compressional forces at plate boundaries, mainly ridge push and continental collision. Intraplate areas of active extension are generally associated with high topography in continents and oceans. The plate driving forces give a general intraplate compressional field, but buoyancy stress related to crustal and lithospheric thickening or thinning can dominate in second-order stress fields which are only five to ten times larger than the thickness of schizosphere, and coincide with specific tectonic features due to buoyancy or other processes such as lithospheric buckling, lateral strength and density contrasts. They induce rotation of the stress field, sometimes dramatic, but generally defined at a regional scale. Plate-wide, first-order, stress fields are renewable but most regional stresses are non-renewable (Engelder 1993).

Plate-wide stresses show a relationship with plate kinematics that we will examine now, and with plate driving forces that we will refer to in Section 3.11.

The parallel between maximum horizontal stress and plate motion is a general rule. In plates not attached to subduction zones such as North America the maximum horizontal stress, S_H, is compressive and parallel to relative plate motion given by transform faults between the ridge and the opposite plate boundary (Zoback 1992; Engelder 1993), both in the oceanic and continental lithosphere. This suggests that ridge push is a major source of this stress field.

It has been noted that in the oceanic lithosphere older than approximately 30 Ma the state of stress is exclusively compressional (Wiens and Stein 1985; Bergman 1986; Stein and Okal 1986) but in young lithosphere the state of stress is much more variable. Tensional mechanisms predominate, but the horizontal component of principal tensile stress can make large angles with spreading direction. This suggests that thermal stresses related to rapid cooling of the oceanic lithosphere near the ridge controls the stress state rather than the plate boundary forces. Normal faulting parallel to the ridge crest shows that the divergent plate boundary is in tension but the push due to the gravitational potential of the ridge overtakes this tension and the stress field becomes compressional

In convergent boundaries, such as subduction zones and collision between continents, there is also a tendency for maximum compressive stress, S_H, to be parallel to absolute and, or relative plate motion; many vast intraplate domains show a nearly uniform-oriented stress field, mostly compressive. Extension in continents is present only in localised domains with high elevation and results in normal faulting perpendicular to the gradient of elevation.

Palaeostress is understandably more difficult to estimate than active stress. In the schizosphere it can be estimated both in orientation and in type by stress inversion of old fault populations, as we will see in Section 2.6. In the plastosphere the orientations and magnitudes of palaeostresses and the strain rates of ductile deformation are usually difficult to determine (Twiss and Moores 1992; Ghosh 1993). The orientations of some microstructures in calcite and quartz can determine the orientations of palaeostresses. Other microstructures such as dislocation density and dynamically recrystallised grain size can be used to infer approximate magnitudes of palaeostresses (Engelder 1993). Twining of calcite can be used to estimate orientation and magnitude of palaeostresses (Groshong et al. 1984). It was inferred (Van der Pluijm et al. 1997) that in the craton of north-west America compressive palaeostresses are perpendicular to Late Cretaceous Early Cenozoic Sevier fold-thrust belt; differential stress decreases from ~100 MPa near the orogenic front to ~20 MPa 2000 km inland. Similar results were obtained for Late Palaeozoic Appalachian belt and foreland in the eastern cratonic area of North America. The decay of differential stress with distance follows a power-law.

These results suggest that intracratonic stresses are unrelated to tectonic characteristics of plate margins and that the deformation of continental interiors follows simple rheology models that we will examine in the next section.

2.6 Distributed Deformation in the Lithosphere

Discontinuous deformation produces fractures and faults in the schizosphere; they pass in a smooth downward transition to shear zones in the plastosphere; so we can consider faults as brittle shear zones to differentiate them from ductile shear zones (Ramsay and Huber 1987).

Brittle deformation of the schizosphere is *distributed*, in the sense that there is a broad region of faulting instead of a discrete fault line. The rigid plate approximation of the lithosphere is inadequate for the description of this deformation, and the usual approach is made through continuum mechanics. The basis for this approach is the observation that despite the discontinuous nature of faulting, the continuum approximation is still valid if the dimensions of the deforming material are much larger than the characteristic spacing of the discontinuities. If this is not the case, a description in terms of microplates should be preferred, but the boundary between the applicability of both descriptions is gradational and the choice of the best description is not always an easy task (Thatcher 1995).

Faults can be interpreted in terms of *stress* or *strain*.

In *stress inversion* methods a regional stress tensor is estimated from a population of faults containing slicken lines (for a synthesis consult Angelier 1994). These methods rely on the assumption that slip on each fault plane occurs in the direction of maximum resolved regional shear stress. It has been shown (Pollard and Saltzer 1993) that fault compliance caused by fault shape and mechanic interaction between faults may perturb, on a local scale, a regional stress field homogeneous in space and time.

When these factors are negligible the stress inversion methods are valid and should be compared with focal mechanisms of earthquakes in populations of active faults.

These restrictions show that in general the relationship between the stress tensor and the strain tensor is complex in discontinuous deformation. This stimulated the development of alternative methods of *strain inversion*. It is assumed that all the strain is accommodated by slip along faults (Reches 1983). It is shown that four sets of faults with orthorhombic symmetry are required to accommodate a general three-dimensional deformation. In special circumstances the fault system can be reduced to three or two conjugate faults, in plane strain, or one single fault system (Reches 1978). So the case of the Anderson model for faulting applies only to plane strain. This generalised theory was confirmed by experiments (Reches and Dieterich 1983); these have shown that maximum and intermediate stresses interchange orientation during yield events, the analogue of earthquakes, whereas the corresponding strains are constant. This should also occur in nature. Source modelling of earthquakes (Lay and Wallace 1995; Yeats et al. 1997) has shown that the complexity of rupture process can induce different fault-slip events in the same fault plane during an earthquake. This should result in a complex array of slicken lines in active faults confirmed by observations, for instance, in the active fault plane of the Spitak earthquake, Armenia, on 7 December 1988 with Ms=6.9.

Molnar (1983) proposed a general method for strain inversion; if a region is cut by faults variably oriented, the average finite strain can be calculated by the sum of asymmetric moment tensor; the rotational component of strain is contained in the finite strain tensor. The boundaries of the sample region should include all the faults. This approach is valid because the strain is caused by displacement of essentially rigid blocks and elastic strains are small. However, in many domains, discontinuous deformation is intimately connected with continuous deformations by folding and bending. Nevertheless, we will see below that the seismic moment tensor contains relevant information for the finite strain estimation.

Field observations and analogue experiments show that distributed deformations can be *coaxial*, when two conjugate sets of faults of equal slip are generated or *non-coaxial*, when only one set of faults accumulates more slip than the other (Cobbold et al. 1989; Fig. 2.7).

Non-coaxial distributed deformation has been modelled by the rotation of rigid blocks in various ways (Fig. 2.8).

In the *domino block model* (Roberts and Yielding 1994), each block behaves like a book in a bookshelf. This model can be further subdivided.

In the *pinned block model* (Mckenzie and Jackson 1983, 1989) the deforming zone is represented by rigid blocks pinned with pivots or screwed to a rigid plate on either side.

In the *floating block model* (McKenzie and Jackson 1983, 1989) the fault-bound blocks are not pinned to the boundaries of the deforming zone but contain smaller blocks floating in it.

In all these models the internal block rotation is antithetic to external block rotation; the external spin is greater than internal spin and of the opposite sense. The internal spin is different for the pinned and floating block models; if it can be measured by tectonic, geodetic or paleomagnetic data it can discriminate between both models (McKenzie and Jackson 1989).

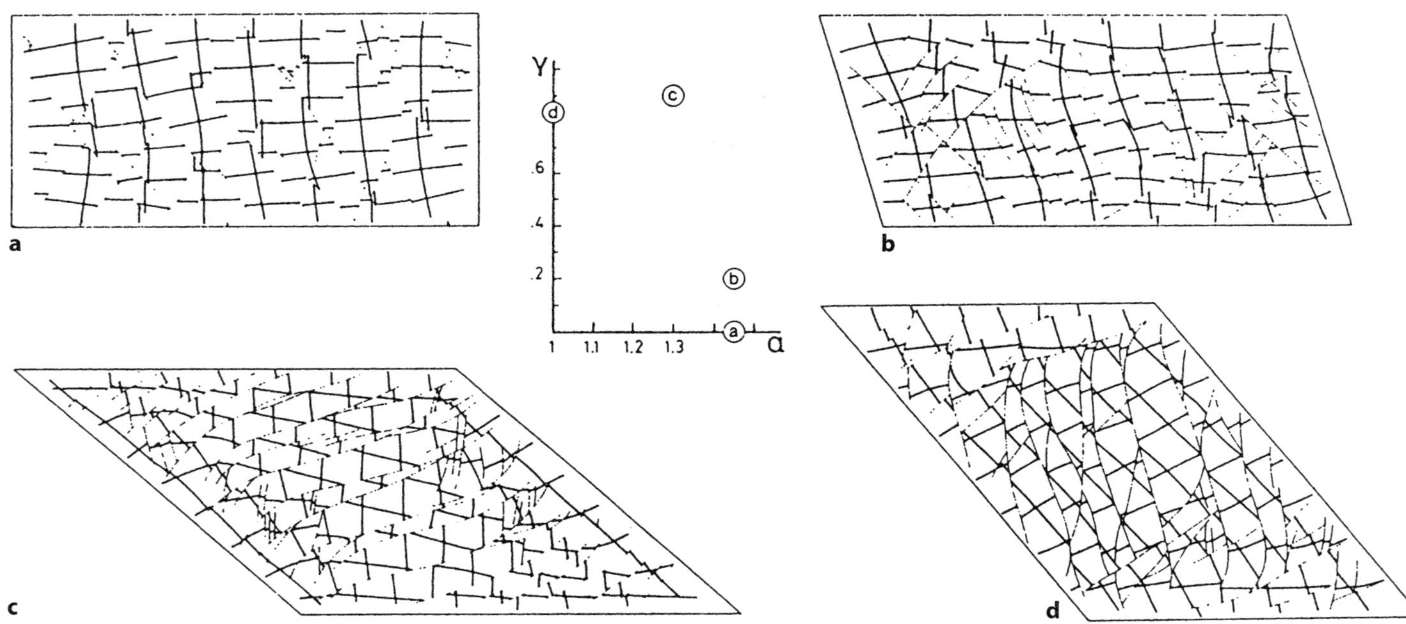

Fig. 2.7. Coaxial and non-coaxial distributed deformation. Strike-slip faults and associated block rotations in four sand packs with differing bulk deformation histories. Outer boundaries of sand packs (*A, B, C, D*) are now parallelograms, but were initially squares. These lines are passive markers originally forming square grids parallel to outer boundaries. *Thin lines* are fault traces. Plane-strain bulk deformation was achieved by combining coaxial stretching along bases of parallelograms with left-lateral simple shearing in same direction. Total amount of basal stretch (σ) and left-lateral shear (γ) are plotted in central diagram for each sandpack. Line drawings (*A, B, C, D*) were traced from photographs. (After Cobbold et al. 1989; with kind permission of Kluwer Academic Publishers)

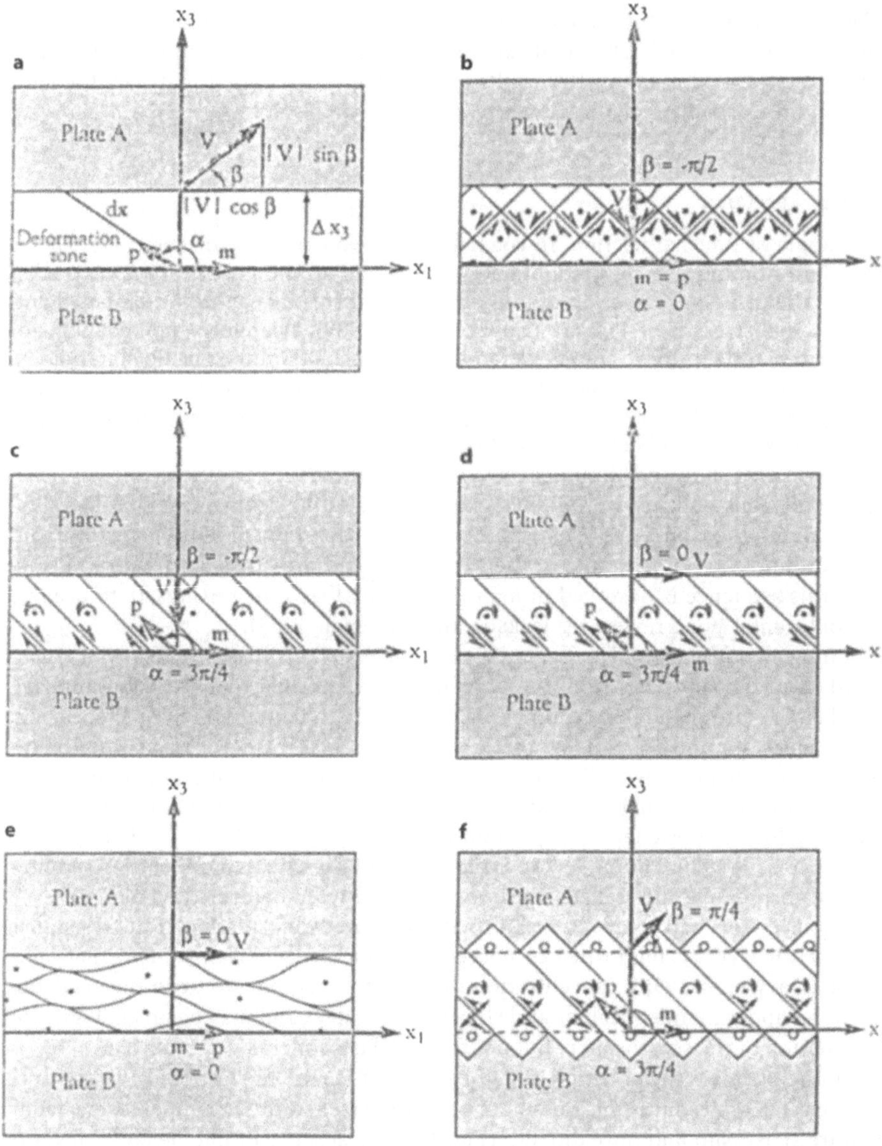

Fig. 2.8. Kinematic model for distributed deformation. **a** General model defining the variables that are adjusted to produce specific models. **b** Shearing on two sets of shear planes symmetrically arrayed about the principal axes of the deformation rate tensor. **c** Shearing on a single set of shear planes that rotate progressively with the deformation. **d** Simple shear with microspin equal to the macrospin. **e** Simple shear with zero microspin. **f** The pinned block model. (After Twiss et al. 1993/copyright 1993 by American Geophysical Union)

Domino domains are frequent in materials that obey the Coulomb criterion for failure (Jaeger and Cook 1981) that explains the Byerlee law for the upper crust. After a uniform strain of less than 10% faults form at the acute angle θ, between the fault plane and the principal compressive stress σ_1 such that

$$\theta = 45° \pm \phi / 2$$

where $\phi = \arctan \mu$

is the angle of internal friction. In coaxial strain two conjugate orientations, $\pm\theta$ are formed. In domino one set predominates: stress and strain tensors are non-coaxial, because the principal direction of bulk incremental strain is at 45° to the faults and the principal stress is at $45° \pm \theta/2$ (Cobbold et al. 1989). Therefore, the behaviour of frictional material is more complex than that of viscous fluids or plastic materials. In experiments dry sand is considered a good analogue for a dry schizosphere, obeying the scale effects; but we are only beginning to understand the behaviour of sand, a point that we will discuss later in this section.

Experiments show that dominos form in some kinematic conditions, such as nearly homogeneous bulk strain. Uniform extension of subhorizontal layers induces extensional dominos; once the layer rotates antithetic dominos predominate, tilting up to 30° around the horizontal axis; then they lock and are cross-cut by new normal faults; this sequence is observed in many extensional tectonic settings, both oceanic and continental (for a review see Jackson and McKenzie 1983).

Homogeneous plane strain on map view induces strike-slip *dominos*; load is transmitted from the sides and from below as in the extensional regime. If the imposed boundary conditions are a combination of coaxial stretching and simple shearing, they produce domino domains synthetic with imposed shearing; external rotation inhibits the antithetic system. Simple shearing alone produces antithetic *dominos* with rotation of up to 30° (Cobbold et al. 1989); this contrasts with simple shear in cohesive materials, where Riedel shear predominates (Cloos 1928; Riedel 1928; Tchalenko and Ambraseys 1970; Freund 1974). Strike-slip and dip-slip components are combined in transpressive and transtensile boundaries between deformed domains.

Various factors can control the nature of distributed deformation. If the resistance of the brittle layers is high when compared with ductile layers below, deformation is diffuse (Davy et al. 1995); concentration of slip on faults may occur in the crust but not in the upper mantle (Molnar and Gipson 1996); dry sand may be an analogue for deformation of the schizosphere but not for the plastosphere.

Another factor is the partition of deformation (Richard and Cobbold 1990); a ductile layer at depth reduces the amount of basal drag transmitted to upper layers; faults are pure strike-slip and pure dip-slip (Fig. 2.9); the absence of ductile layers favours the generation of oblique stresses that can be transmitted as far as the free face and generate oblique-slip faults (Fig. 2.10).

The application of the various models to real tectonic situations is often straightforward at plate boundary zones when there is good evidence that a major fault takes the largest amount of slip and the higher-order faults are congruent with the sense of slip between the two plates. However, in intraplate situations, the hierarchy of fault movement can be ambiguous, with conflicting interpretations in the nature of the deformation between different blocks (McKenzie 1990), such as in Central Asia or in the Iberian Variscides (Fig. 2.11). In active tectonics the evidence from seismotectonics is then cru-

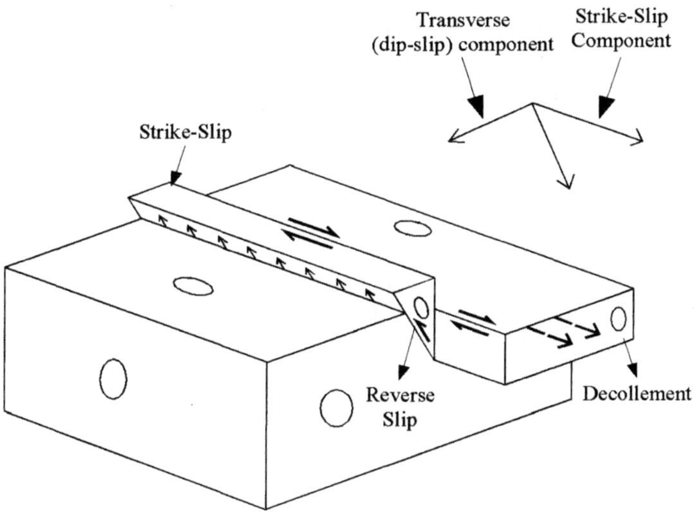

Fig. 2.9. A ductile layer at depth allows decoupling of oblique slip in components of pure strike-slip and pure dip-slip

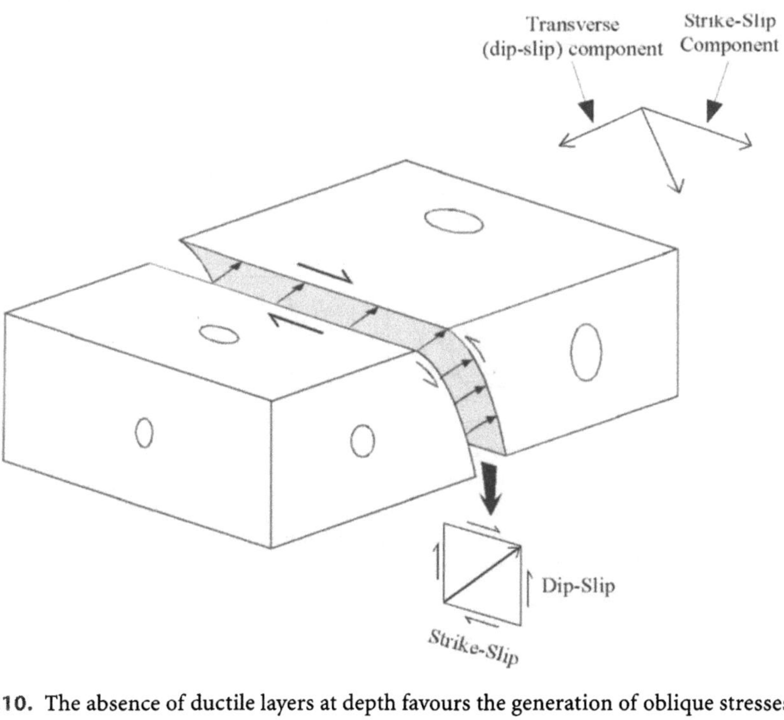

Fig. 2.10. The absence of ductile layers at depth favours the generation of oblique stresses that can be transmitted as far as the free surface and generate oblique-slip faults, in this case dextral-reverse

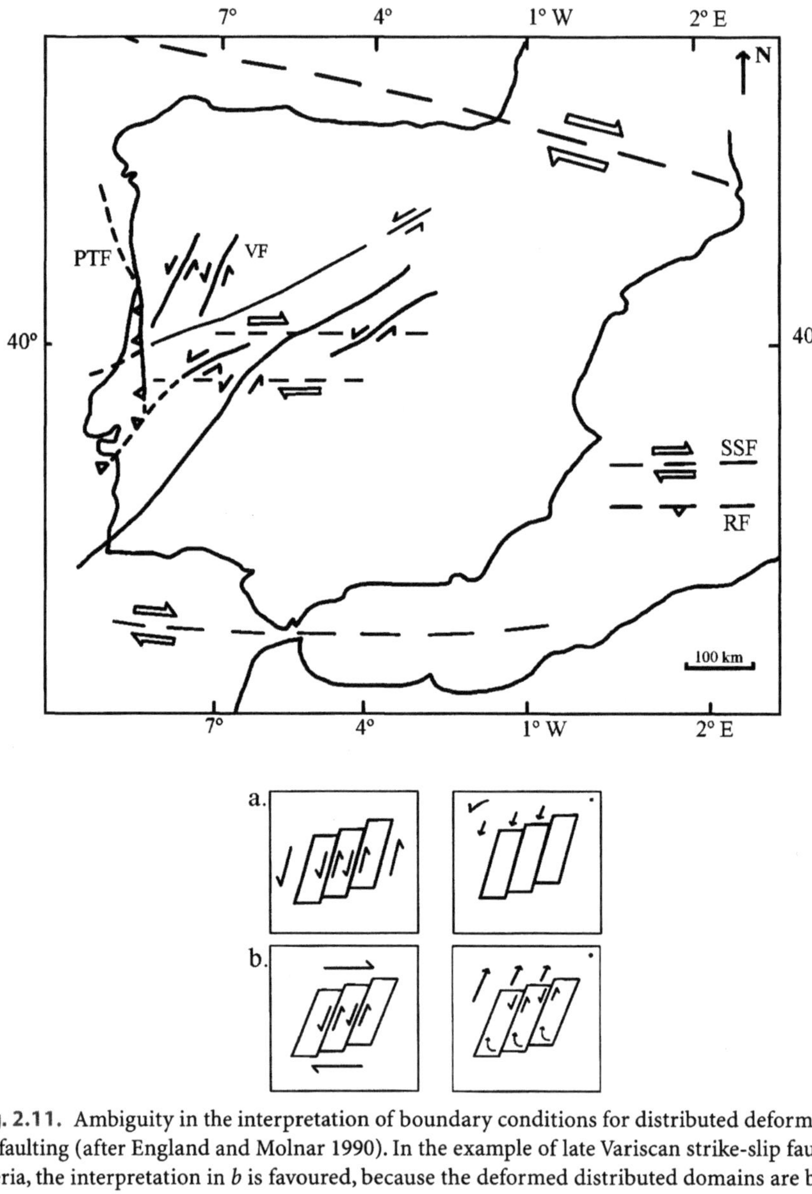

Fig. 2.11. Ambiguity in the interpretation of boundary conditions for distributed deformation by faulting (after England and Molnar 1990). In the example of late Variscan strike-slip faults in Iberia, the interpretation in *b* is favoured, because the deformed distributed domains are bounded by higher order dextral strike-slip faults with the required orientation and compatible with N–S reverse fault at the W boundary of Iberia. *PTF* Porto Tomar Fault; *SSF* Strike Slip Fault; *RF* Reverse Fault; *VF* Vilariça Fault

cial for the definite final interpretation. In this respect, we must notice that the significance of seismic moment tensor for a deformed area has been debated, as follows.

The infinitesimal irrotational strain tensor ε_{ij} can be estimated from a sum of seismic moment tensors, M_{ij}, of (N) different earthquakes

$$\varepsilon_{ij} = \frac{1}{2\mu v} \sum_{x=1}^{N} M_{ij}^{(x)}$$

where μ is the shear modulus, v is the volume of the region, and the seismic moment tensor

$$M_{ij} = M_0 \,(n_i\, b_j + n_j b_i)$$

in which n_j is the normal to fault plane, b_j is the slicken line or slip direction, and $M_0 = \mu A \Delta\bar{u}$ is the scalar seismic moment (see Sect. 3.6.), A is the rupture area and $\Delta\bar{u}$ is the mean displacement.

The asymmetric moment tensor M'_{ij} is

$$M'_{ij} = M_0 b_i n_j$$

and allows the estimation of the rotational strain tensor

$$\varepsilon ij = \delta ui \,/\, \delta xj$$

$$\varepsilon_{ij} = \frac{1}{\mu v} \sum_{x=1}^{N} M'^{(x)}_{ij}$$

according to authors that introduced the concept (Molnar 1983; Molnar and Deng 1984). The physical significance of the asymmetric moment tensor is easily perceived through Table 2.1 and Fig. 2.12. The formal similarity with strain tensor becomes evident.

The fully asymmetric moment tensor (Molnar 1983), as any asymmetric tensor such as the infinitesimal strain tensor, can be expressed as the sum of a symmetric tensor, the stretching, and an orthogonal anti-symmetric tensor, the internal spin, expressing the rate of rigid rotation or vorticity associated with fault slip. However, we must also consider an external spin, due to rigid rotation of the entire subarea with respect to some external co-ordinate frame, including rotation of faults and fault blocks. In coaxial deformation by two conjugate sets of faults with equal displacements the internal spin is zero; if only one set is present the internal spin is at a maximum (Cobbold and Davy 1988). However, the external spin can be greater than the internal spin and of the opposite sense, as in the domino block mechanism. So it was stated that the aseismic moment tensor for a unique seismic event should have the orthorhombic symmetry of the double couple source mechanism of earthquakes; in this manner, earthquake data will not allow the estimation of external spin due to rigid rotation of the whole sub-area, including fault blocks (Cobbold and Davy 1988; Jackson and McKenzie 1988; McKenzie 1990). Other methods must be used for that purpose: palaeomagnetic studies for rotations about vertical and horizontal axes and compatibility equations for velocity gradients, which express the conditions required to maintain material continuity; these methods for integration of finite strain have been developed by some authors (Cobbold 1979).

All the information can be integrated in a complete analysis, using strain rates, crustal thickening, palaeomagnetism, finite strain and fault movements within a de-

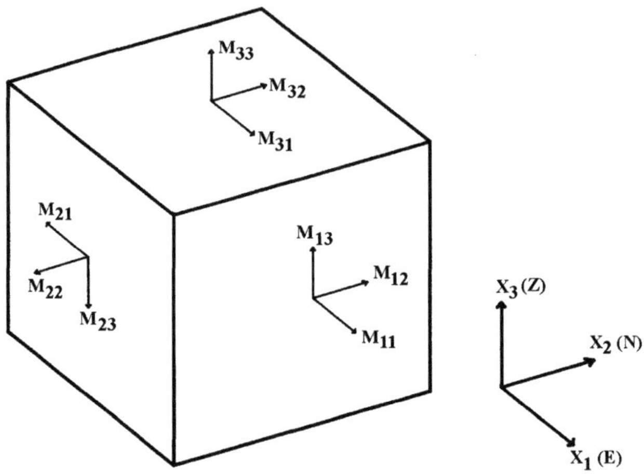

Fig. 2.12. Components of asymmetric seismic moment tensor acting on the faces of a cube. (After Molnar and Deng 1984)

forming zone (McKenzie and Jackson 1983); strain due to fault movements can be estimated independently of seismic moment integration (Molnar 1983); there are scaling laws relating fault displacement, fault length and fault width, generally of fractal nature, that allow the estimation of strain and that will be referred to below in Section 3.7. Finite strain can be estimated from fault-slip data including block rotation (Cladouhos and Allmendinger 1993).

But when seismic moment tensors are calculated for a large deformed area such as central and eastern Asia (Molnar and Deng 1984) an anti-symmetric part is obtained. This suggests that the classical continuum theory is not complete and adequate for specification of distributed deformation.

In fact, the continuum mechanics approach has been challenged by a generalisation of the classical continuum approach by some structural geologists (Twiss and Gefell 1990; Twiss et al. 1991, 1995). The theory on which it is based is the Eringen's continuum micropolar theory; it adequately describes the rotation of rigid blocks that characterises the distributed deformation model.

The kinematics of the *continuum micropolar theory* are distinct from classical continuum mechanics. In simple shear of a continuum there is a unique strain rate and spin for the medium. In continuum micropolar theory we must consider two scales of motion: a large-scale average motion given by a macrostrain rate and a macrospin and a local motion of each rigid grain, composed by a microspin the average rotation rate for all grains; the microspin must be specified because it is not uniquely determined by the large-scale average deformation (Fig. 2.13). It must be stressed that the theory was devised for the brittle field of deformation, however, equivalent concepts exist in the ductile field of deformation. If microspin exceeds the equivalent spin for simple shear, we get super simple shear as a special case for generalised shear (Simpson and de Paor 1993).

Table 2.1. Asymmetric seismic moment tensor. (After Molnar and Deng 1984, JGR p. 6206)

	Description of positive elements of M'ij			Direction
M'ij	Extension on x_1	Top to x_1 around x_3	Top to x_1 around x_2	Row 1: to x_1
	Top to x_2 around x_3	Extension on x_2	Top to x_2 around x_1	Row 2: to x_2
	Top to x_3 around x_2	Top to x3 around x1	Extension on x_3	Row 3: to x_3
	Column 1 [Plane x2 x3]	Column 2 [Plane x3 x1]	Column 3 [Plane x1 x2]	

The consequences of the kinematics are multiple: the seismic moment tensor can be asymmetric (Molnar and Deng 1984); the symmetry of brittle deformation patterns can be orthorhombic or lower, monoclinic and triclinic (Schrader 1988) depending on symmetry particle path fields (Ramberg 1975; Ramsay and Huber 1983). Kinematic models of deformation zones can be generalised to include coaxial pure shear, shearing on a single set of shear planes that rotate progressively with the deformation, also called the domino model, simple shear with microspin equal to macrospin, simple shear with zero microspin and pinned block model.

Focal mechanisms of earthquakes can also be generalised to include non-dipole components; the distribution of P and T axes can have orthorhombic symmetry for pure dipole, plane strain; or a lower symmetry for transtensional, constrictional strain and for transpressional flattening strain; this is because nodal surfaces with zero rate of stretch are neither planes nor cones and are equivalent to surfaces of no finite longitudinal strain (Ramsay 1967). In this theory fault-slip inversions constrain strain rates but not stresses (Twiss and Unruh 1998).

The continuum micropolar theory is a better approximation to schizosphere deformation than classical continuum mechanics. In fact, an analysis of Harvard centroid moment tensor catalogue (Frohlich and Apperson 1992) shows that many seismic events have both dipole and non-dipole source components, instead of the fact that large magnitude earthquakes must occur on planes with larger slip rates (Twiss et al. 1993). Although it is expected that the generalised micropolar theory is more rigorous than the classical continuum mechanics approach, in the specification of schizosphere deformations, most studies have employed the classical approach. This is unfortunate because the micropolar theory can deal more rigorously with the deformable plate real behaviour.

A crucial aspect of the continuum micropolar theory is that it is formally analogue to deformation of a granular material, where fault blocks act like grains. In fact, some specialists on geotechnics (Nascimento 1990) have already tried to apply frictional creep of particulate materials, not only to soils, but also to the rocks that compose the lithosphere (Fig. 2.14).

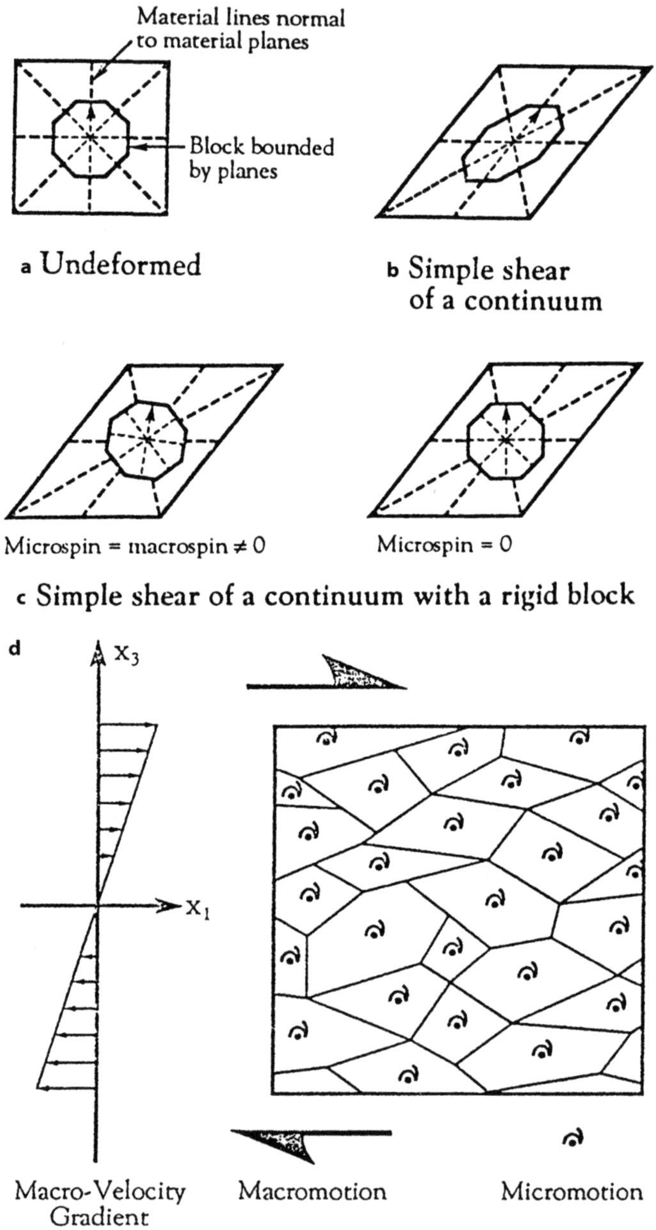

Fig. 2.13. Deformation in continuum micropolar theory. **a** Original state – a circle is enclosed on a continuum. **b** Simple shear of continuum and inclusion. The inclusion deforms like the continuum. **c** Simple shear of a continuum with a rigid block. The inclusion is rigid and deforms with microspin equal to macrospin. **d** Geometry and kinematics of the macromotion and micromotion. The *greater particles* have a low microspin, nearer the macrospin, the *smaller particles* have a higher microspin, much greater than the macrospin. (After Twiss et al. 1993; copyright 1993 by the American Geophysical Union)

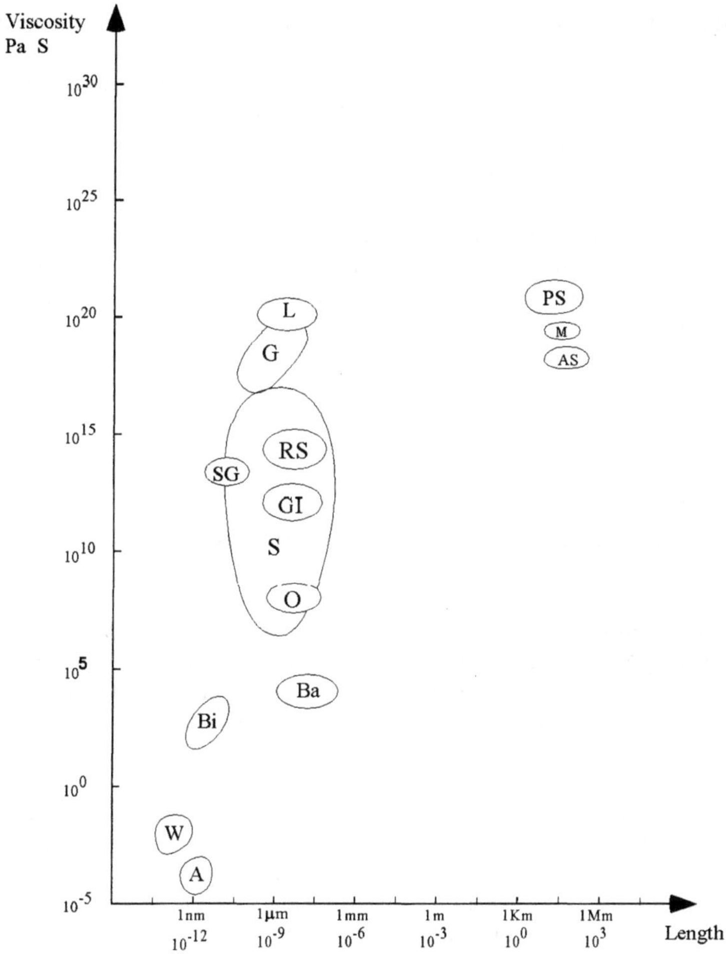

Fig. 2.14. Viscosity of materials and geospheres, *V*, measured in Pa s, as a function of characteristic length, *L*, measured in km, both on a logarithmic scale (after Nascimento 1990; Middleton and Wilcock 1994). *A* Air; *As* asthenosphere; *Ba* basalt magma and lava; *Bi* bitumen; *G* glass; *GI* glacier ice; *L* limestone; *M* mantle; *O* obsidian; *PS* plastosphere; *RS* rock salt; *S* soils; *SG* silica glass; *W* water

The behaviour of granular materials is somehow more complex than that of more usual condensed states of matter such as solids and liquids. One can suppose that granular materials can act *alternatively* as solids – we can construct a sand pile that does not spread like a drop of liquid – and like liquid – the sand pile flows if we add more grains. Recent numerical simulations (Radjai et al. 1996; Radjai 1997) suggest that granular materials can act *simultaneously* as solids and liquids. In fact, in a dry granular material, without cohesion, deviatory stress is not zero, but cannot exceed some threshold that depends on pressure. That is why a solid can sink into sand for some distance, below which the pressure increases and the movement stops: the sand is more solid.

What the numerical simulations suggest is that in a granular material subjected to deviatory stress, the distribution of forces is very inhomogeneous, generating two types of links; hard links from a network that controls the solid behaviour and soft links from a network that controls the liquid behaviour, where the slip is concentrated. We propose that this solid/liquid behaviour is a good analogue for lithosphere deformation; in fact, in linked fault systems we can observe the occurrence of soft and hard links between individual fault segments (Walsh and Watterson 1991; Davison 1994).

We conclude that the behaviour of sand as a good analogue to lithosphere deformation was well known; the heterogeneity of deformation of sand was clearly depicted by photoelastic techniques but we lacked a good theoretical understanding for this heterogeneous behaviour. This is a field where we have achieved recent progress but we need to do much more work along these lines of reasoning.

2.7. Kinematics, Strain and Rheology in Plate Boundaries and Deformed Plate Interiors

We must now synthesise all the data on lithosphere deformation. Kinematics can be obtained from geodetic data at the level of the last century for classical methods or the last decade for modern spatial methods. It must be compared with tectonic and paleomagnetic data for the last Ma. Continuum models of various types will eventually resolve the problem of prescribing the best rheology for plate boundaries and plate interiors

All types of plate boundaries, divergent, transform and convergent have been surveyed by geodetic methods; those which have emerged are the best studied, for obvious operational reasons; the principal examples are: Iceland, Afar and Azores for divergent boundaries; New Zealand and California for transform boundaries; and Japan and eastern Mediterranean for convergent boundaries. In the most active boundaries classical surveys can be used to complete the data set furnished by the more precise spatial geodetic methods. This is important because displacements at the secular level by classic methods can be compared with displacements at the decade level by modern methods. This comparison ensures that transients are of second order, relative to long-term plate motions (Walcott 1984).

We must examine both interplate and intraplate strain; a priori, we do not know if they correspond to different regimes of deformation and how they are connected; *we just know that intraplate strain must proceed at a lower strain rate than interplate strain.* If the rheology is the same in plate interiors as at plate boundaries the information about interplate strain should be extended to include intraplate strain operating at a lower strain rate. However, there is also the possibility that the rheology is not the same at plate interiors and at plate boundaries; continued seismic and tectonic activity can possibly weaken the plate boundaries and confer on them different rheological parameters or even a distinct rheological behaviour. This possibility is suggested namely by numerical models that, in order to simulate natural deformation, must prescribe a different rheology for plate boundary zones.

The methods of finite strain estimation using the classical continuum mechanics approach have been applied to different types of plate boundaries, either divergent, transform and convergent; they must be compared with incremental strain methods for the same domains.

To illustrate deformation at plate boundaries, we must refer to the less complex situation of a transform boundary cutting the continental lithosphere which can be more easily monitored than transforms in oceanic settings. This is namely the cause of the San Andreas Fault Zone in California or the Alpine Fault in New Zealand. The main problem in quantitative modelling is the relationship between plate movement of bounding rigid plates and distribution of the deformation inside the plate boundary of definite width, in the order of 10^2 km, instead of a discrete narrow zone of negligible width. Two models have been advanced for such geodynamic settings.

The deformation in the thin schizosphere (10–20 km) proceeds by seismic slip on active faults; the problem is how deformation is accommodated in the platosphere.

One approach assumes that the platosphere is deformed by stable sliding on the narrow shear zone beneath the seismogenic faults at the surface.

The other approach (Bourne et al. 1998) assumes that the platosphere deforms by inhomogeneous simple shear across the whole width of the deformed zone.

These alternative views are based on different assumptions concerning the driving forces for deformation. In the second model the dominant forces on crustal blocks are tractions at their bases, and the long-term motion of each block of the schizosphere will be given by the average velocity of the underlying platosphere. In the first model viscoelastic coupling between the elastic schizosphere and the viscous platosphere make vertical displacement profiles as a function of depth compatible at both levels with distinct rheology (Savage et al. 1999); this model assumes that lithosphere deformation is being driven from the sides by stress that is comparable in magnitude in schizosphere and platosphere (Haines 1998).

In both cases the slip rates of active faults observed in nature agree with the results of the models within reasonable bounds of uncertainty. This happens because in both cases short and long-term deformation of the schizosphere is driven by steady external forces that result from movement of bounding plates and are transmitted, either from below by a stronger platosphere, or from the sides by coupling of the schizosphere–platosphere of the same strength. To discriminate between both models we need to refine the rheology model for the platosphere, because viscoelastic coupling requires linear, Newtonian rheology, but only power-law rheologies are compatible with a strong localisation of deformation in the schizosphere. Another criterion is the strength of the surrounding schizosphere; the difference must be significant in the first model of weak faults on a strong schizosphere, controlling the stress field, but negligible in the second model of a weak schizosphere driven by a strong platosphere. Concerning this aspect we consider that the models for plate boundary deformation fall into two different classes of interpretation that are mutually exclusive (Tikoff and Teyssier 1994).

In *stress partitioning* models it is proposed that weak faults are embedded in a strong schizosphere; this causes the maximum horizontal compressive stress to be subperpendicular to the plate boundary. So the stress field is controlled by fault properties (Mount and Suppe 1987; Zoback et al. 1987).

Stress partitioning models were introduced to solve a persistent paradox for the San Andreas Fault. If we extrapolate the laboratory friction data on fault strength, we would expect high heat flow by frictional heating during seismic activity across the fault. This was contradicted by measurements along most of the length of the fault (Lachenbruch and Sass 1980). This led to the so-called frictional stress-heat flow paradox (Engelder 1993). One possibility was that the experimental data overestimated the strength of the fault at depth. Other possibilities would be exceptional weak fault gouge or abnormal pore pressures (Hickmann 1991). Stress measurements showed that maximum horizontal compressive stress had a wide range of orientations relative to fault trace, from almost perpendicular to less than 30°. The first situation, with high angles, suggests very low strength in the fault and the second, with low angles predicted by conventional friction data. suggests normal strength in the fault plane.

A recent study (Hardebeck and Hauksson 1999) of the San Andreas Fault showed the systematic variation of the stress field, estimated from an extensive catalogue of earthquakes, across a segment of the San Andreas Fault in southern California. The maximum horizontal stress, σ_H, is typically at a low angle to the fault near it and approximately perpendicular to it at a distance of the order of 50 km. This can be due to three factors. A first possibility is that the fault is inherently weak; a second possibility is that fault zone fluids at high, nearly listostratic, pressures could decrease the effective normal stress on the fault, decreasing its shear strength; a third possibility is the reduction of fault strength during slip by dynamic weakening. The authors exclude the third possibility because the large rotation of σ_H is incompatible with stress data in the field and laboratory coupled with a typical stress drop for earthquakes and cannot explain such rotations of ≥10°, because they require very low background deviatory stress. The first possibility is considered improbable because most candidate minerals responsible for inherently weak fault materials have been eliminated by testing in the laboratory. Therefore, the second possibility remains the most probable explanation. Repeated strain-related fracturing and crack sealing can create low-permeability barriers that seal fluids into the network of currently active fractures and induce stress rotation.

This study was challenged (Scholz 2000) by the view which maintains the San Andreas Fault is as strong as most faults. Stress measurements made in deep (>1 km) boreholes are explained by high coefficients of friction $0.\sigma<\mu<0.7$ and nearly hydrostatic pore pressures, $\lambda\approx0.4$, compatible with Byerlee's law and laboratory experiments. Stress orientation is interpreted by partitioning between the dip-slip component with thrusts and folds that have been rotated 20–300° clockwise since its inception and the strike-slip component in the San Andreas Fault itself. In the more transpressive segments like the Big Bend, shear stress in the fault plane is independent of pore pressure and in the range 100–160 MPa. In the predominant transform segments shear stress diminishes but is well above the 20 MPa required by the weak fault hypothesis. The heat flow paradox is removed assuming that heat transfer is dominated by advective flow allowed by strongly positive scale dependence of permeability. Other authors (Zoback 2000) maintain that fault normal compression remains present in the transform segments within 1–2 km of the fault trace, such as in the segment south of San Francisco, implying a generalised low strength of the main fault.

We must conclude that, in spite of all recent progress in studies of plate bounding fault zones, there are still highly divergent views on their rheology; different specialists prefer models of weak faults in a strong schizosphere or a strong fault in a strong schizosphere; this is due to our inadequate knowledge of the physics of faulting (Zoback 2000).

In *strain partitioning* models plate motion controls the instantaneous strain orientations and finite strain by kinematics response to applied boundary conditions (Oldow et al. 1990, Molnar 1992). The difference in the partitioning results from the system being decoupled in the first case or coupled in the second case, favoured by the presence of a weak zone along the plate boundary or by continuous deformation in the plastosphere that couples the two plates effectively. In strain partitioning, the strain state, strain path or strain components are heterogeneously distributed across the deformed domain. Strain partitioning is the rule in real three-dimensional deformation of geological objects due to dynamic reasons (see revision in Dewey et al. 1998); a simpler explanation might be the fact that, in general, continuous principal surfaces of stress and strain do not exist in three dimensions (Treagus and Lisle 1997).

In *divergent boundaries* the finite plate kinematics can result in a variety of finite strain regimes; this is a consequence of the partitioning between magmatic activity, aseismic fault strain and seismic fault strain (Cowie et al. 1993); the instantaneous displacement measured by geodetic movements on a time scale of 10^1 years can be an order of magnitude above the finite displacements due, on a time scale of 10^6 years, to episodicity in spreading (Delaney et al. 1998).

The recent use of high-resolution sonar allowed the mapping of normal fault populations at mid-ocean ridges (Cowie et al. 1993, 1994; Karson et al. 1996; Lourenço, 1997). These are based on published statistics of the length, scarp heights and spacing of normal faults in the flanks of mid-ocean ridges; other parameters can be used if available. From fault parameter distributions we can estimate strain using scaling laws, as discussed in more detail in Section 3.7.

It was shown that at fast spreading ridges (≥90 mm/year) such as the East Pacific Rise, faulting may account for 5–10% of the total spreading rate. The seismic coupling coefficient is near zero, $\chi<1\%$. In slow spreading ridges such as the Mid Atlantic Ridge faulting represents 10–20% of the total spreading rate and represents less than 85% of the total brittle deformation, $\chi<85\%$.

The difference in behaviour of the two types of ridges is due to stable sliding at fast spreading ridges as opposed to stick-slip at slow spreading ridges which is due to a high and low thermal gradient in each case; another factor could be enhanced hydrothermal effects, due to a higher fault slip rate, which generate alteration products that behave by stable sliding, such as serpentine (Cowie et al. 1993).

Geodetic methods have also been applied to subaerial rifts such as Iceland and the Afar triangle. In these cases active faulting occurs at the divergent boundary and the measured instantaneous rates of displacements are the result of coseismic strains in the very thin schizosphere and not the viscous flow of the asthenosphere. Seismic and volcanic episodes give displacement values that are very different from finite displacement at the scale of 1 Ma. So, seafloor spreading can be episodic (Patriat and Courtillot 1984), a point that we will discuss again in Section 3.8.

Let us now consider *transform boundaries* because the deformation is planar, with no elongation in the vertical axis, if the motion is the pure transform type. However, real transform boundaries can have a component of extension- transtension- or con-

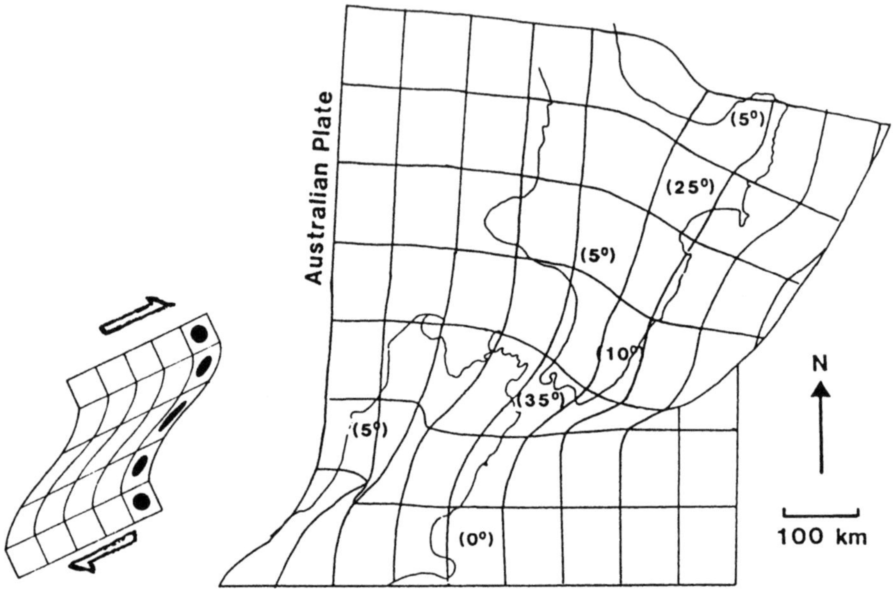

Fig. 2.15. Diagram showing an estimate of the smoothed-out deformation of a 100-longitude/latitude grid system during the last 4 Ma. Grid was orthogonal at 4 Ma. Observed or inferred palaeomagnetic clockwise rotations about vertical axes relative to the Pacific Plate are shown in *brackets*. (After Lamb 1989; with kind permission of Kluwer Academic Publishers)

vergence-transpression (Harland 1971; Sanderson and Marchini 1984). The deformation regimes are well known. If elongation is allowed in the vertical direction transpression will produce oblate ellipsoids and transtension prolate ellipsoids. The amount of partition is highly variable and it depends on relative plate motion and mechanical properties of the boundary zone.

The assumption of rigidity must be abandoned at plate boundary zones; so what rheology fits the data better? A zone where there is reliable data is New Zealand (Walcott 1984, 1987, 1989; Lamb 1989). The Alpine Fault zone is a transform connecting the Tonga-Kermadec Trench, where the Pacific Plate is being subducted, with the Macquarie Trench where the Australian Plate is subducted. Across the Alpine Fault there is oblique collision of two submerged continental fragments. The strain observed by classical and modern geodetic methods corresponds to a dextral shear zone, with the classical geometry of heterogeneous simple shear (Ramsay and Graham 1970); strain in the walls is not specified, because only in the continental segment is the velocity field measured (Fig. 2.15).Therefore, a viscous model can explain the data, even if the schizosphere adapts to plastosphere movement by rotation along vertical axes by a domino model.

An alternative model is proposed on the basis of geodetic data in a very dense network of GPS stations monitoring the deformation in the period 1994–1998 (Beavan et al. 1999). Most of the plate movement, approximately 85%, is observed in a band 5 km NW to 20 km SE of the Alpine Fault, but significant strain is still observed 60 km NE of the main Alpine Fault. The surface displacements are compatible with interseismic strain accumulation in a fault locked 5–8 km above and dipping SE; 60% of the observed strain is stored as elastic and will be released by a future large earthquake and the remainder is dissipated by viscous flow in the wide deforming zone. The deformation at the surface is compatible but cannot resolve the presence of an active flake, with the SE dipping reverse Alpine Fault being antithetic to the Pacific Plate, subducting to NW, below the Australian Plate, suggested by other data.

For the Dead Sea Plate boundary other models have been proposed (Wdowinski 1998). This is a left-lateral transform between the Red Sea in the south and the convergent Arabian-Asian Plate boundary in the north. The present stress field has a maximum compressive stress at nearly 45° to the transform. It is hypothesised that continental transform spirals around Euler Poles of the plates involved because the rheology is elastoplastic. In the interseismic period of the seismic cycle the movement should be parallel to the transform in the rigid plate model; this does not fit the observed data but is compatible with an elastoplastic rheology. In the rigid plate model, Euler Poles of continental plates have no geological significance: they are mere reference points; but in the deformable plate model, Euler Poles become materialised by spiral transforms.

Viscous and elastoplastic models are variants of continuous media rheology. In a different type of model it is assumed that the presence of the plate boundary modifies the rheology of the deformation zone and non-continuous mechanics should be applied. An example is the San Andreas transform Fault Zone, dextral and connecting the spreading centres of Juan de Fuca in the north and Baja California in the south (Mount and Suppe 1987; Zoback et al. 1987). The fault is considered a weak narrow inclusion in a stronger medium; the maximum compressive stress is at ≈81° of the fault zone in the clockwise sense; there is strong decoupling with strike-slip movement in the main faults and thrust movement in secondary faults subparallel to the main faults. This suggests a décollement below the main fault boundary at the surface; shallow seismicity with hypocentres above 15 km confirm this hypothesis and the deeper fault boundary is displaced to the east side of the Transverse Ranges (Hadley and Kanamori 1977; see Sect. 3.7). In this type of model the mechanical properties of the plate boundary are the main control parameters for deformation, according to the concept of *stress partitioning*. This has been challenged by other models.

The efficiency of fault systems in partitioning the strain is high in California but low in the Great Sumatran Fault, a dextral strike slip that accommodates oblique subduction of the Australian Plate in the Java Sunda Trench. This suggests (Tikoff and Teyssier 1994) that this efficiency is a function of *kinematic partitioning* of the displacement field controlled by relative plate motion, in contrast to previous models of stress partitioning (Fig. 2.16).

Theoretical models can represent a wide range of flow laws from Newtonian with $n=1$ to a purely plastic material as $n \rightarrow \infty$. The average value of n in the lithosphere can explain (Sonder et al. 1986) the relation between width, W and length L of a zone of transcurrent deformation

$$W \approx L / \pi \sqrt{n}$$

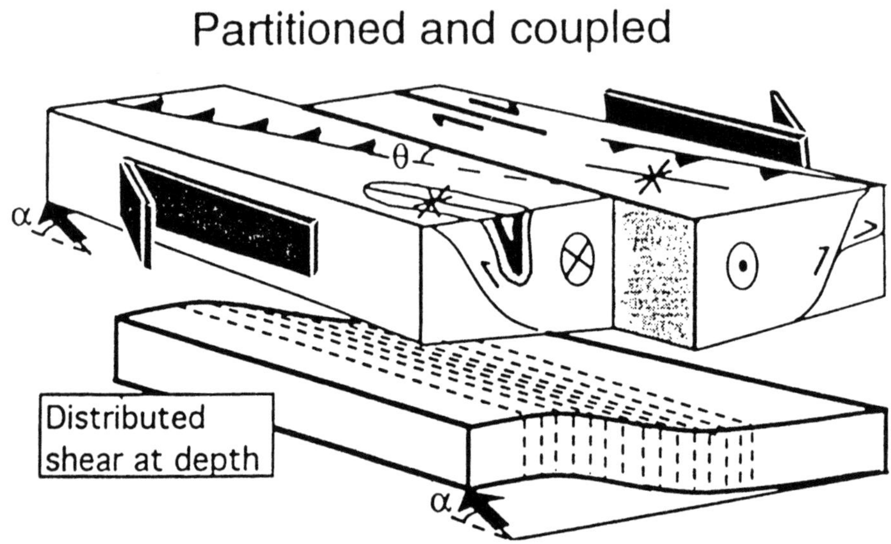

Fig. 2.16. Cartoon of kinematic strain partitioning. Partitioning of discrete slip in the upper crust is controlled by plate motion in a system where the adjacent sides of the main fault are mechanically coupled as a result of distributed shear at depth. (After Tikoff and Teyssier 1994; copyright 1994, with permission from Elsevier Science)

The results for California (England and Molnar 1991) and New Zealand (Walcott 1984) are consistent with an average value of $n=3$.

Another result was obtained when we model the lithosphere as a thin viscous sheet with power-law rheology. The thin *sheet approximation* assumes that stresses and velocities are constant with depth (England and McKenzie 1982). The aspect ratio of the deformed zone, W/L, depends on the tectonic regime. In strike-slip regimes the previous equation is a good approximation. In dip-slip regimes, either compressional or extensional, the approximate solution (England et al. 1985) is

$$W \approx L / \sqrt{n}$$

Therefore, transform plate boundaries are much narrower than convergent or divergent plate boundaries across the continental lithosphere. In all cases the average value for the power-law exponent is $n \approx 3$.

In *convergent boundaries* there is the richest variety of strain regimes (Thatcher 1984; Lambeck 1988). The deformation is more diffuse along ocean/continent convergent zones than in intra-oceanic convergent settings. Geodetic movement can only be carried on the continental side or, rarely, in between the volcanic arc above the trench and islands in the subducting plate.

Japan is the best-studied case, because recent and classical methods are available and big earthquakes have occurred along the plate boundary; but Japan is located in a triple convergent junction and the strain field is complex, for both the horizontal and vertical components. The relationship between plate movement and deformation is much more complex than in simple transform boundaries. The permanent deformation is higher and can account for 20–40% of the plate convergence; the spatial and temporal distribution of strain is also complex during the seismic cycle, with sign reversals in the strain. During the interseismic period of the seismic cycle, the upper plate is being dragged downwards by the subducting plate. Coseismic strain continues with downward movement in the subduction plate but the upper plate rebounds upwards above the trench, and subsides further inland. After coseismic strain in large earthquakes there is propagation of a deformation wave by stress diffusion (Anderson 1975) at a rate of tens of kilometres per year over hundreds of kilometres, showing good evidence for coupling between the schizosphere and the less viscous plastosphere and asthenosphere.

In a recent study in Japan (Shen-Tu and Holt 1995), it has been shown that there is agreement between the direction of the principal strain axes calculated from seismic, neotectonic and geodetic data, but the magnitude of principal strain states from geodetic data is three to five times larger than from seismic or neotectonic data. This suggests that elastic strains transmitted from plate boundaries are responsible for this discrepancy; the agreement between seismic and neotectonic strain rates suggests that a significant percentage of permanent intraplate strain is accumulated by earthquakes. Subduction along the Japan Trench is orthogonal, but in the Nankai Trough there is significant partitioning of oblique subduction of the Philippine Sea Plate under the rigid Eurasian Plate with dextral strike slip across the Median Tectonic Line, accommodating 25% of plate motion velocity. The Sea of Japan moves westward to the Eurasian Plate at a rate of 10±1 mm/year, showing that it is either attached or moving very slowly relative to the North American Plate.

A study in the Andes (Lin et al. 2000) explains the discrepancy between the GPS 30–40 mm/year shortening rate across the deformed belt and the lower tectonic shortening rate below 15 mm/year, concentrated in the sub-Andean thrust belt. Assuming timescale-dependant deformation and a plastic-viscoelastic model for the crust, the instantaneous deformation measured by GPS reflects both permanent deformation and elastic deformation recovered by future earthquakes; the long-term geological rate reflects only the permanent deformation. The latter is concentrated in the sub-Andean thrust belt due to gravitational buoyancy of the high mountain belt. In this model the short-term elastic deformation ($<10^2$ year) implies high-viscous crust (10^{22}–10^{24} PaS). We will reconsider the implications of a viscoelastic lithosphere in Section 3.10.

The examples of New Zealand, Japan and the Andes show that across deformed zones geodetic data sample instantaneous deformation rates, elastic and inelastic; these are above permanent geological strain rates measured by neotectonics; the latter can be above seismic moment release rates if there is significant aseismic deformation.

In continental collision regimes deformation is the most complex due to a multiplicity of controlling factors: width of deformation zone, the largest for all types of boundaries; the absence of plate creation or destruction: the volume is approximately conserved and matter is redistributed by deformation; the complex relationship between relative plate motions and kinematics of the deforming zone, in particular due to gravity effects on the potential energy created by the mountain chain itself (Molnar 1988; England 1992).

The best-studied example is central Asia due to a collision of the Indian with the Eurasian Plates. Rheological models include elastoplastic (Molnar and Tapponier 1975; Tapponier and Molnar 1976, 1977; Tapponier et al. 1982; Cobbold and Davy 1988) and viscous versions (England and McKenzie 1982; Sonder et al. 1986), both numerical and analogous. The first favours localisation of shear in narrow zones by strain softening; the second favours homogeneous deformation. The quantitative parameters that control the behaviour are two non-dimensional members: the Argand number, as we will explain below, measures the relative proportion of gravity and material strength in inhibiting deformation, and the power-law exponent in the dependence of strain rate or stress. This exponent does not represent an individual deformation mechanism (Sonder and England 1986; England and Molnar 1991); it represents the average of deformation by brittleness, with $n \approx \infty$ because stress is independent of strain rate and by creep, with $n \approx 3$. The results of transform boundaries in California and New Zealand, suggesting that $n \approx 3$, explain the relationship between width and length of deforming zones monitored by geodetic methods so deformation is controlled by distributed deformation in the plastosphere and not elastic strain in the schizosphere. However, the brittleness of the schizosphere allows the formation of rotating dominos.

The results of numerical and analogous experiments show a large variation between two end-member tectonic regimes: crustal thickening and lateral expulsion (McKenzie 1990) to the nearest subduction zones; these are in the Pacific for central Asia. Hence, the question is: which model fits the data better?

The amount of rotation can be measured by palaeomagnetism (Kissel and Laj 1989;Molnar and Gipson 1994). It is significant in the extension of the Aegean Sea by dextral transcurrent motion in the North-Anatolian Fault; in the western transverse ranges of California rotation is the superficial response to continuous deformation at depth and is distributed over a broad area. The method cannot be applied to central Asia, because no palaeomagnetic data are available. However, geodetic data confirm the palaeomagnetic data in the Aegean Sea and California; for central Asia the small amount of movement of Shanghai relative to Eurasia, in the order of 8±0.5 mm/year in the direction N 116.5 E (±4.1°) when compared with 50 mm/year of India relative to Eurasia (Molnar and Gipson 1996) favours the thin viscous sheet model with major crustal thickening and minor lateral expulsion.

In tectonically active regions the lithosphere is permanently deformed, over narrow domains in oceans and broad domains in continents. Modelling the lithosphere as a thin viscous sheet above an even more mobile asthenosphere has led to a good approximation of observed structures in nature. In the simple case of a Newtonian fluid the constitutive equation relating stress and strain rate is simply (Ranalli 1995)

$$\sigma'_{ij} = 2\eta\dot{\varepsilon}'ij$$

where σ'_{ij} is the stress tensor, $\dot{\varepsilon}'ij$ is the strain rate tensor and η is the viscosity of the Newtonian fluid; it is a material property, independent of stress, with a negligible variation as a function of pressure but very sensitive to temperature.

The rheology of a power-law fluid is more complex. In a homogeneous and isotropic material the resulting steady-state flow is of the form

$$\dot{\varepsilon}' = A\sigma_s^n \text{ or } \dot{\gamma} = A\sigma_s^n$$

where $\dot{\varepsilon}'_s$ is the tensor shear strain rate and $\dot{\gamma}_s$ its double, A is a constant dependent of pressure, temperature and material parameters and the stress exponent is a constant, $n>1$ for strain-hardening materials, $n=1$ for linear viscous or Newtonian materials and $n<1$ for dilatant materials. In non-linear viscous materials the viscosity is not a material property in the same sense as linear viscosity and must be replaced by *effective viscosity*, specified for a given stress or strain rate, as a function of temperature, pressure and material parameters. The effective viscosity decreases with increasing stress or strain rate and the constitutive equation in tensor form becomes described by an isotropic power-law rheology (England and McKenzie 1982; Ranalli 1995) in terms of invariants of the stress and strain rate deviators as follows.

In tensor calculus we use the invariants of tensors in the establishment of quantitative laws, because invariants are independent of the choice of coordinate systems. For a viscous fluid the flow law relating the deviatory stress tensor to the deviatoric strain rate can be expressed in terms of invariants (Ranalli 1995; p. 74–9). The first invariant of the stress tensor is the mean normal stress; in terms of deviatoric stress it is evidently zero; this expresses the fact that the flow law is not affected by hydrostatic pressure and we can use stress and strain rate deviators. So the flow law can be expressed as a function of the second invariant, which we describe next.

In the thin sheet approximation the components of strain rate and of horizontal velocity are independent of depth (England and McKenzie 1982). It is further assumed that there is a biaxial stress representing one of the three strain rate regimes (Sonder and England 1986).

$$\dot{\varepsilon}_{xx} = -\dot{\varepsilon}_{zz} = \dot{\varepsilon} \qquad \text{(a)}$$

$$\dot{\varepsilon}_{xx} = -\dot{\varepsilon}_{zz} = -\dot{\varepsilon} \qquad \text{(b)}$$

$$\dot{\varepsilon}_{xx} = -\dot{\varepsilon}_{YY} = \dot{\varepsilon} \qquad \text{(c)}$$

In each of these equations the other components of the strain rate tensor are zero because they are independent of depth; $\dot{\varepsilon}$
is positive, x and y are horizontal co-ordinates and z is vertical. Thus Eq. (a) corresponds to horizontal extension, Eq. (b) to horizontal compression and Eq. (c) to vertical strike slip. Therefore

$$\dot{\varepsilon}_{xz} = \dot{\varepsilon}_{zx} = \dot{\varepsilon}_{yz} = \dot{\varepsilon}_{zy} = 0$$

for horizontal compression and the incompressibility of the fluid requires that

$$\dot{\varepsilon}_{zz} = -(\dot{\varepsilon}_{xx} + \dot{\varepsilon}_{YY})$$

The second invariant $\dot{E}'_2$
is

$$\dot{E}'_2 = (\dot{\varepsilon}'_{xx}\dot{\varepsilon}'_{yy} + \dot{\varepsilon}'_{xx}\dot{\varepsilon}'_{zz} + \dot{\varepsilon}'_{yy}\dot{\varepsilon}'_{zz}) + \dot{\varepsilon}'^2_{xy} + \dot{\varepsilon}'^2_{xz} + \dot{\varepsilon}'^2_{yz}$$

and it becomes in this case

$$\dot{E}' = (\dot{\varepsilon}_{ij}\dot{\varepsilon}_{ij})^{1/2} = \dot{E}'_2$$

and the strain rate tensor rather than its deviator may be used. So the isotropic power-law rheology becomes

$$\overline{\tau}_{ij} = B\dot{E}^{1/n}\left(\dot{\varepsilon}_{ij} / \dot{E}\right)$$

where

are the depth-averaged deviatory stresses, $\dot{\varepsilon}_{ij}$ is the strain rate with $\overline{\tau}_{ij}$

$$\dot{E} = (\dot{\varepsilon}_{ij}\dot{\varepsilon}_{ij})^{1/2}$$

where $\dot{E}$ is the second invariant of the strain tensor in a state of biaxial stress and B is a constant; the stress power exponent is usually $n \approx 3$.

The effective viscosity is then

$$\tfrac{1}{2} B\dot{E}'(\tfrac{1}{n} {}^{-1})$$

and it is related to the Argand number, A_r, which is (England and McKenzie 1982)

$$A_r = \frac{g\rho_c L(1 - \rho_c/\rho_m)}{\beta(\mu_0/L)^{1/n}} = \frac{P(L)}{\tau(\dot{\varepsilon}_0)}$$

where $P(L)$ is an estimate of excess pressure arising from a crustal thickness contrast of densities ρ_c, ρ_m; $\tau(\dot{\varepsilon}_0)$ is the stress required to deform the medium at a strain rate characteristic of the system, ($\dot{\varepsilon}_0 = \mu_0/L$)

If A_r is small, the effective viscosity of the medium is large at the imposed strain rate at the boundaries and the flow will be governed by the boundary conditions. Conversely, if A_r is very large the effective viscosity of the medium will not support the forces arising from elevation contrasts due to crustal thickness variations, tending to a state of plane strain.

The thin viscous sheet model for lithosphere was used recently (England and Molnar 1997) to estimate the rheology of the lithosphere in the region of the Tibetan Plateau. The underlying idea is that gradients in stress are balanced by the gravitational body force. The estimation of strain rates from seismic and neotectonic methods allows us to infer variations in viscous stress, assuming a power-law fluid rheology; the stress gradients point uphill toward the centre of Tibet, where gravitational potential energy per unit surface area reaches a maximum. This observation validates the basic hypothesis and allows the estimation of effective viscosity and vertically average deviatory stress for a possible range of strain rates, with the constant B determined from the data and with n=3. The actual values vary from 10^{14} s^{-1}, 1.1×10^{21} Pa s and 22 MPa to 10^{17} s^{-1}, 1.1×10^{23} and 2 MPa. For Tibet the average viscosity is about 10–10^2 times greater than the viscosity of the convecting upper mantle.

A similar study was conducted for the Aegean region (Davies et al. 1997). Geodesic strain was measured in the interval 1892–1992. Strain distribution is compatible with that expected of a sheet of fluid moving towards a low-pressure boundary at the Hellenic Trench relative to stable Eurasia. Strain calculated from seismic moment release is only 20–50% of the geodetic strain. Using a power-law fluid rheology an effective viscosity of ~10^{22}–10^{23} Pa s is estimated for the lithosphere in the studied region, similar to the results for the lithosphere under Tibet. Assuming a slightly above average heat flow, which is consistent with the tectonic setting of the Aegean extensional region behind the Hellenic Trench, it is concluded that the deformation of the lithosphere obeys a rheological law similar to that obtained for olivine on the basis of experimental results in the laboratory. Another study uses the velocity field by GPS repeated measurements in the 1988–1996 period (Kahle et al. 1998). There is a strong right-lateral strike-slip rate to the west of Anatolia across the North Anatolia Fault zone and distributed across the area to the west of it. There is also a smaller N–S extension strain rate behind the Hellenic Trench, with moderate arc-parallel extension and strong compression perpendicular to it. More recent studies (McClusky et al. 2000) have refined these results. This kinematic model is in agreement with restoration in the recent past (Angelier et al. 1982). These studies confirm the view that differentiated topography leads to reduction of lithosphere strength (Burov and Diament 1995).

A review of deformation at plate boundaries suggests a variety of partitioning models, such as strain partitioning and stress partitioning by reduction of shear strength due to high fluid pressures, valid in different situations. Coupling would be the rule and the kinematics would be controlled by plate movements and not by a particular rheology along the plate boundary itself. Decoupling would be only possible along favourable weak horizons across the rheological stratification of the entire lithosphere; a possibility would be the ductile lower continental crust in zones with a favourable geothermal gradient and imposed strain rate.

The study of active (Gordon 1995) and inactive deformed belts shows that strain partitioning follows some general rules. For a mean value of the stress exponent ($n \approx 3$), for power-law rheology, the width of the deforming zone is much larger for the across-strike motion of bounding plates. This can explain the partitioning in a concentrated zone of strike-slip faulting and a wide zone of dip-slip component; in fact, the former component causes no thickening or thinning of the lithosphere, contrary to the latter, which generates buoyancy forces that make further thickening or thinning more difficult. It has also been suggested (Lin et al. 1998) that simple shear is a fabric-weakening process because grain boundaries are attracted to shear plane orientation making grain boundary mechanisms easier, either diffusion-dependent or frictional; pure shear aligns grain boundaries perpendicular to the shortening direction and grain boundary sliding must be taken over by stronger intragranular mechanisms; this leads to localisation of a simple shear boundary-parallel component inside wider zones of boundary-normal component pure shear.

Another factor controls the strain partitioning in the upper crust (Royden 1996). A strong lower crust produces coupling of crust and mantle and stresses are transmitted vertically through the crust. A weak lower crust can exist initially or be developed during crustal thickening leading to decoupling of the upper crust relative to the mantle and stress is transmitted horizontally through the crust.

This can explain a variety of tectonic styles in recent and old orogens (see Sect. 4.10).

Global Tectonics with Deformable Plates

3.1
Introduction

In the previous chapter we analysed the limitations of rigid plate tectonics or standard theory. In this chapter, we will try to prove that these limitations can be overcome if we relax the postulate of plate rigidity for oceanic lithosphere due to hydrothermal seafloor metamorphism. The deformation mechanisms responsible for the strain regime in oceanic lithosphere are revised by A. Mateus in the section 3.12. We will analyse the evidence for aseismic strain in oceanic lithosphere, both by boudinage and homogeneous shortening. We will review the evidence for buckling of both oceanic and continental lithosphere.

We will analyse the evidence from seismic strain both in the generation process by viscoelastic rather than elastic rebound and the significance of fractal dimensions in seismicity and faulting.

We will analyse the methods for estimation of finite strain from finite kinematics, and we will test intraplate strain and paths. We conclude with a distinct view between rigid and soft plate tectonics concerning rheology of oceanic versus continental lithosphere and its implications on the driving mechanism.

3.2
Boudinage of Oceanic Lithosphere: Incremental Aseismic Strain

In the last 20 years global marine geoid data have been obtained by high precision satellite techniques, using sea level as an equipotential surface that images the geoid (for review see Cazenave and Feigl 1994; Sandwell and Smith 1997; Smith and Sandwell 1997).

In the central Pacific regular lineations at the scale of thousands of kilometres with an amplitude of 10–20 cm and wavelength of approximately 200 km were discovered, as well as larger scale lineations with amplitudes of 40–50 cm and wavelengths of 1000–2000 km; “two models have been proposed for these features: small scale convection organised into rolls by shear of the Pacific Plate and lithosphere boudinage resulting from N–S extension of the Pacific Plate” (Dunbar and Sandwell 1988, p. 1429).

The release of previously classified data showed that these lineations are not of local significance and restricted to the Pacific superswell, but should be generalised to the entire oceanic lithosphere (Sandwell and Smith 1995).

A detailed study of gravity lineations was made (Sandwell et al. 1995) in the Pukapuka ridges of the Pacific. They extend for 2600 km north-west of the East Pacific Rise flank to the 40-Ma-old seafloor. These features are a good example of linear volcanic ridges associated with gravity lineations. They are elongated in the direction of present plate motion, forming a band 50–70 km wide in the trough of the more prominent gravity lineations.

The origin of the ridges has been explained by three mechanisms.

One is the action of mini hotspots that move slowly with respect to major hotspots and produce intermittent volcanism (Fleitout and Moriceau 1992). The authors exclude this mechanism, because radiometric dates of the largest ridges show no hotspot age progression incompatible with hotspots models, as we will mention in more detail below (Lynch 1999).

Another is the regime of small-scale convective rolls aligned in the direction of absolute plate motion by shear in the asthenosphere. The authors exclude this mechanism by using a convincing argument: "the occurrence of the ridges in the trough of the gravity lineations is incompatible with the small-scale convection model which would predict increased volcanism above the convective upwelling" (Sandwell et al. 1995, p. 15087; Fig. 5, p.15096; also Fig. 3.1, this Chap.).

The third mechanism is diffuse extension of the lithosphere by boudinage, resulting in lineated zones of extension. The authors favour this mechanism, being more consistent with all of their observations. The stress field that causes this mechanism is due to the geodynamics regime of the Pacific Plate, surrounded by subduction zones on the western side; "if the ridge-push force is much smaller than trench-pull force then near the ridge axis the direction of maximum tensile stress must be perpendicular to the direction of absolute plate motion" (Sandwell et al. 1995, p. 15087).

We propose an alternative explanation for this stress field, which we consider more plausible than the one advanced by the authors. In fact, gravity lineations are also observed in Atlantic-type oceans that are surrounded by passive margins instead of subduction zones as in Pacific-type oceans (Cazenave and Feigl 1994, p. 56); it is probable that they are present in the oceanic lithosphere. Usually, they do not extend to mid-ocean ridges, but they develop rapidly on young seafloor (~3 Ma on the Pacific) and become well developed in the older lithosphere (~6 Ma in the Pacific). These facts prove that ridge push must play a role in the genesis of gravity lineations. Knowing stress field in oceanic lithosphere and postulating realistic strain rates we can predict strain regimes (Ribeiro 1997).

In the oceanic domain we should expect very low strain rates for intraplate deforming because the Wilson cycle is much longer than the episodic orogenic period at convergent boundaries; we should also expect small viscosity contrasts in sediments and hydrothermally altered oceanic crust and mantle in abyssal plains (Ribeiro 1993). We know the state of stress in the oceanic lithosphere (Zoback 1992) at mid-oceanic ridges and young adjacent lithosphere is extensional, but everywhere else is in compression subperpendicular to the ridge axis. If the oceanic lithosphere is deformable or soft, as we claim, we can predict strain regimes if we look at the behaviour of geologic materials under compression, which is well known (Ramsay and Huber 1987;

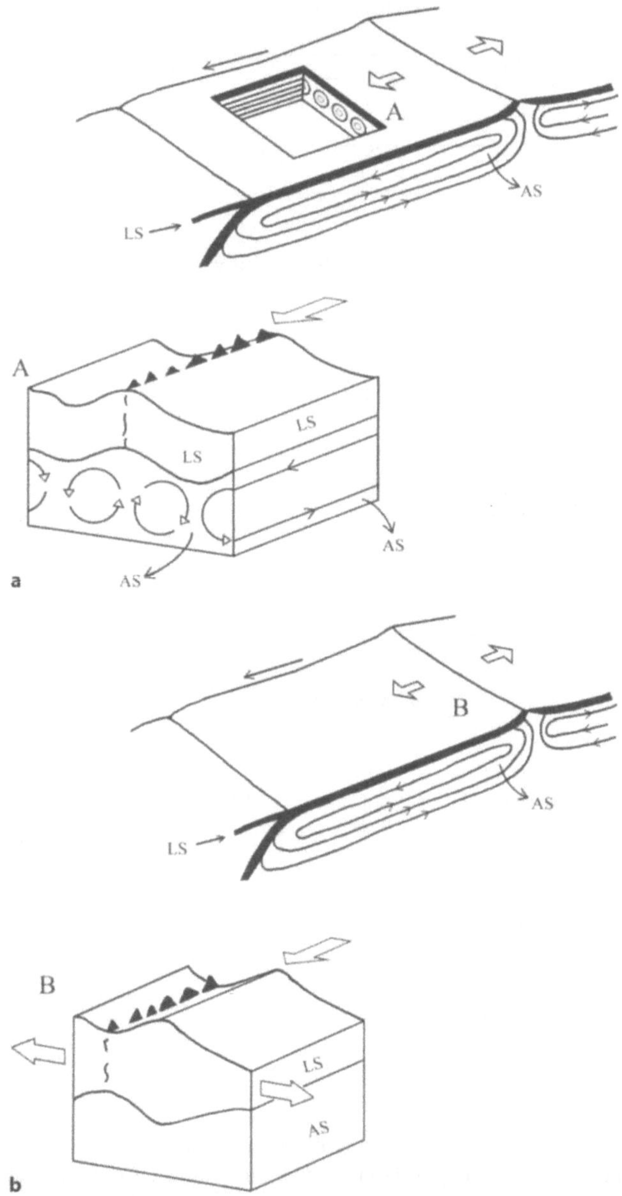

Fig. 3.1. Model for geoid lineations. **a** Secondary convection cells; linear volcanic chain must be above gravity and topography highs on top of upwelling currents, **b** incipient boudinage; linear volcanic chain must be above gravity and topography lows in areas of highest extensional strain. *AS* Asthenosphere, *LS* lithosphere. (Modified after Sandwell et al. 1995)

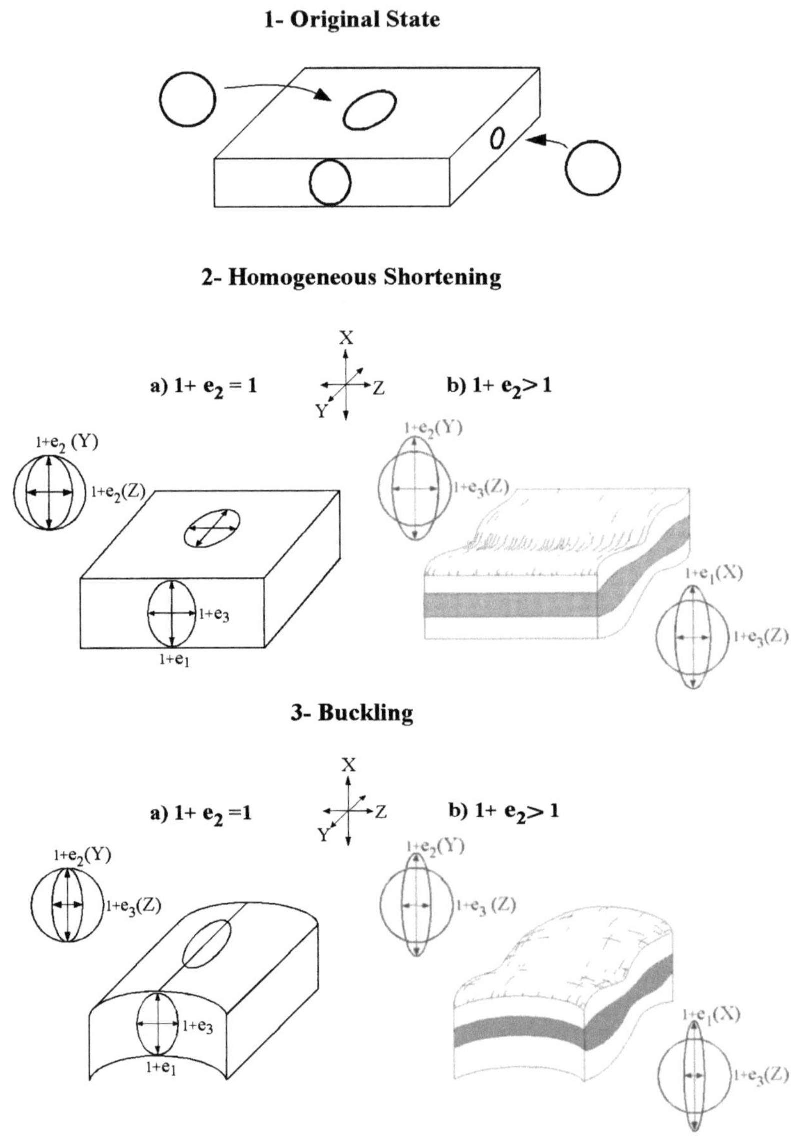

Fig. 3.2. Stages of deformation of layered geological materials. *1* Original state, with reference circle. *2* Homogeneous shortening or layer parallel shortening. *a* Constant volume – the intermediate strain axis, *Y*, preserves the original length, $1+e_2=1$. *b* Volume loss – the intermediate strain axes, *Y*, suffers elongation, $1+e_2>1$, with incipient boudinage – pinch and swell parallel to shortening direction, *Z*. *3* Generation of an instability by buckling. *a* Constant volume – buckle develop with axis of folding parallel to the intermediate axis, *Y*, of constant length and axial plane perpendicular to shortening, *Z*. *b* Volume loss – buckles develop with axes of folding parallel to the intermediate axis, and axial plane perpendicular to shortening, *Z*, and boudinage parallel to shortening direction, $1+e_2>1$. Orientation of strain ellipsoid derived from original sphere: *X* major axis. *Y* intermediate axis. *Z* minor axis. Principal longitudinal strains: $1+e_1$ major; $1+e_2$ intermediate; $1+e_3$ minor

Price and Cosgrove 1990; Ghosh 1993) and the first stage of deformation can be recognised (Fig. 3.2). The material starts to deform by homogeneous shortening or layer parallel shortening with the major axis of strain ellipsoid, X, subvertical; the minor axis, Z, parallel to compression; along the intermediate axis, Y, one can have no deformation (Y=R, the radius of the original sphere) or a slight extension (Y>R), with boudinage at right angles to Y. In oceanic lithosphere we should expect this slight extension because it cools as it moves away from the mid-ocean ridge. The amount of extension will be variable, maximum at fast cooling domains and minimum at slow cooling domains.

In this way, the first response of the oceanic lithosphere to ridge push must be boudinage at right angles to maximum compressive stress. Most materials develop extensional instabilities like boudinage before compressional instabilities like folds, simply because they are stronger in compression than in extension; in the next section we will prove that boudinage is an inevitable consequence of homogeneous shortening of the oceanic lithosphere. This interpretation is reinforced also by theoretical arguments and a general analysis of gravity lineation data.

The interpretation by the small mode of convection has been rejected by more detailed analyses (Davies and Richards 1992; Davies 1999). In fact, one should expect from the fluid dynamics theory and experimentation on rotating Couette flow that Taylor and wavy vortices require a viscous rheology component for the oceanic lithosphere (Tritton 1988). In the Pacific Ocean there was a change in spreading direction 25 Ma ago and the cross-grain lineations are parallel to the present movement of the Pacific Plate and oblique to transform direction older than 25 Ma; however, the lineations parallel to present movement are visible even in the older lithosphere, oblique to old transforms. Maximum compressive stress, inferred from focal mechanisms of earthquakes is parallel to the present direction of movement; hence, it is shown that the elastic memory of the older stress field was erased by younger viscous flow, suggesting a viscoelastic rheology; in this case, we will define a Maxwell relaxation time in Section 3.10, below which the elastic response predominate and, above which the viscous response becomes predominant; Maxwell time is below 25 Ma. Numerical modelling has confirmed the pinch and swell origin for the cross-grain lineations discovered in Seasat data (Dunbar and Sandwell 1988).

The incipient boudinage origin for the geoid lineations is further strengthened by careful analysis of the recent data (Sandwell and Smith 1995 www). In the eastern Pacific when the subduction zone is approached, the incipient pinch-and-well structure can lead to complete boudinage of the young lithosphere. This is happening in the Mendana Fracture Zone (Huchon and Bourgois 1990), which has been opening since 3.5 Ma as a wedge propagating from the Peru-Chile trench in the NE side to the East Pacific ridge in the SW side, fragmenting the NAZCA Plate off Peru (Fig. 3.3). In addition, the geoid lineations are strictly parallel to the Mendana Fracture Zone and perpendicular to the Peru-Chile subduction zone, showing that these geoid lineations are secondary extension structures, parallel to maximum compressive stress in the NAZCA Plate. A more evolved stage of this evolution is represented by the Galapagos Rift, where a former and larger East Pacific Plate was fragmented in the NAZCA and COCOS Plates. Again, the geoid lineations are subparallel to the Galapagos rift trend, confirming their origin by complete boudinage of the oceanic lithosphere. Hotspot activity also favoured this evolution.

Fig. 3.3. Mendana Fracture Zone. Magnetic anomalies numbered after standard nomenclature. Inside the Mendana Fracture Zone that displaced older magnetic anomalies ocean crust has formed since 3.5 Ma by propagation from the Peru Chile Trench to the East-Pacific Rise. *Insert* Location of Mendana Fracture Zone in the plate system of the South-East Pacific. (After Huchon and Bourgois 1990; copyright by the American Geophysical Union)

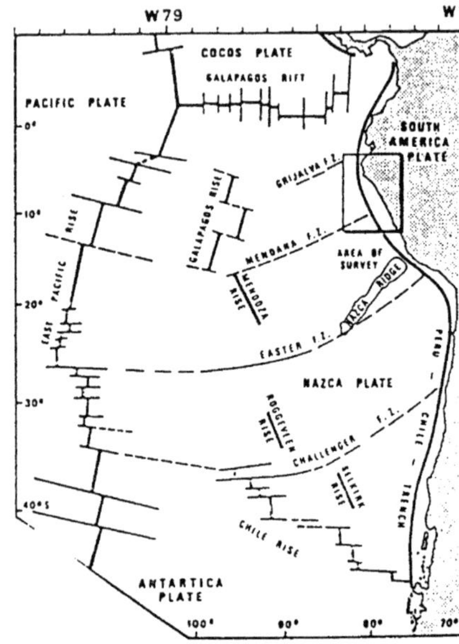

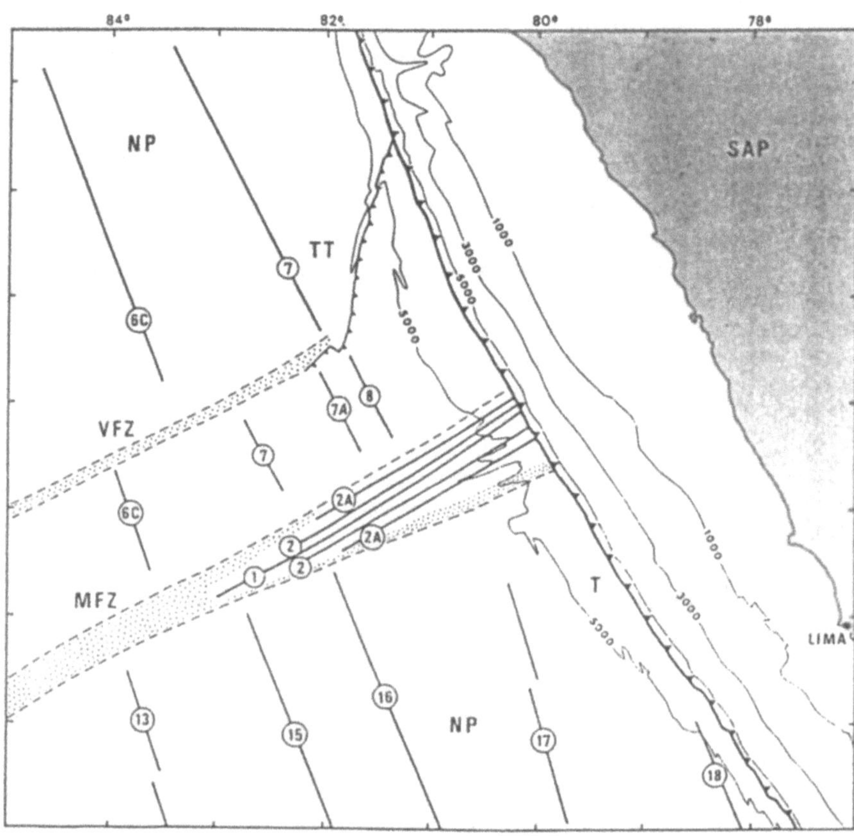

A situation similar to the East Pacific occurs at the intersection of the Louisville seamount chain with the Tonga-Kermadec trench. A large seamount obstructs subduction and induces extensional failure in the subducting plate for 800 km eastward from the intersection between both structural features (Small and Abbott 1998). Again, this is a case of secondary traction in a compressive stress field near the trench; if, however, in the East Pacific the fractured lithosphere is young, less than 40 Ma old, in the West Pacific the fractured lithosphere is older, at least 95 Ma.

In the Western Pacific the geoid lineation pattern becomes fuzzier as the subduction zones are approached. This suggests that slab-pull stretches the subducting plate; thus, the major axis of incremental strain ellipsoid, X, is perpendicular to plate movement given by transform direction near mid-ocean ridges, but becomes parallel to plate movement near subduction zones. This is also confirmed by spatial geodetic methods as we will see in Section 3.3.3.

The mechanism of boudinage of the oceanic lithosphere can be confirmed by geologic observations, theoretical and experimental modelling. Natural occurring boudins at a microscopic scale show that incipient boudinage is expressed by pinch-and-swell structure (Ramsay 1967; Gosh 1993); the profiles of these structures fit the morphology of the geoid lineations described at the plate scale. The mechanics of development of these structures has been analysed by Smith (1975, 1977) and summarised by Gosh (1993). They concluded that well-developed pinch-and-swell structures cannot form in Newtonian materials. The theory also predicts that the ratio of dominant wavelength to thickness of the layer is around four. If we apply this relationship to the geoid data we can perhaps explain the two orders of wavelengths observed in the Pacific data. The shortest wavelength of 200 km suggests a thickness of the competent layer in the order of 50 km. This is compatible with the thickness of mature oceanic schizosphere. The largest wavelength of 1000–2000 km will require a thickness for the boudinaged layer of 250–500 km. The top of low velocity zones (LVZ) in oceans changes from 80–220 km in young domains to 150–400 km in domains aged 135 Ma (Ranalli 1995, p. 138). Therefore, the long wavelength geoid lineations could be due to boudinage of the entire seismic lithosphere above the LVZ.

Another important aspect on the origin of geoid lineations is the mechanism of erasing the old geoid lineations by new geoid lineations when these are not coaxial. This is observed in the Pacific as we stated above. If the lineations were produced by extension we should expect that the old extension structures would be preserved, if the younger stress field were also of a tensional character. However, if, as we claim, the lineations are the result of secondary traction (Tapponier and Molnar 1976), induced by primary compression, the new structures can erase the old ones (Fig. 3.4). In fact, the old structures will be attenuated in the cooling oceanic lithosphere because it is subsiding (Fig. 3.5), and the new orientation of maximum compressive stress will control the orientation of the new geoid lineations parallel to it.

In summary, the pinch-and-swell origin for the geoid lineations suggests that the oceanic plastosphere has a non-Newtonian rheology and that the average Maxwell time for viscoelastic oceanic lithosphere would be below 25 Ma.

Geoid lineations generated by incipient boudinage of oceanic lithosphere are particularly well expressed in the South Pacific Superswell (McNutt 1998); we must analyse the relationship with other geodynamic features of this domain in order to see what conclusions can be drawn at regional and, possibly, at a more global level.

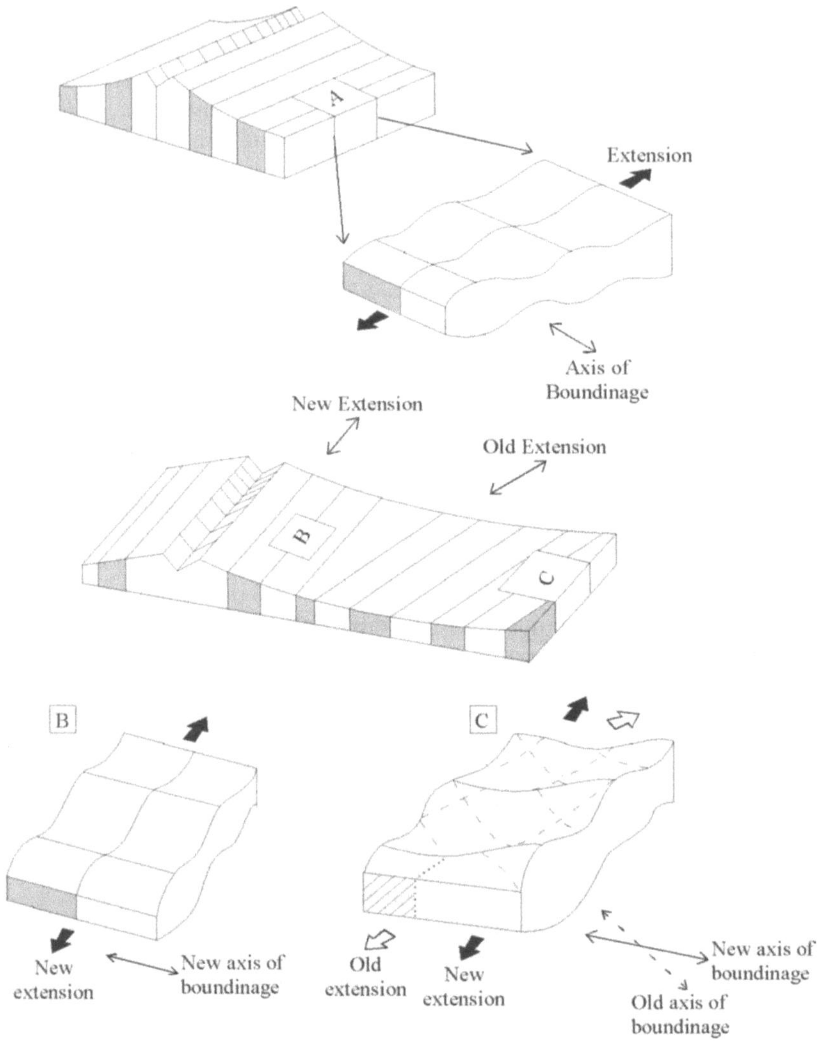

Fig. 3.4. Superposition of old geoid lineation system by a new one; the old extension and old axis of boudinage, orthogonal and the new extension and the new axis of boudinage are shown; the extension is, in both cases, orthogonal to boudinage axis

The superswell is characterised by abnormally shallow topography, negative geoid anomaly, weak flexual rigidity (see Sect. 3.4) and low seismic surface wave velocity, suggesting mantle upwelling (Searle et al. 1995). We observe here the so-called linear ridge groups (LRG) of volcanic origin such as the Pukapuka, Crossgrain and West of Easter Plate (Lynch 1999); these are a new manifestation of intraplate magmatism different from hotspot tracks, but related to them, as we will see below. The LRG propa-

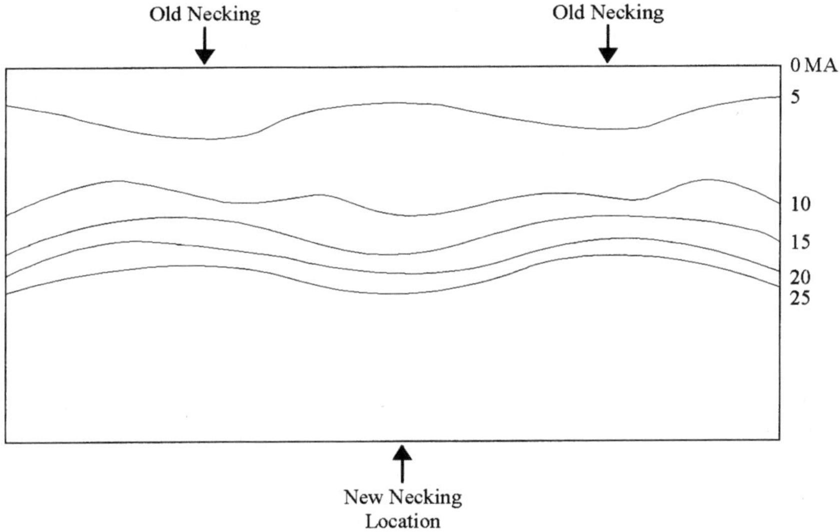

Fig. 3.5. Erasing of old geoid lineation and imprinting of new geoid lineation. The location of necking is different in the old and in the new tectonic setting; the orientation is also different and must be seen at the third dimension

gate from west to east and form on the seafloor between 24 and 1 Ma for Pukapuka and 80–60 Ma for Crossgrain ridges. There is evidence of age propagation for individual ridges on Pukapuka between 27.5±0.4 to 71±0.4 (Lynch 1999) in 2600 km along the ridge. This is inconsistent with the mini hotspot model, because it would be a very fast propagation rate when compared with the Pacific Plate motion above an almost stationary hotspot such as the classic example of the Hawaii-Emperor chain. Nevertheless, the ridge must be genetically related to hotspots because there is a transition from linear ridge to hotspot seamount morphology at the western tip of both the Pukapuka and Crossgrain linear ridges. Magmatism on the eastern tip of linear ridges is of a tholeiitic and alkalic nature (Macdougall et al. 1994) and not of a hotspot nature; this suggests partial melting of the lithosphere with more extensive fractionation in long conduits on colder, thicker lithosphere to the west, where the alkalic volcanic rocks occur. These relationships favour a model of thinning of oceanic lithosphere by heat weakening, related to a hotspot that fed the earlier ridges, but the speed of fracture propagation outpaced that of the hotspot by a self-sustaining mechanism (Lynch 1999). The thinning is reflected by a drastic decrease in effective elastic thikkness (Goodwillie 1995) and implies that thinning of the lithosphere and magma generation are genetically related. Nevertheless, hotspot tracks are not parallel to linear volcanic ridges of the same age (Lynch 1999). It has been suggested (Gordon 2000a) that there were significant movements between hotspots in different oceans and we must conclude that LRG do not record absolute plate motion. We must then ask what is their tectonic control, both at the scale of the entire ridge group and of each ridge element.

The tectonic control for the ridge group can be explained by the global stress field inside the Pacific Plate. The alignment of the ridge group is parallel to the direction of relative plate movement; this is generally the direction of compressive stress, as seen in Section 2.5. The ridge is then due to propagation on a flow perpendicular to the direction of least compressive stress, according to fracture mechanics (Atkinson 1987); this explains the length/width ratio of the linear ridges 2600/60 km for Pukapuka and 1300/75 km for Crossgrain (Lynch 1999). This also explains the situation of Pukapuka at a geoid low, by secondary traction of the Pacific Plate. We must note, however, that Pukapuka, at its eastern tip, formed in a shear floor less than 1 Ma old, given the age of the volcanic ridge 7.1±0.4 Ma and of the seafloor around 7.7 Ma. This means that the crack crossed the boundary between compression parallel to relative plate motion to collinear extension near the ridge at 20–10 Ma, as seen in Section 2.5.

The LRG west of Easter Plate (Searle et al. 1995) coincide with a geoid high, contrary to Pukapuka; its direction is at right angles to older NNW magnetic anomalies, but oblique to younger NNE magnetic anomalies. In this domain, the seafloor fabric oscillates between both directions, suggesting the presence of unstable plate boundaries, probably associated with a succession of propagating rifts of microplates in the last 24 Ma (Searle et al. 1995). It is possible in very young lithosphere, less than 10–20 Ma old, that LRG can develop in contrasting situations because there are highly variable thermal stress fields, as referred to in Section 2.5. Extension parallel to mid-ocean ridges will produce LRG in geoid lows, such as Pukapuka; compression parallel to the mid-ocean will produce LRG in the crest of flexures, perpendicular to the mid-ocean ridge and in geoid highs, such as west of Easter Plate.

The tectonic control of each ridge element is dominated by the stress field at a local level. The individual ridges can show variable arrangements inside each band forming the ridge group, from en echelon to straight cracks; the arrangement depends on the relation between crack-induced stress over remote stresses during propagation; a curving path of overlapping en echelon cracks indicates predominance of local crack induced stresses and straight crack paths indicate predominance of remote compressive crack-parallel differential stress (Olson and Pollard 1989). En echelon single ridges reflect self-organisation, resulting from mechanical interaction of randomly located flaws inside the band, according to numerical experiments (Fletcher and Pollard 1999).

In the continental lithosphere the only evidence for the occurrence of boudinage comes from the interpretation of lineaments as tensile fractures (Nur 1982). The extensional deformation associated with these structures is very low.

3.3 Homogeneous Shortening of Oceanic Lithosphere

3.3.1 Introduction

In continuous media there is a simpler relationship between stress and incremental strain than between stress and finite strain (see, for instance, Ramsay 1967). Therefore, we must look at the state of stress in the oceanic lithosphere; if some symmetry is present we can apply the symmetry principle (Turner and Weiss 1963) to learn something about the symmetry of strain.

If we admit that the oceanic lithosphere can have some viscous component on its long-scale rheological behaviour, we should first consider which stress field acts on it. In Section 2.5 we concluded that maximum horizontal stresses are nearly parallel to plate motion vectors: in young oceanic crust, with ages below 20–30 Ma, the state of stress is mainly extensional and in older oceanic crust, above that age limit, it is compressional.

The resultant stress field is approximately axisymmetric. In a divergent plate boundary, one of the horizontal principal stresses, extensional near the ridge, and compressional further away, is in the direction of parallels around the Eulerian pole of rotation and the other is in the direction of the meridians (Ribeiro 1987).

In continuous media strains there are functions of displacement gradients. In rotation plates angular strains must result from angular velocities. In rigid plate kinematics plate velocities are constant throughout the plate. Therefore, in classic global tectonics the interiors of plates are considered to be rigid on geological time scales: sediments remain horizontal; transform faults remain linear and with constant separation; magnetic lineaments remain linear in oceanic domains (Turcotte and Schubert 1982). The kinematics of rigid plates were considered to have a unique property: great circles perpendicular to transform plates intersect near the Eulerian pole; and spreading rates along the same plate boundary vary as the sinus of the angular distance to the Eulerian pole. This was considered to be a test for rigidity of plates (Le Pichon et al. 1973).

In fact, this is only a trivial solution in plate kinematics: if there is no displacement gradient, there is no strain and the plate is rigid. However, we must look for non-trivial solutions for plate kinematics.

3.3.2 Kinematics and Strain of Deformable Plates in Slow-Spreading Atlantic-Type Oceans

If we admit that oceanic plates are deformable, the deformation field must somehow reflect this axisymmetric pattern of the stress field.

Neglecting the slight deformation due to lithosphere cooling that explains the well-known correlation between depth of the ocean floor and square root of its age that we will refer to in Section 3.3.4, we can try to constrain the deformation regime at the surface of a perfectly spherical earth by searching for a solution that fulfils three conditions:

The first condition states that deformation must be zero at the mid-ocean ridge, implying that in some reference great circle the graticule of spherical co-ordinates project as a square where a reference circle can be inscribed.

The second condition states that transform faults must remain parallel in oceanic domain of the same age, imposing that they must project as parallel straight lines.

The third condition states that the axisymmetry of the deformation must be a consequence of the axisymmetry of the velocity field inside the plate. In rigid plate rotation the vector representing the infinitesimal angular displacement is constant through the entire plate. In non-rigid plate rotation, this infinitesimal angular displacement is the same for great circles around the Euler pole, but varies in the small circles at constant angular distance to the Euler pole. Therefore, magnetic anomalies of the same age will suffer the same homogeneous shortening on the same angular acceleration. Concerning the third dimension, the depth of the plate shell, the boundaries of magnetic stripes with opposed polarity are approximately vertical in the block model (Lowrie 1997) and should remain vertical after deformation; otherwise the original skewness of magnetic anomalies would be affected, even though there are no studies on this topic.

These three conditions ensure that all deformation is in the plane: the plate does not suffer torsion along horizontal and vertical axes: the plate does not lose coherence.

This model can be tested in the case of oblique spreading, where plate movement is parallel to the trend of transform faults and fracture zones, but the spreading direction can make an angle less than 90° and sometimes even less than 45° (Searle 1986) with plate movement. Progressive deformation leads to shortening parallel to the plate movement and the original angle between magnetic anomalies and fracture zones increases as deformation proceeds; this must lead to a systematic increase in this angle as we move into older segments of the spreading ocean. In fact, this is observed in the Atlantic off West Africa between the ridge and the continent. The original trend of magnetic anomalies is preserved near the ocean-continent boundary, highly oblique to fracture zones, and it is due to progressive propagation to SW of the initial rift; this original trend is subparallel to the present trend of magnetic anomalies at the ridge and oblique to transform faults. These geometric relationships exclude horizontal shear which would change the dip of magnetic stripes, but would preserve the original angular relationships observed at the surface of the Earth (Fig. 3.6).

Structural geologists use orthographic projection to estimate strain in geological structures (De Paor 1983, 1986). "Thus, using the orthonet we may measure angles on the deformed objects as if they were in their initial pristine condition. Other nets cannot be used in this fashion because they distort such angles. Only the orthonet conserves vector and tensor operations, and homogeneous strain is one such operation" (De Paor 1986, p. 99). In other words, the orthonet can be used in tensor manipulations such as in the estimation of strain because the great circles of the orthographic projection are ellipses with any axial ratio from 1, a circle, to ∞, a line, as a function of their orientation relative to the plane of projection.

A modified transverse orthographic projection obeys the three conditions referred to above, with the pole around the Eulerian pole of rotation and the mid-ocean ridge projecting as a central reference meridian; in this case, small circles project as straight lines. In the equator of the projection the Tissot Indicatrix in the central reference meridian is a circle, but changes to an ellipse as we approach the Euler pole. However, we can modify the orthographic projection and obtain an orthographic abacus where we trace a circle along the central reference meridian. This was made using the program STEGRAPH (Mendes and Kullberg 1992). We imply that transforms are small circles around the Eulerian pole and they must remain straight during a finite rotation, around the Eulerian pole. The deformation will be the one corresponding to the Tissot Indicatrix for the modified orthographic projection (Fig. 3.7). In fact, if the ellipticity was less than that of the Indicatrix, transforms will converge and if ellipticity

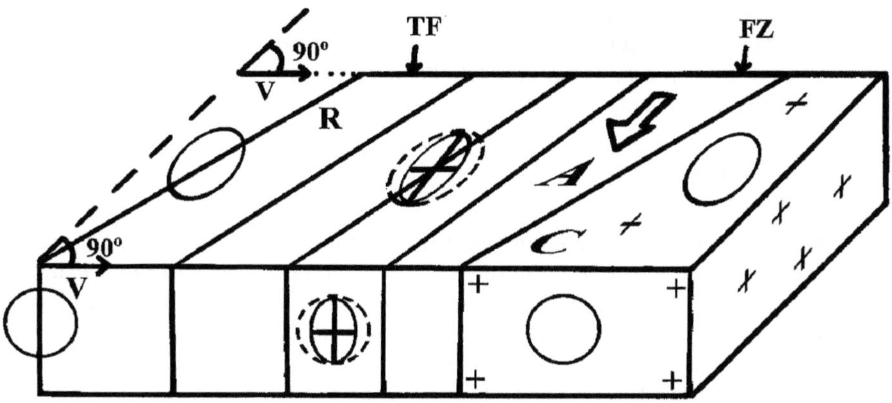

Fig. 3.6. Deformation by time-dependent homogeneous shortening in an ocean with oblique spreading; this is a schematic model for West Africa between the undeformed continent, *C*, and the present Mid-Atlantic Ridge, *R*. plate movement vector, *V*, is parallel to transform faults, *TF*, and fracture zones, *FZ*, and oblique to the trend of magnetic anomalies. With progressive deformation the angle between the magnetic anomalies and TF or FZ increases and approaches 90° because the trend of magnetic anomalies rotates away from the minor axes of strain ellipse at the seafloor; this minor axis is parallel to *V*. The original trend of magnetic anomalies is preserved near the ocean-continent boundary, subparallel to the present trend of magnetic anomalies at the ridge. In the older segment of the ocean, there was propagation of the initial rift in the sense of the arrow *A*

Fig. 3.7. Orthographic tectonic plate deformation abacus, axisymmetric pattern of deformation drawn with STEGRAPH program. (Mendes and Kullberg 1992)

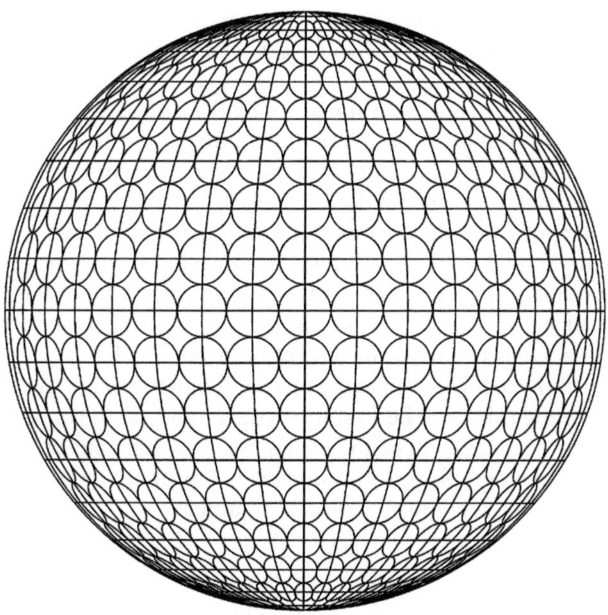

was more than that of the Indicatrix, transforms will diverge. In that case, the axial ratio, R, of a strain ellipse at the surface of the Earth will be only a function of the finite rotation, ω around that Eulerian pole (De Paor 1983), with $1+e_1=e_2<1$, or the strain ratio will be the inverse of the cosine of the finite rotation angle from the ridge crest

$$R=\frac{1+e_1}{1+e_2}==\left(\cos\omega\right)^{-1}$$

We must note that the orthographic projection is not equivalent, because the area is not preserved, or conform, because the angles are altered. In fact, in the oceanic lithosphere, area must not be preserved during deformation because it cools as it moves away from the ridge, as we will consider in Section 3.3.4. Therefore, the orthographic projection is adequate to study deformation of the cooling oceanic lithosphere.

Concerning the fact that orthographic projection is not conform, the following reasoning applies. Rigid rotation of moving plates can be described by oblique Mercator projection around the Euler pole of the two plates involved (Le Pichon et al. 1973); the transforms become parallels in this projection and the magnetic lineations become meridians, if spreading is orthogonal to transforms. This orthogonal relationship is preserved if the Euler pole remains stable. Now, if we orthographically project this oblique Mercator map, we preserve the axisymmetry of the deformation process; the major axis of strain ellipse stays perpendicular to the parallels and the minor axis of strain ellipse stays in the direction of the parallels in the modified oblique Mercator projection. This is due to the fact that the rigid motion of plates has a cylindrical symmetry around the rotation axis of the plates involved; and if the plates deform during motion the symmetry is lower: the circular section of the cylinder becomes an elliptic section due to symmetry breaking by deformation. Therefore, magnetic lineations should stay perpendicular to the transform if they were perpendicular in the original state; the angles are modified by the deformation process,, but not by the projection of the spherical earth surface in the plane of the oblique Mercator projection (Fig. 3.8).

The non-trivial solution given by the orthographic projection must be confirmed by analytical methods. We will use polar co-ordinates due to Earth sphericity.

Consider Fig. 3.9 that represents a cut through rotating spherical plates. The Euler pole is the centre of the circle; angular distances are measured from the ridge. Therefore, this cut can represent what happens at any angular distance from the Euler pole, from 0 to 90°, in a direction parallel to relative plate motion. In this cut *R* is the radium of the Earth and *r* is the radium of the base of the lithosphere. In the undeformed state a plate will move through the angle ω; at the surface the angular distance to the ridge is Lo and at the base of lithosphere is lo.

In the deformed state the plate will move through the angle ω′; the angular distance to the ridge will be L_1 at the surface and l_1, at the base of the lithosphere. These angular distances will be the orthographic projections of L_0 and l_0 respectively; the plane of orthographic projection will be tangent to the Earth at the ridge.

The following relationship is obtained

$$\frac{R}{L_1}=\frac{r}{l_1}$$

because these are corresponding sides of similar triangles.

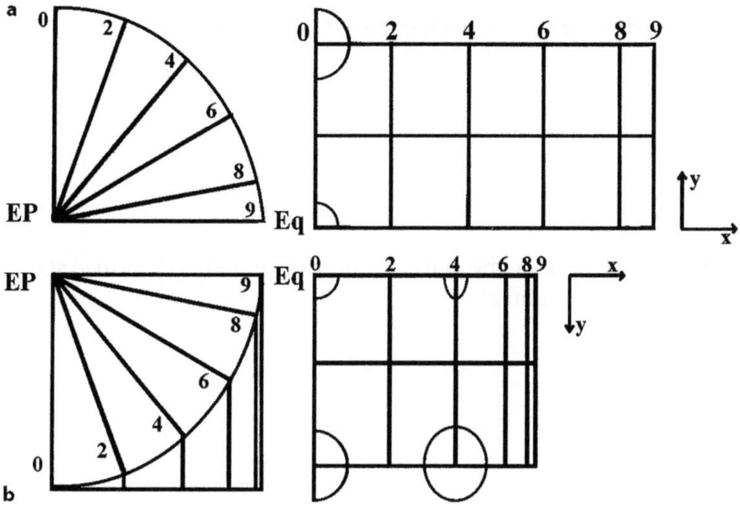

Fig. 3.8. A Polar Mercator projection for a constant spreading-rate ocean; *EP* is Euler pole. Motion is parallel to transforms that project as small *circles*; magnetic lineations project as *meridians*. B Polar orthographic projection for a constant spreading-rate ocean

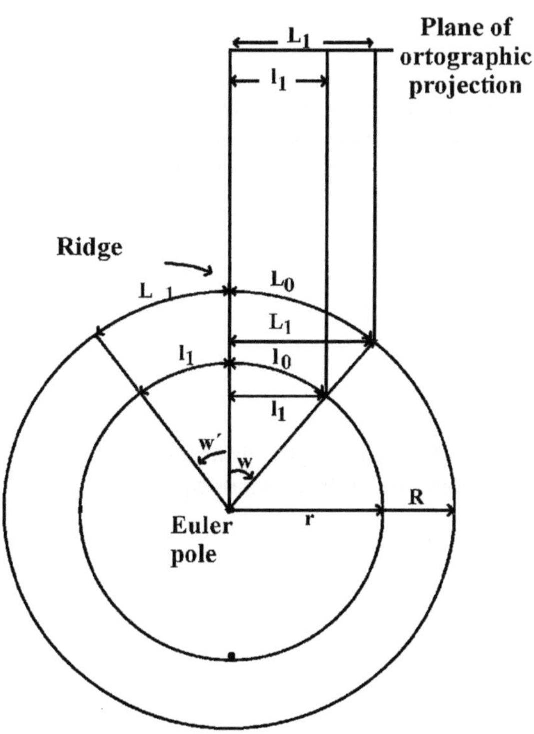

Fig. 3.9. Deformation of a spherical shell, representing the lithosphere that obeys the orthographic projection law

We can calculate the stretches of lines at the base of the lithosphere

$$1+e_r=\frac{l_1}{l_0}=\frac{l_1}{\omega r}$$

and at the surface

$$1+e_R=\frac{L_1}{L_0}=\frac{L_1}{\omega R}$$

These stretches are negative because the lines become shorter in the deformed state.

Dividing Eq. (2) by Eq. (3) we obtain

$$\frac{1+e_r}{1+e_R}=\frac{\frac{l_1}{\omega r}}{\frac{L_1}{\omega R}}=\frac{\frac{l_1}{r}}{\frac{L_1}{R}}$$

However,

$$\frac{\frac{l_1}{r}}{\frac{L_1}{R}}=1$$

by application of Eq. (1).

Therefore

$$1+e_r=1+e_R$$

We prove that the stretch is the same on the top and base of the lithosphere.

We can prove the reciprocal theorem: if the plate deforms according to the orthographic projection rule it will rotate by the angle ω', instead of ω.

Again,

$$\frac{l_1}{r}=\frac{L_1}{R}$$

however,

$$l_1=\omega_r' r$$
$$L_1=\omega_R' R$$

because these are angular distances.

Substituting in the previous equation

$$\frac{\omega_r' r}{r}=\frac{\omega_R' R}{R}$$

therefore

$$\omega_r'=\omega_R'$$

This is a consequence of self-similarity of orthographic projections for different radius at the surface, R, or at the base of lithosphere, r. Without similarity and scaling, quantitative general laws would be impossible (Schroeder 1990, p. 4).

We must also relate finite and incremental strain. For this purpose, we consider a steady-state spreading ocean; in other words, an ocean spreading at a constant rate where the divergent plates will rotate around a fixed Euler pole. We can plot the arc described by the spreading ocean, $\widehat{l}$
as a function of half-spreading rate at the mid-ocean ridge, $\dot{\omega}_0$.

In the undeformed state we verify that

$$\widehat{l} = \dot{\omega}_0 t$$

In the deformed state the stretch is variable, as a function of time.

$$1 + e = \frac{\widehat{l_1}}{\widehat{l_0}}$$

In the ridge there is no deformation. Hence

$$e = 0$$

$$1 + e = 1$$

$$\frac{\widehat{l_1}}{\widehat{l_0}} = 1$$

So we must impose

$$\lim_{\widehat{l_0} \to 0}(1 + e) = \lim_{\widehat{l_0} \to 0} \frac{\widehat{l_1}}{\widehat{l_0}} = \frac{\sin \widehat{l_0}}{\widehat{l_0}}$$

Therefore

$$\widehat{l_1} = \sin \widehat{l_0}$$

The sinus function satisfies the boundary condition at the ridge.

So, the incremental strain is

$$\frac{dl}{dt} = \sin \widehat{l_0}$$

We can estimate the finite strain from incremental strain by integration

$$\int_0^t dl = \int_0^t \sin \widehat{l_0} dt = \cos l_0 = \cos \dot{\omega}_0 t = \cos\theta$$

and

$$\theta = \dot{\omega}_0 t$$

We recover the orthographic rule.

Ocean spreading at different rates will follow different curves in the plot of spreading arc versus time (Fig. 3.10).

We propose a non-trivial solution for kinematics of deformable plates. We must discuss now whether this solution is unique. In fact, solutions such as

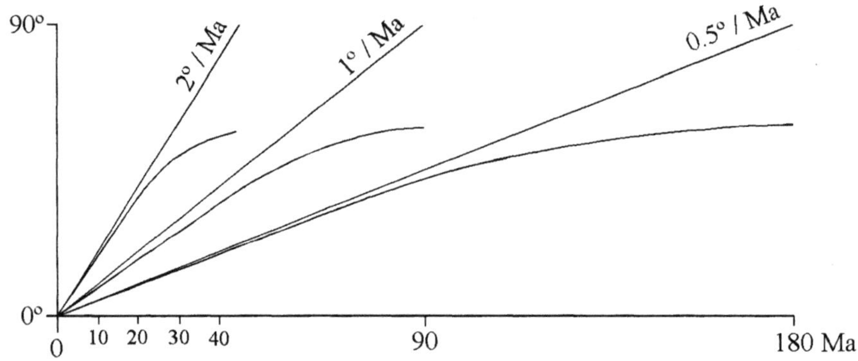

Fig. 3.10. Variation of angular plate velocity, $\dot{\omega}$, as a function of time, T in MA, for oceans spreading at different angular rates, $\dot{\omega}_0$. Curves for 20/MA, 10/MA, 0.50/MA

$$\frac{dl}{dt} = K \sin \frac{\widehat{l_0}}{K}$$

satisfy the boundary condition at the ridge of no deformation, but are scaled to a radius

$$\frac{R}{K}$$

instead of the solution with K=1, which is scaled for R. We will consider the uniqueness of the solution again in Section 3.3.3.

The axisymmetric model of plate deformation has a domain of application limited by the validity of the postulates on which it is based. We considered a symmetric divergent plate boundary around a fixed Euler pole; the spreading rate is constant; the deformation is in the plane and by homogeneous shortening.

We consider first the position of the Euler pole. We recall that we assumed that it is fixed between 3 Ma and present time in the NUVEL-1 model. This is a good approximation for Atlantic-type, slow-spreading oceans where the pole is quite stable during the opening process. However, it is invalid for the large, fast-spreading, Pacific-type oceans where the Euler pole is spinning or wobbling and jumping significantly during the Wilson cycle. We also stress the fact that the NUVEL-1 model mixes spreading rates and transform fault azimuths from 3 Ma ago with slip vectors of earthquakes occurring now. Any misfit between Euler poles for 3 Ma ago and now will introduce a correction that will reduce the strain because it becomes non-coaxial. Also, in the Pacific, changes in plate motion have been recorded in the last Ma as we will see in Section 3.3.3.

The axisymmetric model predicts that plate velocity will drop to zero when

$$\dot{\omega}_0 t = 90°$$

where $\dot{\omega}_0$
is the plate velocity in degrees per Ma and t is the time elapsed since the beginning of spreading. This is a physical impossibility because at 90° from the ridge the strain will be infinite. In an ocean spreading at 2°/Ma this will happen after 90 Ma; in an ocean

spreading at 0.2°/Ma this will happen after 900 Ma. We know that there is no oceanic lithosphere older than approximately 200 Ma; the reason is because old lithosphere is cold and must sink by subduction. There seems to be a contradiction between the axisymmetric model and the real data that we must discuss thoroughly. The axisymmetric model deals only with *homogeneous shortening*; we will see that this deformation regime occurs only for small strains, in the order of <10–20% otherwise instability will develop, namely, the buckling of the entire lithosphere. Above this value of finite strain the axisymmetric model is not applicable.

On the other hand, we know that in slow-spreading oceans the Euler poles are almost stable instead of in fast-spreading oceans where the Euler poles spin at fast rates. Between 162 and 119 Ma the Euler pole for the North America–Africa plate pair stays in Iceland; in contrast, between 77 and 32 Ma, the Euler pole for the plate pair Pacific-Farallon spin very fast in a non-linear manner. The implications of this fact are that in slow-spreading oceans, like the Atlantic, the strain accumulates coaxially and the fixed Euler pole axisymmetric model is a good approximation for the real situation. In fast-spreading oceans like the Pacific, strain is non-coaxial; successive strain increments eliminate each other and the real finite strain is low. In this case, we can still use the axisymmetric model if we go backwards in time with small steps. This reasoning suggests that slow and fast-spreading oceans will converge to a maximum age for the Atlantic phase of the Wilson cycle. After that, buckling and consequent initiation of subduction introduce new constraints that were not considered in the axisymmetric model.

If we apply this model to the North Atlantic, opening at a half-spreading rate of 0.11°/Ma between Europe and North America, according to NUVEL-1, we can estimate a stretch of 0.92, at 90° of the Euler pole, after 200 Ma. This would explain the present situation in the Atlantic, as we will see later on.

If subduction starts in the spreading ocean we should consider a possible modification or generalisation of the axisymmetric model in order to take into account Pacific-type oceans, the subject of Section 3.3.3.

We must now test the axisymmetric model for Atlantic-type oceans.

The obvious idea would be to compare the measured displacements by geodetic methods with estimated velocities from the kinematic model NUVEL-1 based on the magnetic anomaly 2A, approximately 3 Ma old.

There was found to be (Robbins et al. 1993; Fig. 3, p. 28) a high correlation level of 0. 994 with a slope fit of 0.937±0.009 on an analysis of 149 lines spanning 20 stations and crossing 10 plate boundaries. This means that the present rates are 6±1% slower than NUVEL-1 rates.

This discrepancy could be due to different causes (Smith et al. 1990; Robins et al. 1993).

1. Secular or non-linear variations of the plate motions over the 3-Ma time span.
2. Adjustments in geochronological time scales.
3. Non-uniform geographical distribution of the selected stations.
4. Significant intraplate strain (Ribeiro 1993).

Concerning the first factor, we must note that plate motion must be steady because of the convective character of the driving mechanism (Gordon and Stein 1992). Transient motions at plate boundaries are dampened by the viscous asthenosphere and converge to the observed steady motion. Therefore, we should expect accelerations over intervals of Ma if there is a gradual reconfiguration of plate boundaries; but, changes at the scale of years or decades are unexpected, as we stated in Section 2.4. if we stay away from active faults.

Concerning the second factor, we must note that a recalibration of the geomagnetic reversal time scale, based on a new astronomical approach, led to a change in the age of magnetic anomaly 2A. In the improved model NUVEL-1A, the NUVEL-1 angular velocities maintain their direction, but their magnitudes should be multiplied by a constant, α, of 0.9562, that gives 4.4% slower rates. Therefore, the discrepancy between rates measured by geodetic methods and NUVEL-1A is ~2% (De Mets et al. 1994). However, some authors maintain that VLBI data are in better agreement with the predictions of NUVEL-1 than with those of NUVEL-1A; in fact, VLBI gives faster rates +3.4% (±1.2%) than NUVEL-1, instead of -4.4% slower rates than NUVEL-1A (Heki 1996). The astronomical approach has been refuted by some authors, because astronomical perturbations are not strictly periodic, but quasi-periodic in the short term and chaotic in the long term (Lliboutry 1998).

Concerning the third factor, we can minimise its effects by looking at the adjacent pairs of plates that are sampled by more stations; we can hope that in the near future the occupations of new sites will achieve a better distribution with more significance at the global scale.

The previous considerations were the motivation to explore discrepancies between geodetic and theoretical plate motions by intraplate strain.

Let us consider the case of a symmetric slow-spreading ocean around a fixed Euler pole, as in the Atlantic today. We assume that due to inertia of large plates involved, the plate velocity would have been approximately constant in the last 3 Ma and that the discrepancy between the displacement measured along the base line and the predicted displacement by the NUVEL-1 model is due to intraplate deformation along the line in the oceanic lithosphere. In fact, the magnetic anomaly pattern for the last 3 Ma in the Atlantic (Argus et al. 1989) does not show evidence of significant variations in spreading rate during this period (Fig. 3.11). This means that any change in plate motion should be very recent and discontinuous, which seems unlikely.

First, we will consider a line, parallel to plate motion, and at 90° of the Euler pole. The angular displacement, ω, in a time interval t, along the line can be calculated by the rigid plate kinematics theory, from the angular plate velocity, $\dot{\omega}$ around the Euler pole, according for instance to model NUVEL-1 (De Mets et al. 1990).

$$\omega = \dot{\omega}\, t$$

In 1 year this displacement will be D, in cm/year. This value should be corrected by shortening along the length of line in the oceanic lithosphere, L.

On each side of the mid-ocean ridge we can obtain the real displacement, d, from the theoretical displacement, D, by multiplying L by its stretch, 1+e. Thus

$$\frac{d}{2} = \frac{D}{2} + \frac{L}{2}(1+e)$$

However, D is calculated at the geological time-scale – it is in fact the average spreading rate in the last 3 Ma and, on the other hand, d is averaged over a decade; so, the displacement in 1 year, d, is nearly an instantaneous displacement rate. The strain, e,

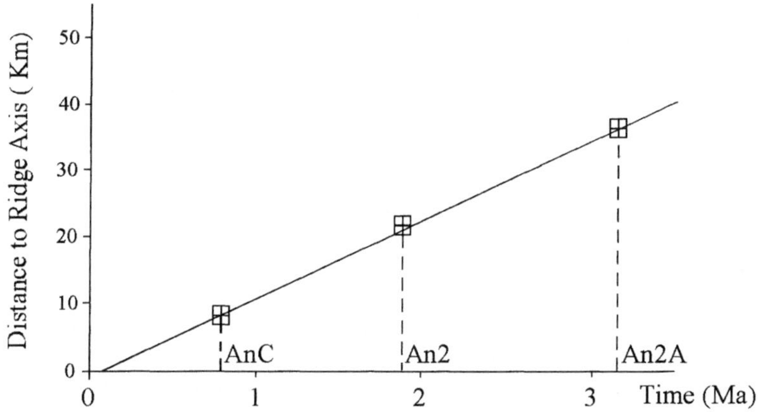

Fig. 3.11. Rate of spreading in the Mid-Atlantic Ridge, north of the Azores based on Argus et al. (1989, Fig. 5, p. 5592); age of magnetic anomalies based on revised magnetostratigraphic scale. (Model NUVEL-1A)

during 1 year should also be considered as a strain rate. It can be estimated from finite strain by differentiation in relation to time:

$$\frac{d(1+e)}{dt}=\frac{d\,(\cos\omega)}{dt}\frac{d\,[\cos(\omega)t]}{dt}=-\sin\dot{\omega}$$

However, $\dot{\omega}$ in radians is a very small angle; with R as radius of the Earth, we obtain

$$\sin\dot{\omega}\approx\dot{\omega}=\frac{D}{R}$$

or finally,

$$d=D-\frac{LD}{2R}$$

This approximation is valid for $L<2R$.

For a line parallel to plate motion at θ degrees from the Euler pole, we will use a co-ordinate frame where the Euler pole is at the North (or South) Pole of the projection, and the ridge is at the central reference meridian. The ocean/continent boundary along the line is specified by its spherical co-ordinates $\hat{\theta}$ angular distance to Euler pole and $\hat{\lambda}$ angular distance to ridge crest, in radians or $\bar{\lambda}$ the equivalent linear distance.

In this case we have

$$D_\theta=D\sin\theta$$

from the plate kinematics theory and

$$\bar{I}=\bar{R}\sin\hat{\theta}\sin\omega$$

from the orthographic projection theory.

The equation relating the measured displacement and the theoretical displacement is easily obtained, by analogy with the previous case

$$\bar{I} = \bar{\lambda} R \sin\theta$$

$$d_\theta = D\sin\theta - \frac{\bar{I}}{2}\frac{D\sin\theta}{R\sin\theta}$$

By substituting Eq. (11) in Eq. (12), we obtain finally

$$d_\theta = D\sin\theta.(1 - \frac{\hat{\lambda}}{2})$$

This Eq. (13) contains Eq. (8) as a special case when θ=90°. Note, however, that in Eq. (13) angular distances are measured along small circles around Euler poles; but, only when θ=90° the small circle becomes a great circle or geodetic (Fig. 3.12).

We applied the last equation to the line Grasse (Stable Eurasia) - Greenbelt (Stable North America), using the plate velocity and Euler pole between the plates, given by NUVEL-1 modified to NUVEL-1A (De Mets et al. 1990, 1994). The line is less than 1° from the plate motion vector, θ=68°, λ=27 or λ=0.471 radians. The theoretical value, $D\sin\theta$ is 2.2 cm/year, the corrected value for intraplate oceanic deformation is 1.9 cm/year, closer to the estimated value of displacement by geodetic satellite methods (Smith et al. 1990), which is 1.8±0.3 cm/year, than the theoretical one.

This analysis does not take into account two factors: slight extension near the ridge and intracontinental deformation across the line; these are of second order relative to intra-oceanic compressive deformation by homogeneous shortening.

The most recent data for the Eurasia–North America plate pair show that we still do not have enough resolution for different methods to prove or disprove intraplate oceanic strain for lines crossing the North Atlantic. The present displacement near the Mid-Atlantic ridge (Bastos et al. 1998) is in the order of 2.5 cm/year for GPS sites in North America (Flores, Azores) and Eurasia (Graciosa, Azores), equivalent to

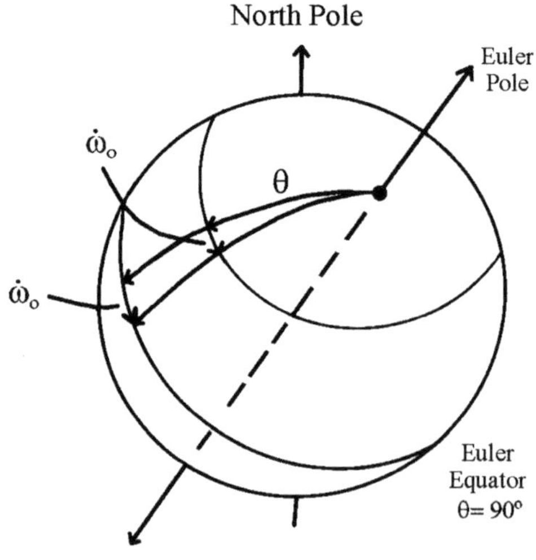

Fig. 3.12. Image of variables for general analytical axisymmetric solution. $\dot{\omega}_0$, angular velocity for spreading at ridge, in degrees/Ma. θ° angular distance to Euler pole varying between 0ϒ, at the Euler pole, to 90°at the Euler equator

Table 3.1. Angular velocity for Europe–North America. (Larson et al. 1997; Bastos et al. 1998)

Method	ω deg/Ma
GPS	0.24
NUVEL-1A	0.21
VLBI	0.26
Goddard	0.22
Jpl-GPS	0.23
GPS (Azores)	0.23

0.230/million years, a line approximately perpendicular to the Euler pole for the plates involved in NUVEL-1A or GPS; this is above NUVEL-1A angular velocity (0.210/million years) and some geodetic data, but below other geodetic data (Table 3.1). We conclude that the geodetic data on the Atlantic-type oceans cannot resolve the question of intraplate oceanic strain due to two factors. First, there are uncertainties in the geomagnetic reversal time-scale for anomaly 3, leading to different plate velocities for models NUVEL-1 and NUVEL-1A; and second, there are uncertainties in the geodetic data because plate velocities show a considerable amount of noise in the database. Therefore, we must enlarge the period of observation, recording longer time series and we hope that the uncertainties will be reduced in a few years. Hence, let us consider the case of fast-spreading Pacific-type oceans.

3.3.3 Kinematics and Strain of Deformable Plates in Fast-Spreading Pacific-Type Oceans

The Pacific Ocean is the best domain to discuss the evidence for plate rigidity or intraplate deformation derived from a comparison between kinematic models like NUVEL-1 and actual space geodetic results.

The Pacific Plate is very wide (allowing very large baselines to be measured), it has a high velocity (implying that the tectonic stresses that drive the plate must also be large) and it is purely oceanic. The Pacific Plate interacts with the Nazca Plate along a large mid-ocean ridge, from where much of the information used to evaluate the Pacific Plate motion comes. This way we can compare the geodetic measurements with the kinematic predictions for the relative Pacific-Nazca motion.

The fast-spreading Pacific Ocean is surrounded by subduction zones. This must somehow affect the intraplate strain, because subduction influences the boundary conditions and the axisymmetric model must be modified or generalised.

The Pacific Ocean is highly asymmetric, with the Mid-Pacific rise situated much nearer to the eastern Pacific subduction zones than to the western Pacific ones. We predict that this asymmetry will influence the strain regime in this ocean, and in fact this is the case, as follows. In the eastern Pacific as the subduction zone is approached the incipient pinch-and-swell structure can lead to complete boudinage of the young oceanic lithosphere as we referred to in Section 3.2. This is happening in the Mendana Fracture Zone and in the Galapagos Rift where a former and larger East Pacific Plate was fragmented in the smaller Nazca and Cocos Plates.

We consider now the existing data on the Pacific Plate, west of the Mid-Pacific Rise. The strain regime corresponds to two different situations.

In the eastern domain of the Pacific Plate, with younger lithosphere, earlier results (Robbins et al. 1993) indicated significant rates of displacement between stations VLBI situated in oceanic islands; some of these results were revised, using mainly VLBI and GPS (Larson et al. 1997), and the discrepancies with the rigid model were attenuated. However, a fact remained, even after revision the discrepancies for lines parallel to plate motion are above the standard deviation of displacement values and indicate shortening; discrepancies for lines at a high angle to plate motion are in the order of error in displacement values (Fig. 3.13). This is explained if lines parallel to plate motion are close to the minor axis of strain ellipse at the surface of the Earth and if lines at a high angle to plate motion are close to lines of no infinitesimal longitudinal strain (Ramsay 1967). If we compare these data with data of independent origin we conclude that shortening is approximately parallel to lineations inferred from geoid data and to maximum compressive horizontal stress where it has been estimated. All these data indicate compression parallel to plate-motion in this sector of the Pacific Plate and negligible extension perpendicular to cross-grain geoid lineations.

In the western domain of the Pacific Plate, with older lithosphere, the data are scarcer. The very recent installation of the VLBI site at Marcus Island (Argus and Gordon 1996) shows, nevertheless, that the line Kwajalein-Marcus Island is lengthening; therefore, there is intraplate deformation because both sites are within the Pacific Plate. This confirms the previous result on the line between Kashima, situated in the North American Plate, and Kwajalein, which converges at a slower rate than predicted by the model NUVEL-1 (Robbins et al. 1993; Ribeiro 1997).

The axial symmetry of the intraplate strain problem suggests a plot to quantify this strain; on the abscissa we plot angular distance to the ridge and on the ordinate, plate velocity. In this way we eliminate the influence of the third factor, the angular distance to the Euler pole. If the plate is rigid, we must obtain a plot of constant angular velocity, a horizontal line with slight dispersion due to measurement errors. If the plate is deformable we must obtain some curve that is a solution to the axisymmetric problem; physically, it signifies that all deformation is in the plane with no torsion around the horizontal or vertical axis; the plate does not loose coherence.

Fig. 3.13. Satellite geodetic stations in the Pacific Plate. Results of the triangle H-Kw-K and H-Kw-M are plotted: the upper value is the displacement measured by geodetic satellite methods and the lower value is the null displacement predicted by NUVEL-1A. *Triangles* are stations inside the Pacific Plate. (Based on Robbins et al. 1993; Argus and Gordon 1996; Boucher 1996; Larson et al. 1997). The *insert* shows geoid lineations inside a quadrangle (after Sandwell and Smith 1997). Band-passed filtered (half amplitudes at 400 and 30 km) *grey shaded* and illuminated from the north to enhance gravity lineations that are parallel to current relative motion between the Pacific and Nazca Plates

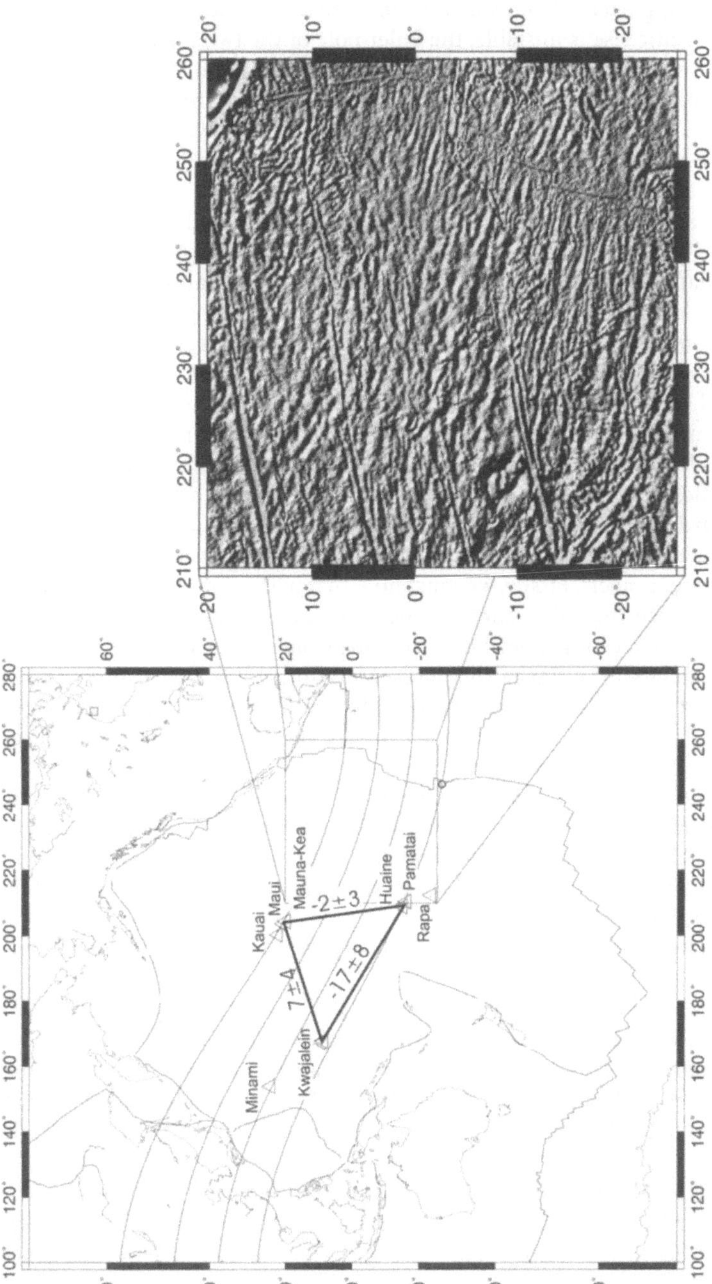
Kauai
Maui
Mauna-Kea
Huaine
Pamatai
Rapa
Minami
Kwajalein
7±4
-2±3
-17±8

In the fast-spreading Pacific Plate, we know that the kinematics of the plates separated by the Mid-Pacific rise is unstable: the Euler pole of the two plates migrates rapidly and there are also variations in the spreading rate. Therefore, when we compare results from the model NUVEL-1A with geodetic satellite data over the last decade we must neglect, in the first approximation, these possible kinematic instabilities; we assume that all plate motions are along small circles around a fixed Euler pole converting a vector problem in a simpler scalar problem.

We plotted the angular velocities relative to the ridge and angular distance to the ridge for stations inside the Pacific Plate (Fig. 3.10), using all the available data and compared these data with results from NUVEL-1A (Robbins et al. 1993; Argus and Gordon 1996; Boucher 1996; Larson et al. 1997; Ribeiro et al., in press).

For some authors (Argus and Gordon 1996), the comparison between plate kinematics model NUVEL-1A and space geodetic results confirms plate rigidity, because the relative speed of intraplate sites far from plate boundaries is below an upper boundary of 2 mm/year. For plates, in general, and particularly for the Pacific Plate, minor differences between both methods have been justified by minor variations in rigid plate velocities; if we consider error bars within 2σ values, where σ is one-dimensional standard deviation, we obtain a plot of constant angular velocity within ±10% error, compatible with plate rigidity. However, this analysis does not take into account the following aspects. Firstly, we must note that with a longer time series, the displacements converge towards a mean value contained in the 1σ error interval. Some authors, in more recent papers, have used this smaller interval to demonstrate changes in plate velocity in the last 3.16 Ma (De Mets and Dixon 1999; Norabuena et al. 1999), supported by longer time series. Secondly, we must note that mean values are not distributed randomly, as will happen if they resulted from error measurements, but follow a definite coherent pattern.

The stations situated near the Mid-Pacific Rise, such as Patamai and Huhaine, show plate deceleration relative to the angular velocity at the Rise and the stations in the older western Pacific such as Hawaii, Kwajalein and Marcus show plate acceleration relative to the previous ones. We will try to show that those facts are compatible with a deformable plate model.

A review of the most recent data (Ribeiro et al., in press) can be summarised as follows.

A global solution (Sillard et al. 1998) integrates all the data from SLR, VLBI, GPS and Doris in the International Terrestrial Frame, ITRF/96. The authors consider that data lose some internal consistency, but are more reliable because systematic errors are minimised by cross-checking the data. Using this data, we obtain the solution in the plot of angular velocity versus angular distance to the ridge in the Pacific Plate shown in Fig. 3.14.

For our analysis we chose sites in the Pacific Plate surveyed by space geodetic techniques which are far from plate boundaries and sample plate deformation along a wide band to the PCFC-NAZC relative motion (Fig. 3.13). A previous set of baselines (Robbins et al. 1993) also shown in Fig. 3.13 already suggested that significant intraplate deformation might occur along this band, while at high angles to the relative velocity, relative geodetic rates are negligible. The space geodetic data used in this paper (Table 3.2) are defined in the ITRF97 report (Boucher et al. 1999). Site velocities relative to ITRF97 were converted to local co-ordinates (north, east and down) according

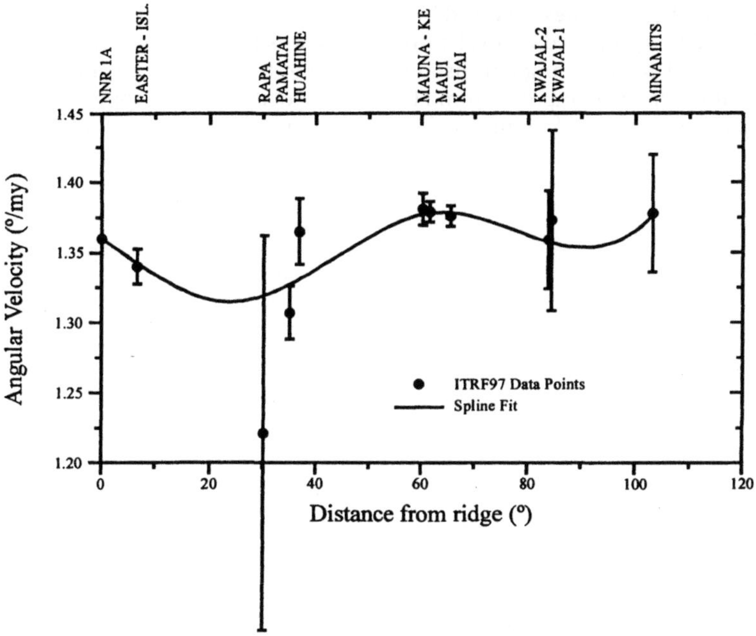

Fig. 3.14. Plot of angular velocity relative to ridge, W, versus angular distance to ridge, θ. The stations are listed in Table 3.3, from the Mid Pacific Rise to the west

to the matrix transformation given by Cox and Hart (1986). For our analysis we transformed the horizontal components of the velocity into relative angular velocities, measured along the direction of the PCFC-NAZC motion (Table 3.2) assuming that the difference between ITRF97 and NNR1A is negligible for this kind of study. In fact, a rotation between these two frames only affects the absolute values for the angular velocities, but leaves the site-to-site differences unchanged. The reference error used is the spherical error (Sillard et al. 1998) computed as the square root of the quadratic sum of the standard errors of the Cartesian velocities.

The angular velocities (with 1σ error estimates) for each site were then plotted as a function of the approximate angular distance to the mid-oceanic ridge (Fig. 3.14). The reference velocity for zero offset was taken from the NNR1A model. If plates have been accelerating since 3 Ma, then the offset velocity would be higher than plotted.

Using data isolated from VLBI or GPS, we obtain a different plot. Therefore, we must discuss the validity of each solution.

The ITRF is compatible with the inferences from geoid data on boudinage of the oceanic lithosphere referred to above and in Section 3.2. There is a negative acceleration near the ridge and a positive acceleration as we approach the subduction zones of the West Pacific. The negative acceleration can explain boudinage at right angles to the ridge, parallel to plate movement; the positive acceleration will cause its disappearance further away.

Table 3.2. Pacific Plate sites surveyed by space geodetic techniques VN and VE are the ITRF97 velocities projected to local co-ordinates. w is the angular velocity relative to stable Nazca. ADM is the angular distance to Pacific-Nazca mid-ocean ridge. Error is the radial uncertainty computed from ITRF97 1s velocity errors. Capital letters indicate the space geodetic technique used to survey the site: V-VLBL, S-SLR, D. Doris, G-GPS

Site name	Lat. (°N)	Long. (°E)	VN (Mm /year)	VE (Mm /year)	Error (%)	w±s (°/Ma)	ADM (0)
Rapra (D)	-23.35	212.40	28.1	-45.9	29.	1.22±0.14	30
Patamai (G)	-16.43	210.43	29.87	-57.29	3.2	1.307±0.019	35
Huahine (S/D)	-15.27	208.97	29.23	-65.11	3.6	1.365±0.023	37
Mauna-Kea (V/G)	19.80	204.55	31.31	-61.61	1.6	1.381±0.011	60
Maui (V/S)	20.72	203.73	32.48	-60.32	1.0	1.379±0.007	62
Kauai (V/D)	22.13	200.33	31.20	-61.08	1.0	1.376±0.007	65
Kwajalein-1 (V/S)	9.40	167.48	23.83	-69.93	9.7	1.373±0.065	84
Kwajalein-2 (G)	8.71	167.72	24.90	-67.39	5.4	1.360±0.035	84
Minami Tori Sima (V)	24.28	153.68	24.12	-39.07	6.3	1.378±0.042	103

In contrast, the data from individual methods show a positive acceleration of the plate between the ridge and subduction zones. If this is the case, the Pacific Plate should be extended everywhere parallel to plate movement and should show boudinage parallel to the ridge. This is not the case and we prefer the ITRF solution, which is free from any geological assumption, but is consistent with independent evidence, as just seen. Moreover, this solution will be compatible with a dynamic interpretation of the role of ridge push and slab pull in the driving mechanism of the plates(see Sect. 3.11).

In the analysis of the data, angular velocity at the ridge is a crucial parameter, because we know that the direction of the plate motion has changed in the last 1 Ma (Searle 1983); on the other hand, it cannot be measured directly by satellite geodetic data because the Mid-Pacific Rise is immersed. A possible kinematic change between 3.16 Ma, the age of magnetic anomaly 2A on which NUVEL-1A is based, and present time will be undetected. However, we think that this will not affect our conclusions concerning plate rigidity versus deformability. We will see later (see Sect. 3.10) that the rheological properties of oceanic plates will show viscous response predominant over elastic response for forcing in the order of 20 Ma, hence the long-term deformation of the plate will depend more on integrated kinematics at the scale of 20 Ma than on short-term to instantaneous changes in the last 3 Ma.

Therefore, for the moment, we will only discuss the internal coherence of spatial geodetic data for the Pacific in the last decade; we leave the accurate comparison between this data and kinematic model NUVEL-1A for Sections 3.10 and 3.11 because it

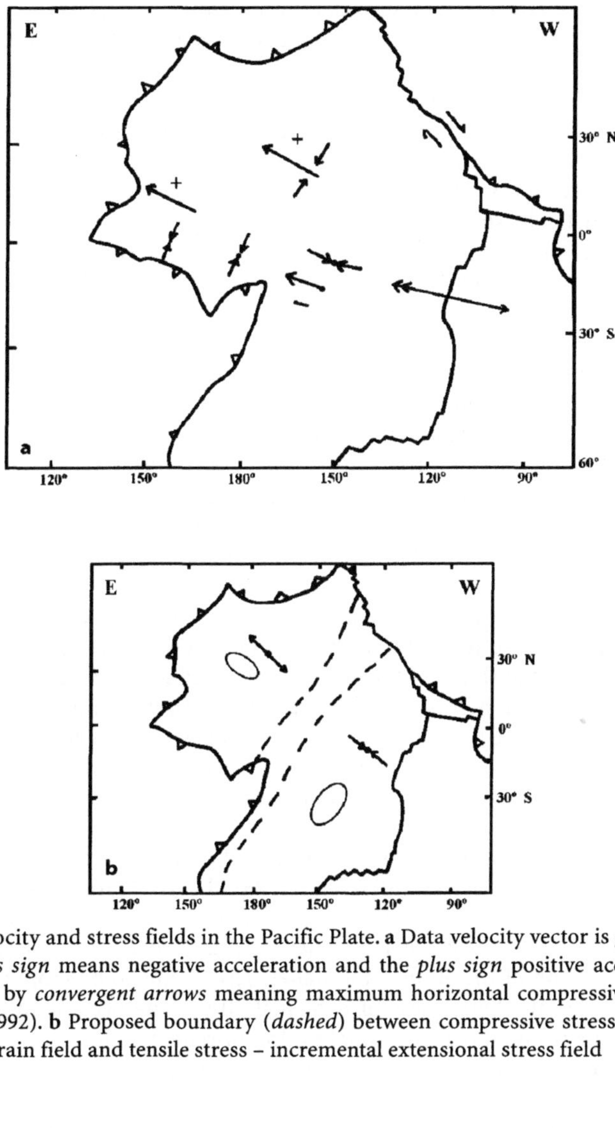

Fig. 3.15. Velocity and stress fields in the Pacific Plate. **a** Data velocity vector is given by an *arrow*; the *minus sign* means negative acceleration and the *plus sign* positive acceleration. The stress is given by *convergent arrows* meaning maximum horizontal compressive stress (after M.L. Zoback 1992). **b** Proposed boundary (*dashed*) between compressive stress - incremental compressive strain field and tensile stress - incremental extensional stress field

is the basis for discussion of steadiness or unsteadiness of plate kinematics, with implications for both the rheology of the lithosphere and the driving mechanism of plate motion.

If the Pacific Plate deforms we must look at the relationship between the state of stress and the incremental strain inside this plate as follows.

The maximum horizontal compressive stress is oriented parallel to plate motion in younger oceanic lithosphere and becomes perpendicular to plate motion at the longitude of Hawaii (Zoback et al. 1989; Zoback 1992). In Hawaii itself we could attribute the permutation of stress to second-order patterns due to local causes such as flexural stresses due to loading by the Hawaii hotspot volcano. However, the maximum compressive stress stays oriented NE–SW to the west of Hawaii, so it must reflect a real first-order pattern related to plate motion (Fig. 3.15).

An explanation has been advanced for the permutation of stress (Zoback et al. 1989, p. 295) in the following way: "this change in Shmax orientations coincide roughly with the 80 Ma isochron, the point at which the $t^{1/2}$ subsidence of cooling oceanic lithosphere dramatically slows down from $t^{1/2}$ to an approximately exponential decay", t being the age of the isochron. If this is the cause of the permutation why does it not occur in the Atlantic, where the subsidence law is the same as that in the Pacific? We will return to this problem in Section 3.3.4.

If we consider the orientation of the stress tensor and the orientation of the incremental strain tensor at the seafloor together, inferred from geodetic satellite data, we note that maximum horizontal compressive stress and the major axis of incremental strain ellipse remain approximately orthogonal in the entire Pacific Plate; there is no sufficient information about stress and strain states to decide if the permutation of stress coincides with the permutation of incremental strain axes; in the case of viscoelastic rheology, we should expect some delay to occur, due to stress relaxation, before permutation of incremental strain. This is compatible with the existing data, although not proven by it. However, in any case, the permutation of incremental strain as stress permutes requires a viscous component for the rheology of the Pacific lithosphere. Therefore, we suggest that the flow near Hawaii changes because it becomes constrained by the shape of the West Pacific subduction zones, especially as the corner of the Tonga Trench is approached. Therefore, the strain inside the Pacific Plate is controlled by the flow patterns: the major axis of strain ellipse perpendicular to plate motion in the young lithosphere of the East Pacific and parallel to plate motion in the older lithosphere of the Pacific west of Hawaii, due to convergent flow (Ribeiro 1997; Ribeiro et al., in press).

Another indication of strain inside the Pacific Plate comes from elongation of residual depth anomalies in Hawaii (McKenzie 1983); the axial ratio of the elliptic shape of the anomalies is around 1.5 with the major axis oriented N60°W. The maximum compressive horizontal stress is oriented here N45°E (Zoback et al. 1989) at a high angle to the elongation of the elliptic anomalies. This seems to be more than a coincidence and favours the hypothesis of deformability of the oceanic lithosphere. It is possible that the residual depth anomalies do not reflect total strain suffered by the Pacific Plate, but only the strain accumulated since the less viscous magma produced by hotspot activity recently reached the surface layers of the more viscous, colder lithosphere around Hawaii. Thus, the axial ratio of this ellipse is increased by the ductility contrast between a less viscous inclusion and the more viscous medium where it is contained.

We conclude that the existing data point to shortening parallel to plate motion in the eastern domain of the Pacific Plate and to elongation parallel to plate motion in the western domain of the Pacific Plate. An infinitesimal neutral zone must exist between both domains. This strain distribution is compatible with orthogonal flexure (Twiss and Moores 1992) of the whole Pacific lithosphere around a vertical axis and convexity to the East, the same sense as the West Pacific subduction zones (Fig. 3.16). A similar model has been invoked for arcuate deformed belts in the continental li-

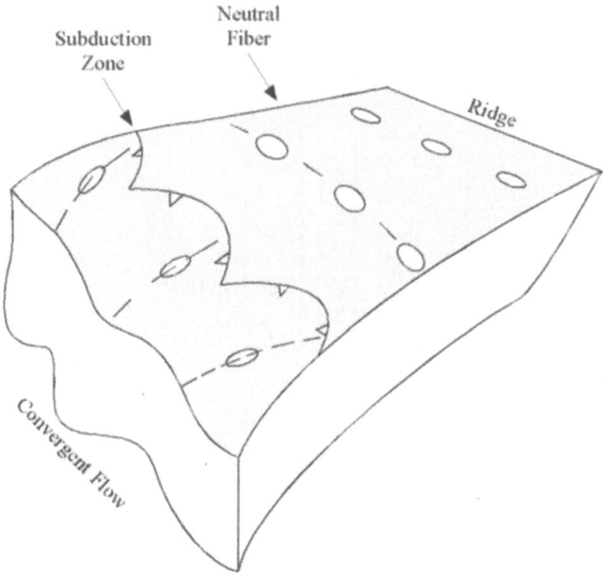

Fig. 3.16. Scheme of strain distribution in a Pacific-type ocean: compression perpendicular to strain ridge; orthogonal flexure in a neutral fiber and convergent flow in downdip subduction zone

thosphere, such as the Ibero-Armorican Arc as will be seen in Section 4.10 (Ribeiro 1974; Ries and Shackleton 1976; Ribeiro et al. 1995). This behaviour would be expected if the oceanic lithosphere were highly viscous. A possible effect would be dispersion of Euler-stage poles with age in this plate. This effect would be particularly notable in periods of avalanches of upper-mantle material into the lower mantle and related superfast subduction (Tackley et al. 1993; see Sect. 3.11).

We can infer that plate kinematics of the Pacific Plate confirm the soft plate model and that this kinematic behaviour is compatible with what we know from plate dynamics. To the east of the Mid-Pacific rise, in young oceanic lithosphere, ridge push is stronger than slab pull from the East Pacific subduction zones and can induce whole lithosphere failure subperpendicular to the Rise in the Galapagos Rift and Mendana Fracture Zone. To the west of the Pacific rise ridge push is dissipated in older lithosphere and slab pull increases as we approach the West Pacific subduction zones. We will return to this subject in Section 3.11.

We know from the plate tectonics theory that an Atlantic-type ocean will become, after approximately 200 Ma of spreading history, a Pacific-type ocean, surrounded by subduction zones. In the light of the soft plate theory, a link must exist between deformation of the oceanic lithosphere during the Atlantic and Pacific stages of the Wilson cycle. What the existing data suggest is that the oceanic lithosphere can accommodate nearly 10% of homogeneous shortening, then it becomes unstable and starts to buckle (see Sect. 3.4). Finally, whole-lithosphere failure will lead to subduction initiation(discussed in more detail in Sect. 4.6). Once subduction starts, sprea-

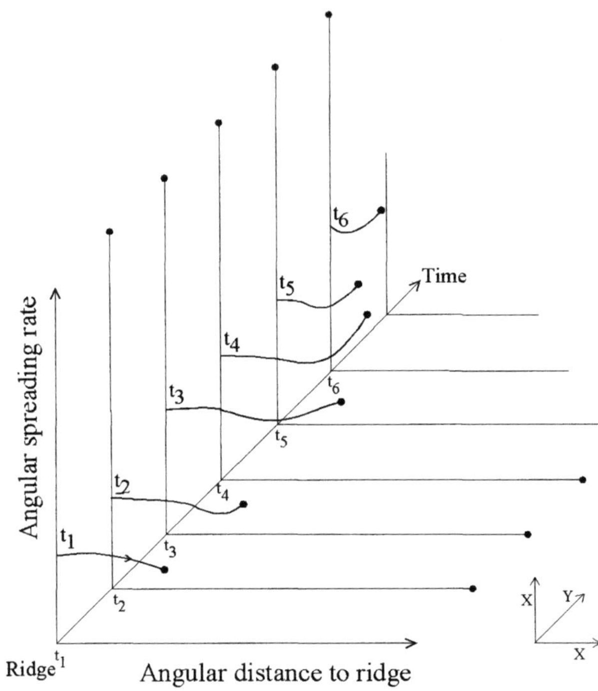

Fig. 3.17. Life cycle and deformation history of an ocean. A three-dimensional plot with co-ordinates. *x* angular distance to ridge. *y* time. *z* angular spreading rate. At t_1 subduction starts at S; the plates decelerate according to path to S. At t_2 slab pull induces extension near the subduction zone, but the ocean is widening. At t_3 spreading is accelerating, but extension by slab pull becomes increasingly important. At t_4 the subduction rate supersedes the spreading rate. At t_5 the subduction rate accelerates and the ocean is closing rapidly. At t_6 the ocean is almost closed

ding can accelerate because the ocean has the largest width during its life cycle. If the spreading rate is larger than the subduction rate, the ocean will continue to open. However, subduction will change the boundary conditions for deformation, because slab pull will reverse compression perpendicular to the ridge axis and will extend the previously compressed domain. The curve for deformation will change from monotonically decelerating with a minimum in the old oceanic lithosphere to a more complex and asymmetric periodic curve with a minimum inside the middle-aged oceanic lithosphere. Eventually, subduction will become faster than spreading and slab pull will completely supersede ridge push. The ocean will close, ending its life time (Fig. 3.17).

We can conclude by saying that during the Atlantic stage a strain wave propagates through the oceanic lithosphere as it spreads. The minimum of strain as a function of distance will be at the ridge and the maximum at the ocean-continent boundary at all times of the spreading history; this will be equivalent to a freely supported sinusoidal oscillation, slightly damped by density variation by cooling as we will see in

Section 3.3.4; only one quarter of the wavelength of this strain wave will occur during the Atlantic stage. When subduction starts, the oscillation becomes forced because the free harmonic introduced by ridge push will be modified by a forced oscillation derived from slab pull; the net result will be a more complex sinusoidal curve that we observe in the Pacific stage.

We can use the analogy of a deformable plate model with an harmonic oscillator to put the previous quantitative laws, established by a purely kinematic approach, in a more robust dynamic basis.

We will now consider a very simple model of an Atlantic-type ocean, spreading at a constant steady rate. We will neglect variations in topography and in lithosphere thickness, considered negligible because they are in the order of 10^1 and 10^2 km, respectively, when compared with the Earth radius 6.33×10^3 km (Fig. 3.18).

We trace a free-body diagram composed of the oceanic and continental lithosphere; the oceanic lithosphere is thinner ≈50 km, than the continental lithosphere ≈100 km. We will consider a driving force, F_D, for plate motion which acts against a resistive force, F_R; x and w are a linear and arc distance to a ridge crest. The driving force will supersede the resistive forces, otherwise plate movement will stop. We postulate that the driving force, F_D, is approximately constant across the length of the entire plate, x, because any acceleration of the plate is of second-order magnitude relative to plate velocity $\dot{x} = \frac{dx}{dt}$.

The resistive force, F_R, will be zero at ridge-crest and will increase with distance to ridge crest, x; its value will be stabilised inside the continental domain, quasi-rigid. In the oceanic lithosphere we hypothesise that F_R will vary with distance in a linear way such as

$$F_R(x) = -Kx$$

In fact, the variations of F_R parallel to plate motion x, are small, otherwise we would produce significant intraplate deformation; the same hypothesis was formulated to explain variation of topography with age in an ocean opening at a steady spreading rate; driving and resistive forces are balanced with linear increase with age of push from the ridge (Lister 1975; Wiens and Stein 1985; Stein and Okal 1986).

For a constant mass we can write a linear differential equation with constant coefficients such as

$$m\frac{d^2x}{dt^2} = -Kx$$

because

$$\frac{dx}{dt} = \dot{\omega}_0 \cos\dot{\omega}_0 t \text{ and at } t = 0$$

$$\frac{dx}{dt} = \dot{\omega}_0$$

The formal analogy with a free harmonic oscillator is complete if we equate the spreading velocity at the ridge, $\dot{\omega}_0$ with the natural period of the oscillator. If k=0, we have the trivial solution with no variation in plate velocity or $\dot{x} = \frac{dx}{dt}$ will be constant; in other words, the plate will be moving, but rigid; this happens in the continental lithosphere.

We can make this formal analogy explicit in the following way:

Let us consider a co-ordinate axis system such as x perpendicular to ridge crest; y parallel to ridge crest and z vertical; the mass created at ridge crest per unit time is:

$$m=\rho x z_0 y.$$

where ρ is the mean density of lithosphere at ridge crest and z_0 is the lithosphere thickness at the ridge crest.

However, $x=w_0\, t$
where w_0 is half spreading velocity. Therefore, if t=1 then

$$x=w_0$$

and

$$m=\rho w_0 z_0 y_0$$

but

$$\frac{d^2x}{dt^2} = -\omega_0^2 \sin\omega_0 t = -w_0^2 x$$

therefore,

$$m\frac{d^2x}{dt^2} = -kx$$

is equivalent to

$$m\left(-\omega_0^2\right) = -kx$$

or

$$m\omega_0^2 = k,$$

analogous to equivalent formula for harmonic oscillator, and finally

$$k = \rho\omega_0^3 z_0 y$$

the force per unit length of the ridge F/Y and per unit time is

$$\frac{F}{y} = -\frac{k}{y}x$$

with

$$-\frac{k}{y} = \rho\omega_0^3 z_0$$

Therefore, k will depend on ρ and z, approximately constant, and on the variable w_0, half-spreading rate.

This theory considers only first-order linear effects; but, non-linear effects are considered negligible for the reasons exposed above. In fact, in the asthenosphere, the

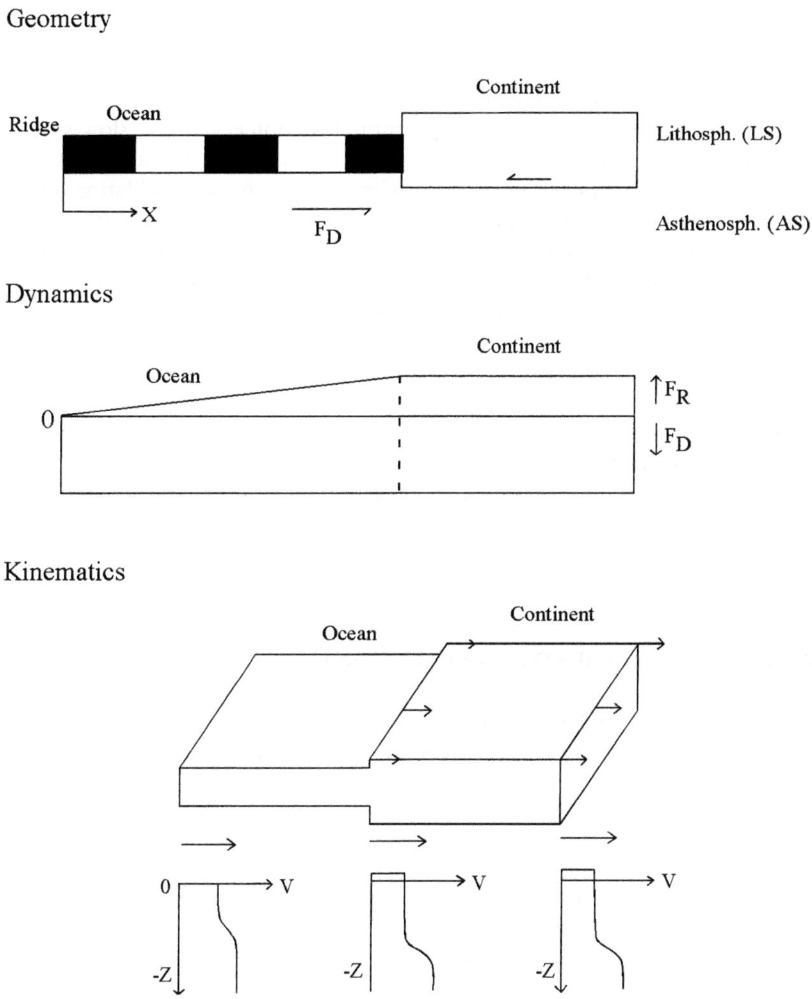

Fig. 3.18. Geometry, dynamics and kinematics of Atlantic-type ocean, spreading at a steady rate. F_D Driving force for plate motion. F_R Resistive force for plate motion. *X* distance to the ridge crest. *Z/V* Depth velocity profiles

steady slow viscous flow implies that the fluid is not accelerating (Davies 1999); the lithosphere is viscoelastic in a general sense, so the local acceleration varies in space, but also in time because there is a "memory" of the past in their actual rheological behaviour (Lliboutry 1992). These are the only rheology constraints in the model.

By analogy with free oscillation for Atlantic-type oceans, for Pacific-type oceans we should try the linear differential equation governing the dynamics of a forced oscillation, which is

$$m\frac{d^2x}{dt^2} = -kx + F(t)$$

where $F(t)$ is slab pull variable with time, t.

We have seen in Section 3.3.2 that the linear approximation is valid for slow-spreading Atlantic-type oceans; in this case, the resistive force is proportional to distance to the ridge. In the Pacific-type the linear approximation to a forced oscillation becomes invalid; perhaps for large speed plates the drag becomes proportional to a power of plate speed. Ridge push and slab pull extends for angular distances above a quarter of a circle and the axisymmetric solution is no longer applicable. Slab pull must follow a complicated function for two reasons: it must integrate the effects over the trace of sinuous subduction zones at highly variable distances to mid-ocean ridges and it should extend to considerable depth; converting the two-dimensional problem is a much more complex three-dimensional problem. The change in dynamics from linear to non-linear expresses the change in tectonic regime from stable homogeneous shortening to unstable buckling and faulting that we will refer to in Section 3.4.

All the oscillations are slightly damped, because there is a slight variation of density inside the plate as it moves away from the ridge crest, because the plate cools. We will consider this aspect in the next section.

3.3.4 Cooling and Thickening in the Oceanic Lithosphere

There is a well-known correlation between the depth of the ocean floor and the square root of its age (Parsons and Sclater 1977), for ages confined between 0 and 70 Ma, of the form

$$d(t)=2500+350\ t^{1/2}$$

with depth, d, in meters and age, t, in Ma. This correlation is explained by cooling of an elastic half-space representing the oceanic lithosphere in isostatic equilibrium (Turcotte and Schubert 1982). For the older ocean floor the previous correlation is no longer valid, but the following expression

$$d(t)=6400-3200\ \exp^{(-t/62.8)}$$

is a good approximation for $t>20$ Ma. This means that for older parts of the oceanic lithosphere additional heat transfer to the base of the oceanic lithosphere is needed in a pure cooling model. The empirical curves were fitted to the very similar North Pacific and North Atlantic data (Parsons and Sclater 1977). Hence, the curves reflect mainly the fact that the ocean is in isostatic equilibrium and are independent of the rate of spreading and of the path followed by the plate.

However, a recent synthesis of revised data (Davies and Richards 1992; Davies 1999) maintains that the single square root-age trend can explain bathymetry even for older ocean floor. In the same way the ocean lithosphere follows this simple square root of age versus thickness law (Turcotte and Schubert 1982).

The combination of satellite altimetry and ship depth soundings gives more accurate results than classical bathymetric maps (Sandwell and Smith 1997). The data can be interpreted by subsidence due to lithospheric cooling in the 055-Ma age range.

Beyond that limit the depth pattern flattens abruptly and there is increased variability of depth at older ages; these facts can be explained by a stochastic model in which randomly distributed reheating events caused by intraplate magmatism such as hotspot activity and raise of the ocean floor (Smith and Sandwell 1997).

The viscous component of deformation in the oceanic lithosphere proceeds under conditions of slight volume change, given density increase by cooling. This means that the very viscous fluid is not incompressible, but must be slightly compressible. This is also expressed by a very small thickening of oceanic crust with age, proved in the Pacific (McClain and Atallah 1986): crust younger than 30 Ma has a mean thikkness of 5.67 km, whereas crust between 30 and 100 Ma has a mean thickness of 6.01 km. This thickening is, in our interpretation, a consequence of homogeneous compression. In standard theory this effect will be a consequence of deepening of seismic Moho by serpentinisation of the upper mantle as a function of the increasing age of the oceanic lithosphere. The validity of each of the alternatives could be tested by comparing the thickening of crust in two oceans spreading at different rates, such as the fast Pacific and the slow Atlantic. If thickening is a result of deformation it should be larger in the Pacific than in the Atlantic. If it is a consequence of serpentinisation it should be age-dependent. Unfortunately, no data are available for the Atlantic, to the best of our knowledge.

3.3.5 Transform-Rift Relationships

Since the beginning of plate tectonics transform faults are a requirement of plate kinematics (Wilson 1965) in the sphere (Morgan 1968). However, their dynamic significance is still poorly understood; do they control or simply record plate motions (Cox and Hart 1986)? If the more usual relationship is orthogonality between ridge and transform, a requirement of minimum energy conditions (Lachenbruch 1973), why are there so many exceptions to this rule and, in fact, a complete spectrum of oblique rifting to leaky transforms (Talyor et al. 1994)?

In soft plate theory transforms are interpreted as discontinuities (Thom 1982) in the velocity field of a very slow-flow solid (Lliboutry 1992) which coincide with discontinuities on thermal structure of the oceanic lithosphere. The best analogy is the plate tectonics-like structures found in a crust-veneered lava column during an eruption at Kilauea Volcano, Hawaii (Duffield 1972). Rises, sink transform faults, and triple junctions were produced and dimension analysis even provides an estimate of viscosity for the convecting fluid in the asthenosphere of 10^{21} Pa. The most obvious difference between plate tectonics and the lava structures is the asymmetrical subduction in the first and symmetrical downwelling in the second.

Once we accept that transforms are discontinuities in the velocity field that can exist in very slow solids, but also in rapidly moving fluids, such as in the atmosphere and hydrosphere (Pedlosky 1987), many puzzling aspects of transform faults in standard plate theory can be explained by assuming a viscous component in plate rheology instead of a rigid rheology.

In rigid plate theory the jagged shape of a mid-ocean ridge system is an inherited feature from the geometry of the initial breaking up of the continent. However, this can be challenged by another point of view where the "mid-ocean ridge is a dynamic system which evolves to adjust to changes in plate motion" (Sandwell and Smith 1995). From the dynamic view, we can explain many puzzling aspects of the transform-rift system much better.

Velocity discontinuities can be strictly parallel, but with opposite signs of movement; this happens with transform faults and transfer faults; they differ because there is area preservation in transfer faults, but not in transforms. Both are compatible with rigid or soft rheologies, producing both hard and soft links (Roberts and Yielding 1994). Transform faults can accommodate the breaking of a random shape inside the continent, but in a fixed direction; the geometry is sensitive to initial conditions, but the kinematics are completely deterministic.

If transform faults are velocity discontinuities they must not end abruptly against rift segments as the rigid theory requires, where the active transform becomes an inactive fracture zone inside the divergent plates. These have topographic expressions that are explained by different cooling histories on each side. Thermal stresses must induce along them a dip-slip fault mechanism with the young side subsiding relative to the old side, shown by focal mechanisms of earthquakes.

In the soft plate model velocity discontinuities must cross plate interiors, in fracture zones. These must show strike-slip focal mechanisms caused by plate tectonic stresses, besides the dip-slip mechanisms caused by the thermal stresses (Fig. 3.19); the seismic moment release rate should be lower along fracture zones than in active transforms, because the displacement rates are one order of magnitude lower than in active segments. It is even predicted that fracture zones should show reversed polarity to transform segments, because the younger block near to the rift move at a faster velocity than the older block in the other side of the fracture zone (Fig. 3.20). This is, in fact, observed in a study of the Hayes transform in the Atlantic (Engeln et al. 1986) and in other cases (De Long et al. 1977). As we move away from the rift, the velocity field becomes more uniform and the heterogeneities caused by the presence of transform faults become smoothed out inside each plate; so this effect should be maximum near the mid-ocean ridge and fade away from it. In the rigid plate theory, the reverse polarity of the strike-slip component relative to the transform fault segment (De Long et al. 1977) is attributed to thermal contraction in the horizontal direction perpendicular to the ridge crest, at a higher rate in the young crustal and elevated side of the fracture zone. In this case, fracture zones should show coupled oblique slip, because the forcing mechanism is thermoelastic for both components, with the dip-slip component larger than the strike-slip component; as referred to in Section 2.5, this is valid for near ridge domains. However, in fracture zones, the focal mechanisms with reverse strike-slip components are largely predominant over almost pure dip-slip events. This can be explained by the soft plate model, with decoupling between the dip-slip component, thermoelastic in origin and largely aseismic and the seismic strike-slip component; the latter is due to discontinuities in the time of ridge push attenuated by the intraplate viscous component of deformation. This analysis applies to young oceanic lithosphere cooling fast; in slow cooling older oceanic lithosphere the situation is different and will be analysed in Section 3.9.

Transform faults are an essential feature of plate tectonics. Rift-transform interplate systems differ from graben-strike-slip intraplate fault systems in their boundary conditions (Fig. 3.21). In intraplate deformation area is preserved and a quasi-homo-

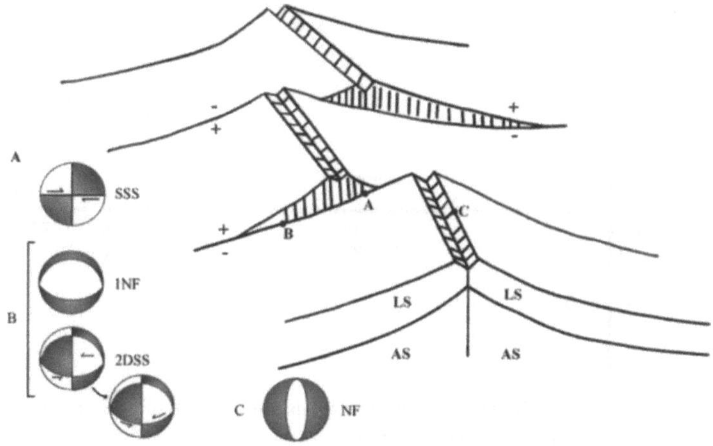

Fig. 3.19. Reversal of polarity in the strike-slip sense at fracture zones, shown by focal mechanism of earthquakes. *A* Transform fault with sinistral strike-slip motion; the ridge is displaced dextrally, the reverse sense, as is always the rule in transforms. *B1* Fracture zone is activated as normal fault with downthrow of the hanging wall that is cooling more rapidly according to the rigid plate theory. *B2* Fracture zone is activated with dextral strike-slip motion, opposite to transform from which it derives. This is compatible with soft plate, but not rigid plate theory. *C* Normal fault is parallel to the ridge

geneous stress field can explain both the normal faults and conjugate strike-slip faults. In intraplate deformation, the area is increased at rifts and the tensional stress field near the rift is deflected near transforms where there is torsion of the plates across the width of the transform fault zone, eventually with an opposite sense of movement between plates involved. This situation is analogous to the torsion of a circular cylinder whose axis passes through the Euler pole of rotation of the two plates involved. Maximum compressive stresses make angles of ±45° with the plane of torsion, which is, in this case, the transform fault plane and the intermediate stress axis is radial or vertical in this case. It has been demonstrated (Mandl 1987; Treagus and Lisle 1997) that two of the three principal planes of stress cannot be described as continuous surfaces. The torsion of an upright circular cylinder will produce principal σ_1 and σ_3 trajectories winding around the cylinder and making angles of ±45° with the horizontal torsion plane; σ_1, σ_2 or σ_2 σ_3 surfaces cannot be defined because σ_2 must be simultaneously radial at the upper and lower sections of the upright cylinder and also contain its axis; the surfaces in question must then be broken and no definable continuous surfaces are defined. It has also been demonstrated that continuous principal surfaces of stress or strain will not be defined in general for three dimensionally varying stress or strain fields, contrary to what happens in cases of plane stress or plane strain that can be analysed in two dimensions. It is meaningful that continuum mechanics defines its own limits; although this was realised more than one century ago, we are rediscovering its implications just now. It follows that transform faults are an excellent illustration of non-linear science.

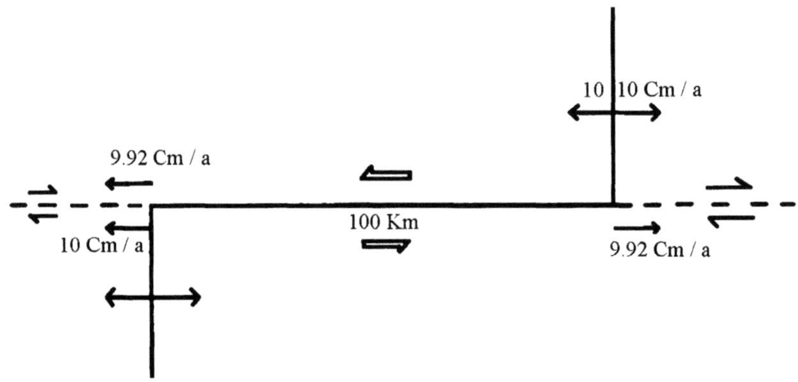

Fig. 3.20. Velocity field in soft plate theory can explain the reversal of polarity between fracture zone and transform fault from where it derives

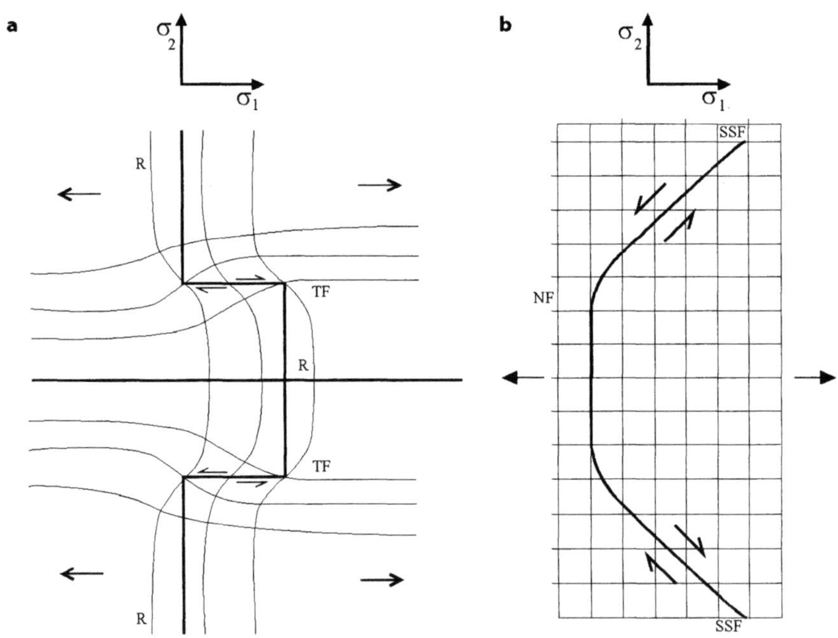

Fig. 3.21. Two-dimensional stress fields in inter- and intraplate fault systems. **a** Rift transform intraplate system. *R* Rift; *TF* transform fault; σ_1 least compressive stress; σ_2 intermediate compressive stress. **b** Graben – strike-slip intraplate system. *NF* Normal fault; *SSF* strike-slip fault; σ_1 and σ_2 as in **a**

Other anomalies are described at the junction of transforms and rift segments due to kinematic constraints, but cannot be used to differentiate between the rigid and soft plate models (Engeln et al. 1986).

3.4 Buckling of the Lithosphere

Geological observation supported by analytical solutions (Ramsay and Huber 1987; Price and Cosgrove 1990; Gosh 1993; Ramsay and Lisle 2000), physical analogue modelling (Shemenda 1994) and numerical modelling (Martinod and Davy 1994) suggest that after some homogeneous shortening geological materials tend to buckle. Both continental and oceanic lithospheres develop buckles, but many questions remain: what threshold triggers buckling? What is the most adequate rheology to explain the data? What factors control the mechanism of buckling?

We should expect that homogeneous shortening is easier in ocean than in continental lithosphere because the chemical environment is more favourable and volume loss is possible. Therefore, the transition from one tectonic regime to the other should occur at a higher lever of strain in the oceanic lithosphere; we have no data for the continental lithosphere, but we have some estimations for the oceanic lithosphere in the central Indian Ocean (Chamot Rooke et al. 1993; Martinod and Molnar 1995); the shortening is 3.4±0.9% with only 0.1% causing long-wavelength undulations (Gordon et al. 1998). It seems that in this case, a shortening of only a few percent can induce buckling. We will see that this threshold should depend on parameters that can vary between large values.

In the continental lithosphere (Burg and Ford 1997) the problem is more complex than in the oceanic domain; two facts control the behaviour in this case (Stephenson and Cloetingh 1991; Martinod and Davy 1992; Nikishin et al. 1993; Burov and Diament 1995): geothermal gradient and strength of the ductile lower crust.

If the geothermal gradient is high, leading to a temperature above 600 °C for the Moho, the upper crust is the only plastic layer; the models predict crustal buckling with wavelengths about four times the thickness of the upper brittle crust, in the order of 40 km; the model is extremely sensitive to the rheology of the lower crust.

However, for geothermal gradients leading to Moho temperatures between 450 and 600 °C, there are two plastic layers: the upper crust and the uppermost mantle. Therefore, two instabilities must develop: a long wavelength involving the whole lithosphere in a coupling mode with four times the thickness of the active lithosphere and a short wavelength involving only the upper crust. In Tibet (Jin et al. 1994) two wavelengths develop: a shorter one in the order of 150 km and a longer one, in the order of 500 km, on the basis of gravity and topography data. This suggests the presence of a weak decoupling zone at the Moho with 60 km depth in this region.

Flexure of the lithosphere has been modelled extensively (for a revision see Turcotte and Schubert 1982; Fowler 1990; Middleton and Wilcock 1994; Burov and Diament 1995; Ranalli 1995). Various rheologies have been assumed for the flexible plate and it is still being discussed which rheological model fits the existing data better.

If a body is stressed by some load it will react by a slight change of shape, either by elastic, viscous or viscoelastic small deformation. A significant change occurs when the components of change in shape are altered radically, by *breaking* and, or, by *buckling*; in this way buckling can be considered unstable rheological behaviour, inherently non-linear. The initiation of buckling must, in fact, be approached by the Catastrophe theory, a precursor to the Dynamic system theory (Poston and Stewart 1978).

In the simplest form, the deflection of a thin elastic plate is governed by the fourth-order differential equation of equilibrium

$$D\frac{d^4w}{dx^4} = V(x) - H\frac{d^2w}{dx^2}$$

where x is horizontal distance, $w(x)$ is the deflection of the plate, $v(x)$ the vertical force per unit length applied to the plate and H a constant horizontal force per unit length applied to the plate and is the flexural rigidity of the plate

$$D = \frac{Eh^3}{12\left(1-\sigma^2\right)}$$

where E is the Young modulus, h is the thickness of the plate and σ the Poisson's ratio.

In elastic flexure many parameters control the response of the plate, such as the load, which controls the curvature of the plate; the presence of intraplate stresses that change the geometry of the flexure and the wavelength of the load, that can isostatically compensate in variable degrees; for λ~100 km there is no isostatic compensation and for λ~1000 km the compensation is complete. Therefore, at long wave lengths the lithosphere has little flexural strength (Allen 1997).

The depth-dependent rheology is replaced by a uniform elastic plate analogue with an effective elastic thickness (EET; Cloetingh and Banda 1992); in continents, this EET varies strongly as a function of geothermal gradient with larger values in cratons (100 km) and lower values in young mobile belts and rifts (25 km).

There is also a strong correlation with the thickness of the schizosphere inferred from hypocentres of intraplate earthquakes. In the continental lithosphere it must be noted that estimates of EET have little correlation with any geological boundary within the lithosphere; they provide only a convenient basis for comparison between continental tectonic settings (Burov and Diament 1995, pp. 3906–3907); and the coupled/decoupled nature of crust-mantle relationship is expressed by a bimodal distribution of EET as a function of age; the lithosphere is decoupled when the crust is thicker than a critical value that increases with age from 35 to 40±5 km, close to the average thickness of the continental crust; below this critical value the EET is 60–75 km and above it can reach 110 km.

In oceans elastic rheology gives good results for vertical loads with durations of 10 Ma (Lambeck 1988, p. 500); for horizontal loads at a geological scale the elastic rheological model fails because it predicts huge compressive stresses much larger than lithospheric rocks can suffer (Martinod and Davy 1992).

We can address a general critique to modelling of the oceanic lithosphere buckling with elastic rheology; in the proposed models it is considered that the vertical load is applied in the absence of horizontal force. This would be valid if there was only nonrenewable stress in the oceanic lithosphere, but we know that this is not the case. This implies that we must look for other rheological models than a simple elastic lithosphere (Burov and Diament 1995, pp. 3906–3907).

The viscoelastic rheology has also been proposed; in this case, the duration of the load is fundamental. The shape of flexure changes with time, with an increase in amplitude and a decrease of the wavelength with increasing duration of loading. The viscoelastic rheological models allow the estimation of Maxwell relation time for the lithosphere as a function of the geometry of the flexure. Viscoelastic models have also been used for modelling the outer rise morphology in subducting plates (De Bremaecker 1977; Melosh 1978).

The choice between elastic and viscoelastic rheology is difficult to resolve; in fact, geodetic data have not sufficient resolution to detect changes of the geometry of buckles, both in space and in time. These are marked mostly in the stratigraphic record of sedimentary basins (Quinlan and Beaunmont 1984). However, this interpretation with viscous relaxation effects is difficult to separate from effects due to a change in the elastic constants which is due to thermal effects in the lithosphere under sedimentary basins (Middleton and Wilcock 1994).

More elaborate models remove the assumption of uniform lithosphere and the viscosity becomes temperature-dependent; plastic effects can be included (Ranalli 1995) with a constitutive equation of work-hardening plasticity. The plastic rheology fits the morphology of the outer rise better than elastic rheology (Turcotte and Schubert 1982). Some authors (Burov and Diament 1995) have studied plate equilibrium in a rheology-independent form, where moments of forces are function of the stress tensor; if the medium is inelastic (plastic, viscoelastic), it can be described as deformation of an elastic medium with variable elastic properties in space and time. The EET estimated is only valid for instantaneous or static plate geometry and may provide only the real strains, but not stresses; the EET will vary with changes in the plate geometry or in the distribution of loads.

We can summarise the rheology of buckled lithosphere (Gerbault et al. 1999) in the following way: if the lithosphere buckles prior to its brittle failure, the stress needed for buckling supersedes rock strength; if faulting occurs before buckling, then the stresses are too low to nucleate folding. A numerical model assumes folding as a mode of deformation and simultaneous faulting as a mechanism of brittle deformation (Gerbault et al. 1999); it has been shown that oceanic lithosphere with a thermal age of 60 Ma possesses a 40-km-thick competent layer and develops single layer faulting and folding with 6% shortening after 7 Ma of compression. In continental lithosphere with a thermal age greater than 400 Ma two modes can develop, depending on strength and thickness of the lower crust: a thin strong lower crust favours a coupled mode for crust and mantle folding; a thick weak lower crust favours a biharmonic folding with two superimposed wavelengths for crust and for mantle. It is concluded that the presence of faults weakens the lithosphere, both in compression and in tension (Wees and Cloetingh 1994)

These results stimulated some authors to apply a viscoplastic rheology for the most well-studied case of lithospheric folding in the Indian Ocean. This is a diffuse plate boundary (Bull and Scrutton 1992; Chamot-Rooke et al. 1993) where the Indian Plate, which is moving north, is slowed by the India-Asia collision in the Himalayas whereas the Australian Plate is subducted under Asia and moves more rapidly to the north. This causes high magnitude seismicity, geoid anomalies, large gravity anomalies and active compression giving large undulations in the oceanic basement; they are perpendicular to the shortening direction with an average wavelength of 200 km and amplitudes of 1–2 km. The active tectonics started 7 Ma ago, possibly in relation to uplift of the Tibetan plateau at that time (Molnar et al. 1993).

The kinematics of these structures have been attributed to buckling (McAdoo and Sandwell 1985; Karner and Weisel 1990; Bull et al. 1992; Beckman et al. 1996) or to inverse boudinage (Smith 1977) of the lithosphere (Zuber 1987). Seismic imaging (Loudon 1995) shows higher Moho below topographic highs, favouring the first deformation regime. Numerical modelling confirms this view (Martinod and Davy 1992). On the other hand, the wavelength of the folds is about four times the thickness of the plastic plate, in accordance with the observed wavelength of 200 km in a 40-km-thick schizosphere.

Martinod and Molnar (1995) modelled the rheology of the oceanic lithosphere as follows: the schizosphere obeys a modified Byerlee law which supposes that fluid pressure increases proportionally with the lithosphere pressure to a depth of 38±8 km for the 60-Ma old lithosphere, where seismicity stops; this is the brittle layer being effectively plastic; the plastosphere obeys the flow law for creep of olivine. This strength envelope implies a differential stress, σ_d, of 35–140 MPa during compression of the lithosphere, equivalent to a force F per unit length, L, of $3.5\text{–}14\times10^{12}$ N/m depending on the thickness of the lithosphere (H) because $FL^{-1}=\sigma_d H$. This is much less than the Byerlee law (1978) in the absence of significant fluid pressure; the plastosphere is weak, allowing a rapid growth of the dominant wavelength.

The authors study the growth of folding by analytic calculations based on a perturbation method; assuming that lithospheric buckling is not isostatically compensated and lithosphere strength must support the stresses arising from gravity acting on horizontal density variations. It is concluded that the thickness of sediments controls the amplitude of undulations of basement, which develop only under the sediments of the Bengal deep sea fan.

Physical modelling (Shemenda 1994) is also successful in the simulation of lithosphere deformation in the central Indian Ocean. It must be remarked that this success of physical modelling is a consequence of simulating the continental lithosphere with a greater thickness of the same material than the oceanic lithosphere; this suggests that the idea of a stiffer oceanic lithosphere required by the standard theory is not supported in these models.

Recently, it was shown (Royer and Gordon 1997) that if the Indo-Australian Plate is subdivided into three subplates we obtain a better approximation to the rules of plate kinematics. The three subplates are from N to S, the India, Capricorn and Australia subplates. The three subplates become divided by two diffuse plate boundaries: near the ridge system of the Indian Ocean there is diffuse divergence in two triple junctions: one between the Somalia, India and Capricorn plate; the other between the Australia, Capricorn and Antarctica Plates. These diffuse divergent boundaries pass laterally to two diffuse convergent boundaries between India and Capricorn and between Capricorn and Australia; the two convergent boundaries verge to a unique plate boundary zone SW of the Sumatra–Java trench. The rate of convergence and divergence along these plate boundary zones is very slow, below 20 and 10 mm/year respectively, suggesting that this rate controls, at least in part, the behaviour of the oceanic lithosphere when compared with the continental lithosphere. As the subplates Australia, Capricorn and Antarctica are spinning quite fast relative to each other, as shown by the fact that respective Euler poles of rotation are inside each subplate, the angular strain rates involved in the deformation of these diffuse plate boundary zones are quite high. This suggests that buckling is initiated after some finite strain has accumulated by homogeneous shortening or if the strain rate rises above some threshold. Across the Capricorn-India subplate the shortening is 3.4±0.9% since 7 Ma (Martinod and Molnar 1995) with only 0.1% caused by long-wavelength buckling. A more recent study (Gordon et al. 1998) suggests that the motion between the three plates in the central Indian Ocean started earlier, at ca. 20 Ma, and kinematics changed between 20 and 11 Ma; the convergence estimates from seismic profiles are valid only for the last 7.5–8.0 Ma. In this model the rate of convergence was slow (~1 mm/year) from 18 to 11 Ma, increased to ~4 mm/year between 11 and 3 Ma and further increased to ~6 mm/year since 3 Ma; this is only half the spreading rate in the slowest mid-ocean ridge.

These previous results cannot be extrapolated to situations where the rotations are at much larger angular distances to the Euler pole; at Tore Seamount, which we will refer to in more detail in Section 5.3, we estimated an homogeneous shortening of 6±1% from 95 to 5 Ma and buckling of 15±3% since 5 Ma. In fact, in other zones active buckling must exist in the oceanic lithosphere, such as the diffuse boundary Eurasia-Africa east of Madeira-Tore rise and to the west of Gibraltar. They are clearly shown in satellite altimeter maps of the oceans (Sandwell and Smith 1995 www); they are more frequent in older oceanic lithosphere. In fact, the signals for boudinage and for buckling are entirely different; boudinage produces a fine tectonic fabric parallel to plate motion, given by transforms, in younger domains; buckling produces a coarse tectonic fabric perpendicular to plate motion in the oldest oceanic domains near passive margins, or as we approach subduction zones. This confirms that there is a succession of deformation regimes in the oceanic lithosphere (Fig. 3.22).

In the first stage the material starts to deform by homogeneous shortening or layer-parallel shortening with the major axis of strain ellipsoid, X, subvertical; the minor axis, Z, parallel to compression; along the intermediate axis, Y, one can have no deformation (Y=R, the radius of the original sphere) or a slight extension (Y>R), with boudinage at right angles to Y. In the oceanic lithosphere we should expect this slight extension because it cools as it move away from the mid-oceanic ridge.

In the second stage, instabilities are amplified and the material starts to buckle and fault; the fold axis will coincide with Y, a direction of continued extension because of continued cooling; these buckles will amplify and will accommodate increasing movement in faults; in the oceanic lithosphere this happens in regions of high intraplate deformation in incipient or well-established converging plate boundaries (Stein 1993).

In the last stage, the material eventually fails along surfaces at approximate right angles to Z; in the oceanic lithosphere this corresponds to initiation of subduction, in the oldest parts of the oceanic lithosphere, where significant strain has accumulated (Ribeiro 1994, 1998b; Shemenda 1994). A turning point on the Wilson cycle has been reached and a passive margin will evolve to an active margin(see Sect. 4.6).

Proof that the oceanic lithosphere of young to mature age can be shortened without buckling was recently published (Chu and Gordon 1999). A finer plate kinema-

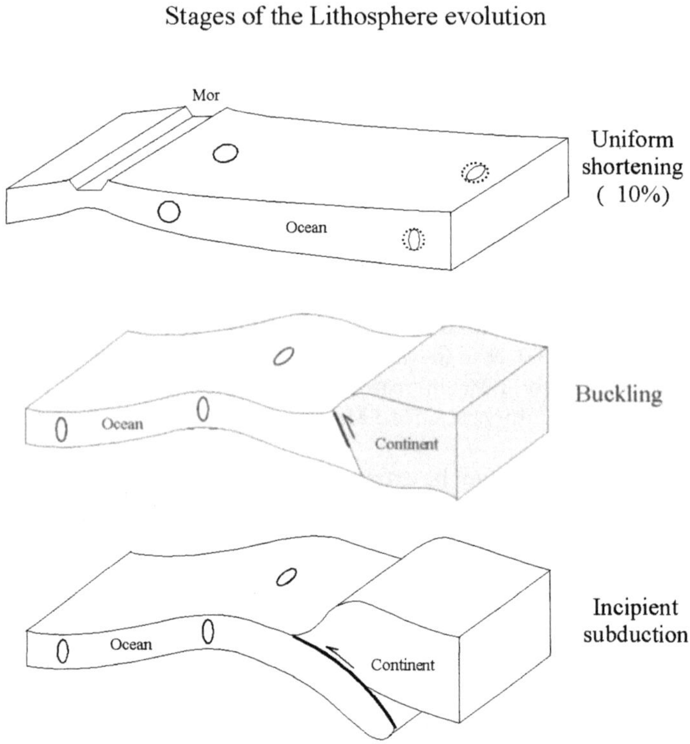

Fig. 3.22. Strain regime stages of evolution of oceanic lithosphere. In the *uniform shortening* stage there is homogeneous deformation by layer parallel shortening. In the *buckling* stage an instability starts to develop in the old oceanic lithosphere with synchronous faulting. Finally, whole lithosphere failure leads to *incipient subduction* at the contact of the oceanic and continental lithosphere

tic model divides Africa into two plates: Nubia (West Africa) and Somalia (East Africa), as already proposed in older kinematic models (Le Pichon et al. 1973); a Nubia Somalia Euler pole is located SW of Madagascar. It explains the divergence across the East African Rift at approximately 6 mm/year to the north of the pole and convergence at 2.4±0.6 mm/year in a diffuse zone between the Euler pole and the SW Indian Ridge. This shortening has no expression in topography, or gravity and no narrow band of earthquakes. This contrasts with the behaviour seen in the central Indian Ocean and shows that the tectonic regime is not buckling, but homogeneous shortening of the oceanic lithosphere by aseismic processes proceeds at a very low strain rate. This is exactly the behaviour that we admitted in Section 3.3. and previous publications (Ribeiro 1993, 1996). The dynamic modelling suggests a force per unit length of less than $\sim 1.1\times 10^{13}$ Nm^{-1} for contraction of the oceanic lithosphere with ages from 0 Ma at the ridge, to 100 Ma near the continental margin of Africa. The width of the deformed zone is considered to be around 2000 km.

An implication of the splitting of Africa into the Nubia and Somalia Plates is that on the intersection of segments of SW Indian Ocean ridge with the diffuse plate boundary zone between Nubia and Somalia Plates there is a very peculiar strain state with extension nearly perpendicular to the ridge, but with compression parallel to it; this is proved by the focal mechanism of an earthquake on 26 February 1988, with a moment magnitude of 6.8, near the ridge.

In conclusion, we must notice that the EET of the lithosphere calculated from the shape of buckle and associated Bouger anomaly is always less than the thickness of the lithosphere inferred from rheology profiles. The concept of EET applies to the response to vertical loads on a 10^3-year scale; and rheology profiles intended to explain the response to horizontal loads on a 1-Ma scale. In fact, the response of the plastosphere to short load can be by reversible transient creep, with no effect on flexural rigidity; but, on a 1-Ma scale, the plastosphere must be ductile (Lliboutry 1998).

3.5
Incremental Seismic Strain: Viscoelastic Rebound

During seismic cycle both seismic and aseismic strain can occur. The seismic coupling coefficient χ is defined as (Scholz 1990, p.284)

$$x = \frac{\dot{M}_o^s}{\dot{M}_o^g}$$

where $\dot{M}_0^s$ is the seismic moment release rate and $\dot{M}_0^g$ is the moment rate calculated from fault slip rates or plate kinematics at plate boundaries. This allows an estimation of the degree of instabilities in faults and hence, of their rheological properties, in both interplate and intraplate environments. If aseismic permanent strain occurs, we infer that elastic rebound is only a first approximation to the rheology of the schizosphere layer; as an alternative we should consider viscoelastic rebound as a better solution. The brittleness of the schizosphere does not preclude that its underlying mechanism is compatible with a viscoelastic rheology as a model macroscopic stress-strain-time behaviour (Scholz 1990, p. 34); this behaviour can be expressed in various temporal and spatial patterns of seismicity well described by fractal concepts (Turcotte 1997), as we will see in Section 3.6.

A pure elastic schizosphere would require a pure elastic rebound as a mechanism for generating earthquakes. The real seismic cycle is more complex than a simple step of instantaneous displacement in an elastic medium. Geodetic methods show the presence of preseismic, coseismic, postseismic and interseismic slip affecting large areas around active faults that generate the seismic events. Both spatial and temporal distributions of strain indicate that the lithosphere should have a viscous component on its rheology giving a response which is viscoelastic sensu lato. This can be explained by aseismic slip in the plastosphere or by coupling to a viscous asthenosphere (for a review see Scholz 1990).

In all phases of seismic cycle there are features that cannot be explained by elastic fracture mechanics. In compound earthquakes (Scholz 1990, pp. 210–221) there is a delayed rupture in different active faults which is too long to result from elastic processes; this is due to viscoelastic relaxation in the immediate postseismic period. This explains the fact that active fault modelling leads to an apparent thickness of 4 km, much lower than the thickness of 10–15 km of the schizosphere (Stein et al. 1988), showing that it weakens due to fault activity. The viscoelastic response can be due to coupling to asthenosphere relaxation or pore fluid coupling within the schizosphere itself.

Aftershocks obey the Omori law, with hyperbolic decay. They relax the stress concentration produced by dynamic rupture of the main shock, with a time-dependent strength that can be caused by static fatigue – a process of failure after a characteristic time under the application of load below the instantaneous breaking strength or by fluid diffusion induced by pore-pressure changes at the tip of earthquake rupture (Scholz 1990). Afterslip, sometimes episodic, accompanies aftershock, but is aseismic and occurs in weak rocks, such as sediments, but also in strong rocks by viscoelastic relaxation. A recent study (Heki et al. 1997) has shown that after an interplate thrust earthquake at the Japan Trench, on 28 December 1994, with a moment magnitude M_w=7.6 afterslip releasing an equivalent energy of M_w=7.7 occurred over the period of 1 year , according to inversion of GPS measurements. This demonstrated that afterslip can contribute to the small seismic coupling coefficient found world-wide in deep sea trenches due to the presence of weakly-coupled patches along subduction zones. Aftershocks and afterslip can be explained by a quasi-ductile continuation of the coseismic deformation or, as a relaxation in the schizosphere of stresses accumulated in the plathosphere after the main shock. In both cases the schizosphere suffers quasi-ductile flow and is a brittle-elastic material and not a strictly elastic one (Unruh et al. 1997).

Aseismic slip is not restricted to the plastosphere and is known to occur also in the schizosphere. Active faults can exhibit stable sliding, stick-slip, or both (Scholz 1998). Aseismic slip cannot be measured by seismic activity and the seismic coupling factor is then below a value of one.

Aseismic slip depends on lithology, tectonic setting and environmental conditions, namely pore fluid pressure. The frictional stability conditions are only known in very few rock types; serpentine, for instance, can stabilise sliding; ductile faults occur in weakly consolidated sediments. Dêcollements seem to be aseismic, especially if they are lubricated with weak materials such as salt. Layered rocks suffer deformation by folding related to active faulting and earthquakes (Stein and Yeats 1989).

Fault creep as stable aseismic fault slip is rare (Scholz 1990; Yeats et al. 1997) and seems to be confined to specific segments of active faults in California, probably due to the presence of serpentine in fault gouge, and to the North Anatolia fault.

A record of seismic coupling in different tectonic environments shows some regularities (Scholz 1990). Aseismic fault slip is rare in the continental schizosphere, except in areas with active dêcollement.

In oceanic ridges most spreading is aseismic, and the main process is magmatic accretion, probably by dyke intrusion along the central active volcanic zone.

In oceanic transform faults the seismic coupling factor is negatively correlated with the spreading rate and positively correlated with $\sqrt{T}$ where T is the average age of the oceanic crust. This suggests that the schizosphere boundary is controlled by temperature, in particular the 600–800° C isotherm. The negative correlation with the spreading rate probably depends on the amount of serpentinisation of the peridotitic upper mantle; we recall that the presence of serpentinite, even in small quantities, may stabilise sliding.

In subduction zones there is an evident lack of complete seismic coupling (Scholz 1990); we can synthesise by saying that "seismic coupling correlates positively with convergence rate and negatively with slab age, taken together" (Scholz 1990; p. 319).

The existence of intraplate earthquakes, representing something like 5% of the global seismic moment release (Scholz 1990) is a clear proof that plates are not rigid, but deformable. Some intraplate earthquakes occur in broad zones near and tectonically related to plate boundaries, showing that these are more or less diffuse; others occur in mid-plate regions and are unrelated to plate boundaries. These mid-plate earthquakes are generated by consistent stress direction over large areas defining *stress provinces* (Zoback and Zoback 1980; Zoback 1992). This suggests that they reflect plate tectonic driving forces reactivating previous inherited heterogeneities inside the plates. The mid-plate earthquakes are more frequent in the continental than the oceanic lithosphere; the standard interpretation is that the oceanic lithosphere is stronger than the continental lithosphere. However, another possibility exists: aseismic strain is higher in the oceanic than in the continental domain, caused by different rheologies if the oceanic lithosphere is viscoplastic and the continental lithosphere is elastoplastic.

The complexities in the relationship between seismic, and aseismic slip are best illustrated by the study of the Landers earthquake, 28 June 1992, the largest in southern California in the last 40 years, with M_L=7.4. The rupture is 70 km on a NNW fault, with a maximum displacement of 6 m, right-lateral. Within a few source dimensions of the earthquake rupture many aftershocks occurred for some 48 days, and a major triggered earthquake, Big Bear, with M_L=6.4 occurred the same day. This can be explained by linear elasticity (Jaumé and Sykes 1992; Stein et al. 1992), assuming static stress changes caused by the main shock. Stress is transferred to neighbouring zones, increasing in stress triggering zones and decreasing in stress shadow zones (Stein 1999) in alternate quadrant lobes. However, within minutes, triggered earthquakes of a magnitude below 5 occurred up to 1250 km from the source in areas of strike-slip or normal faulting with geothermal and recent volcanic activity. This is explained by a non-linear interaction between large dynamic strains by passing seismic waves from the main shock and crustal fluids and perhaps magma. (Hill et al. 1993; Scholz 1998). All the deformed area around the Landers main shock was studied by radar interferometry (Massonnet et al. 1993, 1994), a recently developed satellite geodetic technique (Massonnet 1997). The surface deformation could be modelled by a synthetic interferogram calculated with an elastic half-space dislocation model. However, the more interesting result is the behaviour of active faults located where the static stress changes are vanishing rapidly, but dynamic stress is still quite significant. A slip of 19±10 mm occurred during or soon after the main shock on the left-lateral ENE Garlock fault, 175 km NNW of the main shock. The segment which was activated is at least 100 km long (Cazenave and Feigl 1994, p. 108) and coincides with an alignment of aftershocks; to generate a surface rupture 100 km long by coseismic slip a M_L=7.5 earthquake should have occurred on this fault, which was not the case. We should infer that the triggered afterslip of Garlock fault occurred by a combination of seismic failure caused by small static stress changes and aseismic slip during the post-seismic period of the seismic cycle with a characteristic exponential decay time of ~48 days; this aseismic slip was triggered by acceleration of displacement of the plastosphere following the main shock (Anderson 1975) that diffuses stress rapidly(see Sect. 2.7); this contrasts with static stress changes that are only reduced by the next earthquake in the sequence; the aseismic slip expresses minor transient ductile displacements in the shear zones of the plastosphere situated below the active brittle faults of the schizosphere. (Fig. 3.23).

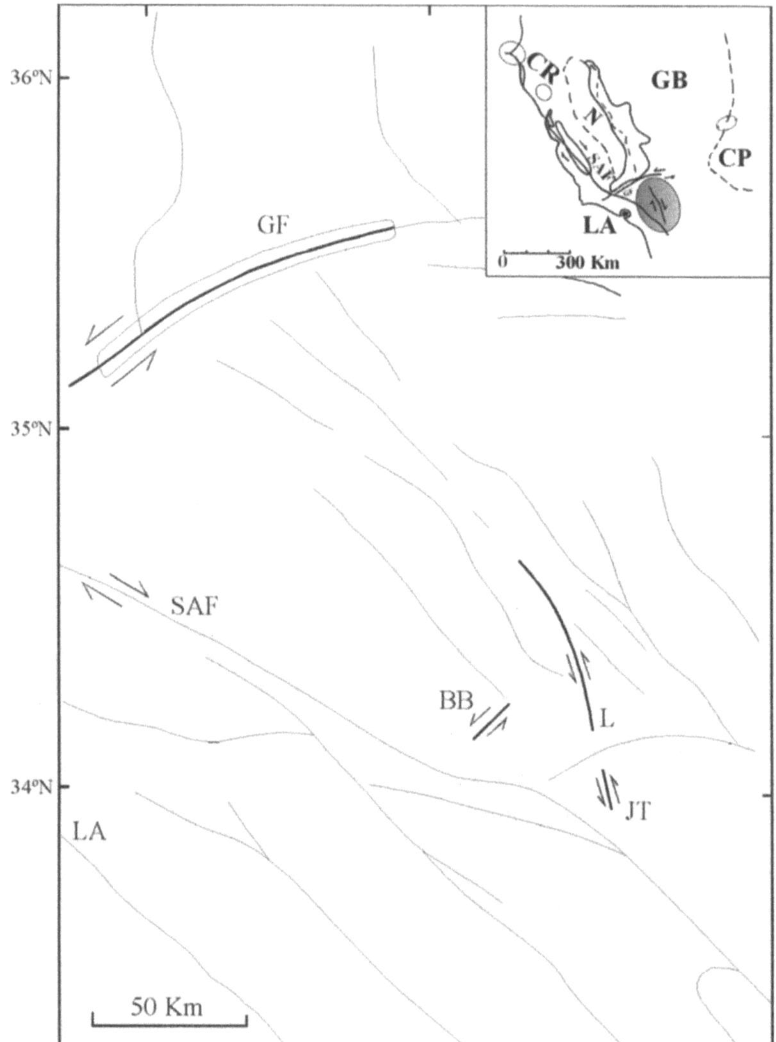

Fig. 3.23. Active faulting around Landers epicentre (28 June 1992). *Thick lines* are main active faults. Surface ruptures are thick lines associated with the following strike-slip earthquakes: *JT* Joshua Tree, M=6.1, April 23; *L* Landers, M=7.5, June 28; *BB* Big Bear, M=6.4, June 28; sense of slip is shown by *arrows*. The 100-km segment of Garlock Fault that moved aseismically is delineated. (Based on Jaumé and Sykes 1992; Stein et al. 1992; Massonet et al. 1994). *Insert* Seismic activity within 10 days after the Landers earthquake. Zone of static stress changes with aftershocks is shown by *ruling*. Zones with distant and small aftershocks are *delineated*. Structural provinces are *CP* Colorado Plateau; *CR* coast ranges; *GB* Great Basin and *SN* Sierra Nevada. *SAF* San Andreas Fault and Garlock Fault are shown; *LA* Los Angeles and *SF* San Francisco, are indicated. (Adapted from Hill et al. 1993)

Therefore, we propose a model of triggering of aseismic slip in active faults located beyond more than one or two source dimensions of the main shock by transient stress changes due to main shock. As other earthquakes do not showed this triggered seismicity and deformation features there is a probably a threshold of $M_l \approx 7$ below which the effects become negligible. This model can explain the puzzling fact that although many faults must move by aseismic slip, fault creep remains a very exceptional feature of active faults.

3.6
Tectonic Significance of Fractal Dimensions in Seismicity

3.6.1
The Gutenberg-Richter Law as a Fractal Distribution

Seismologists established (Gutenberg and Richter 1954) a correlation between the frequency of occurrence of earthquakes to their magnitude, m, such as

$$\log \dot{N} = -bm + \log_{10} \dot{a}$$

where b and $\dot{a}$ are constants and $\dot{N}$ is the number of earthquakes per unit time with a magnitude greater than m occurring in a specified area.

The Gutenberg-Richter law is just an example of a law obeyed by many variables in nature, where a characteristic dimension is not specified, but instead varies along many orders of magnitude; the earliest example is personal income in economics (Mandelbrot 1982; Schroeder 1991); many examples exist in earth sciences, such as distribution of faults, earthquakes, rivers, relief, meteorites, orebodies and many others (Middleton 1991; Barton et al. 1992; Turcotte 1997). Power laws are equivalent to fractal statistical distributions.

Another measure of earthquake size with better physical significance is the scalar seismic moment, M with

$$M = \mu A\bar{D}$$

where μ is the shear modulus of rock, usually taken as 3×10^{11} dyn/cm^2, affected by the active rupture area, A and mean displacement $\bar{D}$; M is measured in dyne-cm.

From scalar seismic moment M we can calculate the moment magnitude M_w such as:

$$M_w = \frac{2}{3} x \log M - 10.73$$

$$M_w = \frac{2}{3} x \log M - 10.73$$

The moment M and magnitude are related by

$$M_W = LogM = cm + d = 1.5M_s + 16.1$$

where c and d are constants and M_W is the moment magnitude.

The static stress drop, Δ_σ, during an earthquake is the difference between initial and final shear stress on the fault surface; it is given by:

$$\Delta_\sigma = C\mu(\frac{\bar{D}}{L})$$

where μ is the rigidity, L is the length of rupture, and C is a non-dimensional shape factor depending on the type of fault, but always near 1, so $C\approx 1$. This is simply the application of the theory of elasticity, with shear stress proportional to shear strain change,

$$\Delta_e = \bar{D}/L.$$

For a circular fault of radius R

$$\Delta_\sigma = C\mu\left(\frac{\bar{D}}{R}\right)$$

$$S = \pi R^2$$

hence,

$$M = \frac{\pi\Delta_\sigma}{C} A^{3/2}$$

and

$$M_w = \log M = 3/2.m + d = 1.5M_s + 16.1$$

with C=1.5 if we assume an approximately constant stress-drop model, valid for large inter-plate earthquakes (Kanamori and Anderson 1975). In fact, if L is a length dimension, seismic moment scales with the cube of it,

$$M_0 \alpha L^3$$

but magnitude scales with the square of it

$$m \alpha L^2$$

so moment and magnitude should be scaled by a power exponent of 3/2

$$M \alpha m^{3/2}$$

It was demonstrated (Turcotte 1997) that fractal dimension of distributed seismicity is

$$D = \frac{3b}{c}$$

and taking the theoretical value c=1.5, valid for events with intermediate duration compared to the natural period of the recording instrument, we obtain

$$D = 2b$$

So the empirical frequency-magnitude given by the Gutenberg-Richter relationship is equivalent to a fractal distribution (Aki 1981) with a fractal dimension of regional or global seismicity twice the b-value, the power law exponent. We must discuss now whether the Gutenberg-Richter relation is the result of a fractal network of faults, each producing characteristic events that are a function of fault length or if it results from power-law earthquake distributions on individual faults or some contribution of both (Turcotte 1997).

3.6.2
Earthquakes as Self-Organised Criticality

Earthquakes are an example of *dynamic systems*, whose state changes with time (Middleton 1991). They can be *conservative* if energy, mass or momentum are conserved; and *non-conservative or dissipative* if energy is lost; this is the general case in mechanical systems because of friction effects. The behaviour of dynamic systems depends on the *Lyapunov exponent*, a measure of the instability of the trajectory towards an *attractor* in *phase space* of a dissipative dynamic system. If a system diverges after n iterations from two slightly differing initial states, x and x+ε such as that $\varepsilon_{(n)} \approx \varepsilon e^{ny}$ λ is the Lyapunov exponent (Baker and Gollub 1990, p. 85).

A *fixed point attractor* implies that the system becomes fixed with time; a *limit cycle* implies the system becomes periodic; a *strange attractor* requires more than two dimensions and has a *fractal* structure, self-similar at all scales; it can explain *deterministic chaos* with sensitivity to initial conditions.

The Lyapunov exponent, λ, can be negative, if the system becomes periodic; it can be positive, $\lambda>0$, in chaotic systems; between both cases λ can be zero at the "border of chaos" (Bak 1990).

Cellular Automata (Wolfram 1984) can be used to simulate computationally dynamic systems; they were classified according to their asymptotic behaviour into class I, homogeneous and equivalent to systems with fixed point attractors; class II, periodic and equivalent to systems with limit cycle attractors; class III, chaotic and equivalent to systems with strange attractors. Class IV are complex and undecidable.

An analogy has been suggested with rheology behaviour (Langton 1990): class II would be equivalent to crystalline solids; class III to gases, completely disordered, and class IV to liquids, not completely ordered nor disordered, between classes II and III. It has also been suggested that class IV is just a transient between classes II and III (Bak 1996).

The fractal character of the seismicity process is the fingerprint, both in space and time, of self-organised critically (SOC). According to this theory, "large interactive systems naturally evolve toward a critical state in which a minor event can lead to be a catastrophe" (Bak and Chen 1991); the analogy used is a sand pile that, once the critical state is reached, can start an avalanche of any size, even catastrophic, when a grain of sand is added to the pile (Bak et al. 1988; Bak and Tang 1989; Ito and Matsuzaki 1990). This is considered to be weak chaotic behaviour, where uncertainty increases according to a power law as opposed to fully chaotic behaviour, where uncertainty increases according to an exponential law.

Some authors (Turcotte 1997), using the analogy of slider-block models, have stated that the interaction of segments in active faults produces windows of fully chaotic behaviour in the sequence of failures that causes seismic events along the faults. The differences between both models are important because long-term prediction is possible in weakly chaotic systems, but not in fully chaotic ones (Bak and Chen 1991, 1995; Bak 1996; Ito 1995).

We can state that in fully chaotic behaviour, the Lyapunov exponent is positive, $\lambda>0$; in weak chaotic behaviour, $\lambda=0$. An example of fully chaotic behaviour is the evolution of the atmosphere, where the weather prediction above a period of some days becomes impossible; this was in fact, the first example of deterministic chaos with sensitivity to initial conditions (Lorenz 1963): we think that the lithosphere behaves differently from the atmosphere. That is what makes earthquake prediction undecidable in the short term and with increasing accuracy in the long term; this is exactly the reverse of weather prediction. Therefore, we have to substantiate why we prefer the weak chaos concept in the case of earthquake faulting rather than the full chaos. In fact, the lithosphere is dragged by the asthenosphere; and the schizosphere is dragged by the plastosphere, a conclusion that we will refer to on various occasions throughout this book. Therefore, there is a continuous driving force, applied steadily in the base of the lithosphere, as plate tectonics tell us. This situation is maintained if the stage Euler pole does not change; and the sense of drag is only reversed at a turning point of the Wilson cycle, when subduction starts. This is quite different from the atmosphere where turbulence is fully developed. It is possible that the situation is unsteadier in the case of intraplate earthquakes than in the case of interplate earthquakes; but, even active intraplate faults do not change the sense of movement at random. Therefore, we think that self-organised criticality is the best theoretical model so far for earthquake faulting generation. It is also the best theoretical model so far to generate fractal objects in space such as active faults, and in time the so-called 1/f noise such as the earthquake time series, a point strongly emphasised by the author of the model (Bak 1996). The Gutenberg-Richter law is one of the best illustrations of the concept of self-organised criticality because it extends over 16.5 orders of magnitude (−2 to +9) "Earthquakes may be the clearest and most direct example of a self-organised critical phenomenon in nature" (Bak 1996, p. 85).

The experiments with real sand piles showed a very rich variety of behaviours, a strict power law, but also more complex laws, with power law small avalanches, no intermediate avalanches and quasi periodic big avalanches; the last case has some similarities with the non-linear characteristic earthquake concept. These different types of behaviour seem to be controlled by friction effects; low friction favours strict SOC behaviour (Bak 1996).

3.6.3 Self-Similarity and Characteristic Earthquakes

The generality of the Gutenberg-Richter Law (Gutenberg and Richter 1954) tends to prove that earthquakes are self-similar at all magnitudes because the b value is assumed to be constant and near one. In fact, the Gutenberg-Richter law is only approximate and the b value changes in space, according to tectonic regime and also in time, during the seismic cycle.

The Gutenberg-Richter law introduces scaling laws for earthquakes. In fact, the energy released during an earthquake, E, is simply

$$E = A\sigma D$$

where A is the rupture area, D the displacement and σ the mean value $(\sigma f+\sigma i)/2$ of the shear stress that acts before, σi, and after, σf, the earthquake on the active fault plane. From the previous equation the well-known relationships (Slemmons 1977; Wyss 1979; Wells and Coppersmith 1994) between magnitude, and rupture area or length, L, and displacement follow.

$M = \log E \propto \log A$ or $M = \log E \propto \log L$

$M = \log E \propto \log D$

For some faults the characteristic earthquake is well defined and the most stable parameter is the rupture area, A, because during successive earthquake cycles, barriers and asperities can survive (Aki 1984) and they control the geometry of the rupture area irrespective of fault type (Wyss 1979). Shear stress is obviously largely heterogeneous because the heterogeneity of the active fault rupture area must be high; in fact, fault planes are intrinsically heterogeneous (Aviles et al. 1987; Okubo and Aki 1987; Ribeiro et al. 1991) for purely geometric reasons and also because of heterogeneities in strength.

In conclusion, one must expect that the rupture area will be the most stable parameter because it only depends on the large-scale heterogeneity of fault plane, expressed for instance as the Barrier Interval (Aki 1984). In contrast, displacement per event and shear stress must be unstable because they ultimately depend on the "mismatch" between geometry and strength on the two walls of the fault. This incites us to investigate which parameters could control the rupture area of "characteristic" earthquakes. The generality of the Gutenberg-Richter law does not preclude deviations from strict self-similarity. A better approximation is found if we differentiate small earthquakes that can grow in both length (L) and width (W) from large ones, where L is unbounded, but W is limited by the thickness of the schizosphere (Pacheco et al. 1992). Databases and theoretical arguments (Rundle 1989) show that b should be 1.0 and 1.5 for small to large earthquakes respectively; small earthquakes affect an essentially two dimensional domain while large earthquakes affect a three-dimensional domain. A non-linear plot in frequency size at the logarithmic scale results from these considerations. A corollary will be that in the same seismogenic zone large and small seismic events belong to different fractal sets (Scholz and Aviles 1986) and do not obey the same scaling relationship (Fig. 3.24).

The same conclusion emerges if we plot in the same frequency size logarithmic plot data from instrumental seismicity and from palaeoseismology (Schwartz and Coppersmith 1984) for the same active fault. Different models were introduced to explain the frequency-size plots for each fault (Fig. 3.25).

In the *variable slip model* the displacement per event varies along the fault, but the slip rate is constant. A linear frequency-size will result in this model.

The non-linear plot can be explained by two models. In the *uniform slip model* the displacement per event is constant for each segment of the fault and the slip rate along the fault is constant. In the *characteristic earthquake model* the first condition also holds, but the slip is variable along the fault. These are probably end-member cases in a continuum of possibilities with the uniform slip model more probable in interplate fault zones and the characteristic earthquake model probable in intraplate faults because larger elastic strains can accumulate more easily in the segment boundaries (Scholz 1990). In both models some segments of the active faults produce constant size large earthquakes which do not scale with smaller earthquakes (Cluff 1978) that have been termed "characteristic earthquakes" in a loose broad sense with no specification of the underlying model (uniform slip versus characteristic earthquake). In the following we will use "characteristic" without specification in sensu lato

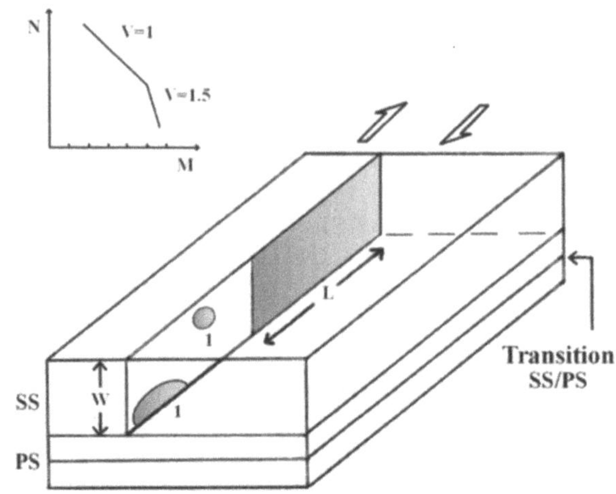

Fig. 3.24. Scaling laws for large and small earthquakes. The strike-slip active fault can rupture in large earthquakes that cut the entire thickness of the schizosphere (*SS*); the rupture is approximately rectangular with area A=LxW; this seismic moment is proportional to the square of length dimension $M \alpha L^2$; there are also smaller earthquakes, with circular or ellipsoid areas with diameter *l*, where the seismic moment is proportional to the square of the *l* and to displacement *d*, so $M \alpha L^3$. The two types of earthquakes lead to a non-linear distribution of frequency-magnitude, with $b \approx 1$ for small earthquakes and b reaching 1.5 for large earthquakes. (Pacheco et al. 1992)

and "characteristic earthquake model" in the precise sense that refers to the underlying model.

Some authors have discarded the concept of "characteristic" earthquakes on the basis of too short a period of sampling of data. However, examples of very distinct seismotectonic areas show repetition in the historical record of "characteristic" earthquakes (Schwartz and Coppersmith 1986), such as Alaska, Mexico, Greece, Japan, Turkey. The characteristic earthquake concept was confirmed by a study (Stirling et al. 1996) of 22 strike-slip faults associated with frequent and well-monitored instrumental seismicity, from California, Mexico, Japan, New Zealand, China and Turkey. Only in four of those faults does the extrapolation of instrumentally derived curves to larger magnitudes agree with geological estimates on the basis of palaeoseismological studies. Therefore, the characteristic earthquake concept is the rule and the linear magnitude-log frequency Gutenberg-Richter law for each fault is the exception if it really occurs.

The physical basis of the "characteristic" earthquake consists in the permanence of stable barriers in successive seismic cycles along the same active fault, allowing repeated rupture of an entire segment (Aki 1984). In addition, the tectonic evidence for fault segmentation is compulsive and confirmed by palaeoseismologic studies in the better studied active faults.

In fact, despite statements to the contrary, the characteristic earthquake concept does not contradict the self-organised critical model for earthquake faulting, but rather confirms and extends it. To prove this we will consider a model of an active boun-

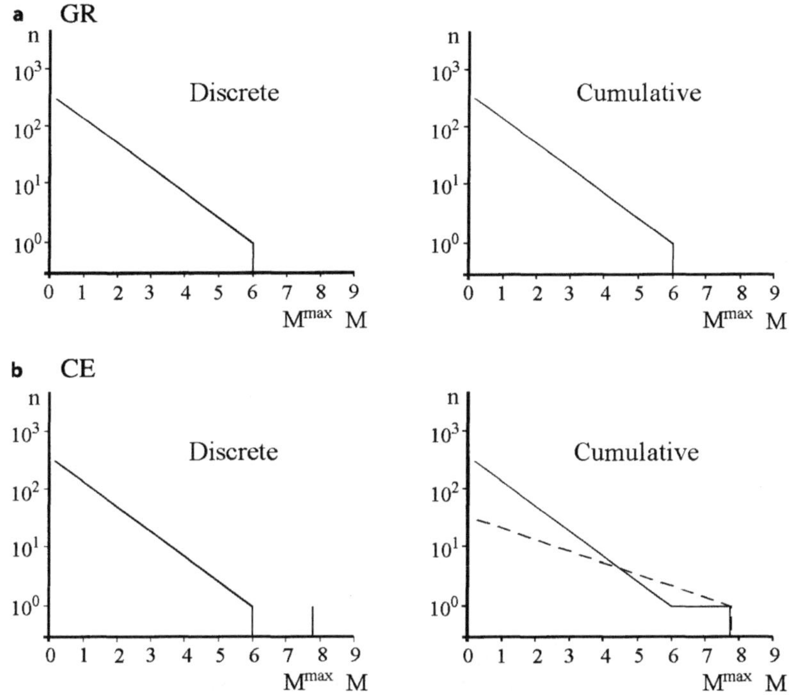

Fig. 3.25. Models for frequency-magnitude distribution. **a** *GR* Gutenberg-Richter relationship. In discrete and cumulative versions *M* is magnitude, *n* is the discrete number of events of a given magnitude per year and *N* is the cumulative number of events greater than or equal to a given magnitude, *n* and *N* are on log scale and *M* is the log of energy. **b** *CE* Characteristic earthquake model. The frequency-magnitude relationship is non-linear in a log-log plot. M^{max} is a the maximum magnitude. The *dashed line* is the distribution for asperities with low anomalous *b* values ($b \approx 0.5$) and M^{max} is more reliably estimated from extrapolation from the frequency-magnitude relationship for asperities. (After Schwarz and Coppersmith 1984; Stirling et al. 1996; Wiemer and Wyss 1997)

dary such as the San Andreas Fault (Fig. 3.26). In the plastosphere the boundary is represented by a broad shear zone, ≈80 km wide; it must be entirely ductile, even if strain is heterogeneously distributed; hence, no localisation of slip in narrow zones should occur in the plastosphere (Molnar and Gipson 1996).

In the schizosphere slip is concentrated on active faults; these show a very clear hierarchy; the first order is the San Andreas Fault itself; the second order is represented by shorter faults such as the Hayward fault; the third order corresponds to secondary branches to the second order ones; and so on down to minor branches with lengths in the order of 1–100 m. Active fault systems are similar to river systems that also follow a power law or fractal distribution, the Horton law (Horton 1945). Each order follows a non-linear Gutenberg-Richter law, with a “characteristic earthquake”; the seismogenic source is finite and so magnitude or seismic moment, correlated with the area of rupture, must also be finite, because in a critical system there is a finite si-

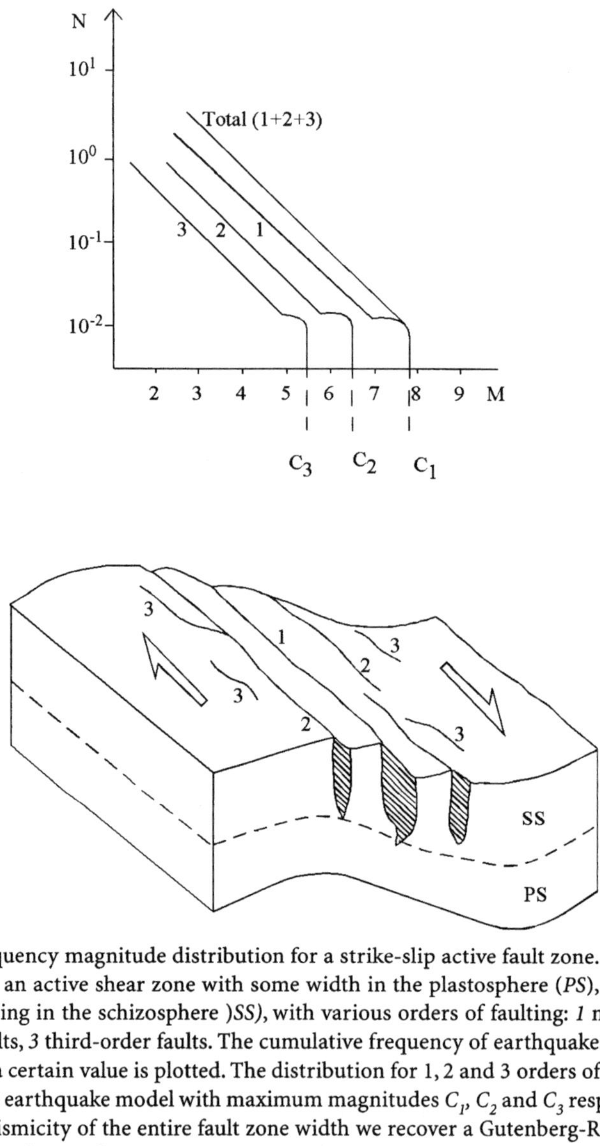

Fig. 3.26. Frequency magnitude distribution for a strike-slip active fault zone. The active zone corresponds to an active shear zone with some width in the plastosphere (*PS*), and distributed strike-slip faulting in the schizosphere)*SS)*, with various orders of faulting: *1* major fault, *2* second-order faults, *3* third-order faults. The cumulative frequency of earthquakes *N* with magnitude *M*, above a certain value is plotted. The distribution for 1, 2 and 3 orders of faulting follows a characteristic earthquake model with maximum magnitudes C_1, C_2 and C_3 respectively. Taking together the seismicity of the entire fault zone width we recover a Gutenberg-Richter relationship with b value of around 1

ze scaling effect (Olami et al. 1992). However, if we take the whole San Andreas zone we must also obtain a Gutenberg-Richter linear law for lower magnitudes and an upper cut-off characteristic earthquake generated in the first order San Andreas Fault. The limitation in this approach is that in the idealised model we suppose there is no deformation outside the shear zone ≈80 km wide; but we cannot isolate this fault zone from other intraplate faults adjacent to the plate boundary; and we cannot isolate

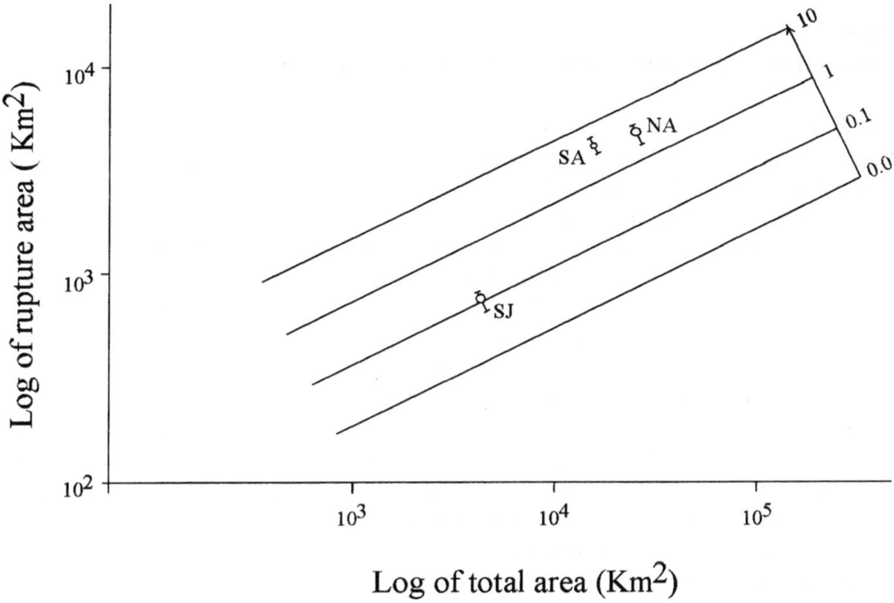

Fig. 3.27. Relationships between maximum rupture area *R*, total area *A* and slip rate *S* of active strike-slip faults. All variables are in log scale; *R* and *A* are in km^2 and *S* in cm year^{-1}. This is the graphic expression of the fractal law of the form: $R=a[A]^b[S]^c$ where *a*, *b* and *c* are constants (After Ribeiro 1986; 1990). *NA* North Anatolian Fault; *SA* San Andreas Fault; *SJ* San Jacinto Fault

the San Andreas fault zone from other plate boundaries connected by it, the Juan de Fuca ridge in the north and the East Pacific ridge in the south. Nevertheless, stress accumulation that causes seismicity must have tectonic natural boundaries rather than artificial boundaries, completely unrelated to the seismotectonics of the interplate fault zone until the depth of the schizosphere.

It has been suggested that the characteristic earthquake magnitude or seismic moment is a fractal function of total area of the specific fault and its slip rate (Smith 1976; Slemmons 1982; Ribeiro 1986, 1990, in press). The dependence on slip rate suggests that time effects are compatible with viscoelastic rebound, but not elastic rebound (Fig. 3.27).

A maximum credible earthquake for the entire schizosphere must also exist, above a measured moment magnitude of 9.5, and below a moment magnitude of 10.6, for rupture of an entire great circle (Yeats et al. 1997). This does not mean that other geological processes, such as meteorite megaimpact, could not reach a magnitude of 13, such as Chicxulub at Cretaceous-Tertiary boundary, 65 Ma ago (Hsu 1994) equivalent to approximately 30^4 times the energy of the biggest tectonic earthquake.

3.6.4
Regional Variations of b-Value as a Function of Tectonic Setting

The b-value established above is b≈1 at global scale, expressing strict scale invariance, but with variations in different tectonic environments. It was proposed (Miyamura 1962) that very low b-values (0.4–0.6) occur in old cratons, low values (0.6–0.7) in continental rift zones, medium values (0.7–1.0) for the Circum-Pacific and Alpine orogenic belts and high values of 1.0–1.8 for oceanic regions (Scheidegger 1982).

The mechanical interpretation for these values, in nature and in brittlely deformed rocks in the laboratory, measured by acoustic emissions (Main et al. 1990; Meredith et al. 1990) assumes a scale invariance of flaws with a fractal-dimension $1<D<3$. If the fracture system is dominated by one main fracture D tends to 1; if there is a volume filled with many small fractures D tends to 3; in the middle there are diffused systems, but with a hierarchy of fracture with D≈2. Hence, the b values should vary from 0.5 to 1.5 (Aki 1981). As there are also variations of the b value during the seismic cycle (Aki 1981), the b-values estimated by Myamura above 1.5, implying $D>3$ suggest that the sample regions were not stationary in time. In fact, some data (Evernden 1970; Fig. 4381) are better fitted by non-linear theoretical curves with a variable b-value. In general, there are many possibilities for relationships between fractal dimension and scale (Korvin 1992): D is independent of scale; D has two values as a function of scale in the so-called bifractals, or more than two values, or D is a gradually changing function of scale in multifractals. The bifractal model has been adopted by some authors (Pacheco et al. 1992): seismic moment scales as L^3 for small events, but L^2 for large events that rupture the entire schizosphere with constant thickness; in this model the b value can change from 1 to 1.5 at the global scale with an increase in seismic moment.

At the regional scale the variations in b value persist and its causes must be investigated, because there is the interesting possibility that they could be related to the rheology of the lithosphere in the different domains.

A recent study (Frohlich and Davis 1993) has showed systematic variations in the b value as a function of the same factor; in one of the catalogues b varies with the focal mechanism of seismic events: the mean value is 1.06 for normal faults, 0.86 for shallow thrusts and 0.77 for strike-slip faults. This suggests that a higher b value for the normal fault earthquake is a consequence of conditions of lower stress drop; hence, the b-value would be negatively correlated with the strength of the faulted media. The variations of the b value with tectonic environment are more problematic than in the previous study because of the geographic division adopted by the authors in the one proposed by Flinn, Engdahl and Hill (1974), which mixes different tectonic settings in the same region. Using the greatest range of magnitude, the b value varies from 0.73 to 1.16 and increases from low activity to high activity areas. A more detailed study with a regionalisation according to tectonic settings is needed in order to confirm the earliest findings by Miyamura (1962). We must refer to a study where a good correlation was found between the b value and age of the oceanic lithosphere in the Azores-Gibraltar plate boundary between Eurasia and Africa (Buforn et al. 1988). In the youngest sector near the Azores, $b=1.3\pm0.2$; in the central sector, $b=0.71\pm0.13$ and in the eastern sector, with the older lithosphere, $b=0.75\pm0.2$. In this case, the b value decreases with ageing of the oceanic lithosphere, suggesting a correlation between the b value and the lithosphere rheology: a lower b value implies a stronger lithosphere. What is the physical basis for b-value variations?

3.6.5 Rheological Significance of the b-Value

The b-value is routinely estimated in seismic studies. It is known that variations in the frequency-magnitude must be due to different physical processes; the parameters that control the b-value are material properties, such as heterogeneity; magnitude of applied shear stress or effective stress and temperature gradient (Wiemer and McNutt 1997).

The effect of heterogeneity is demonstrated in experiments of rock fracture (Mogi 1962). In experiments the b-value and hence the fractal dimension D can be estimated by monitoring acoustic emission (Main et al. 1990); temporal variations in D can be explained by fracture mechanics criteria for failure; they can explain the occurrence of various types of precursors in the intermediate and short terms, important in earthquake prediction. The spatial variations of the b-value can be interpreted from both experiments and field observations of seismicity. They reflect fractal flaw spatial distributions. $D=1$ ($b=0.5$) reflects critical stress concentration, with the probability of repeated fracture at a particular site at a maximum; this is equivalent to critical coalescence of microcracks during dynamic failure in experiment and foreshock in nature. With $D=2$ ($b=1$) the probability is independent of previous events, corresponding to random processes equivalent to background seismicity; with $D<2$ ($b<1$) there is stress concentration leading to localisation of deformation, with positive feedback process due to high stress intensities. With $D>2$ ($b>1$) we have negative feedback at low stress intensities, such as dilatancy, leading to highly diffuse fracture systems.

We conclude that fractal dimension of flaws increases as stress intensities decrease because length distribution of cracks becomes more heterogeneous. Stress intensity depends on applied stress and flaw length according to linear fracture mechanics (see Sect. 3.7). However, conditions in the laboratory and in the field are obviously different; the schizosphere is more heterogeneous than intact rock samples used in laboratory experiments. Hence, in the beginning of the fracture process $b=1.5$, $D=3$ in the laboratory and $b=1$, $D=2$ in the field, because the schizosphere already contains major faults and sutures.

The effects of stress are demonstrated by many observations (Wiener and Wyss 1997) : b decreases when ambient stress is increased in the laboratory; b is proportional to pore pressure in cases of fluid-induced seismicity; b decreases during mine bursts; b is lower in compressive than in extensional tectonic regimes, as we saw above; and b increases with depth along strike-slip faults in California.

The effect of temperature gradient is shown by an increase in b-value from regional values of $b\approx0.7$ to values of $b>1.3$ or even $b=1.5$ in volcanic regions as we approach active magma chambers (Wiemer and McNutt 1997; Wyss et al. 1997). This change in b-value can be explained if the fractal dimension D, which we demonstrated is twice the b-value, changes from 1.5 to 3, or in other words, the crack system changes from the tendency to fill a plane with topological dimension 2 to the tendency to fill a volume with topological dimension 3. Theoretically, the distribution of earthquakes has five dimensions, corresponding to three spatial co-ordinates, one time co-ordinate and a size co-ordinate, magnitude or moment (Goltz 1997).

It is possible that the three proposed physical mechanisms for the variation in b-value play in the same sense, for instance in volcanic areas (Wiemer and McNutt 1997): the vesiculation of ascending magma can induce an increase in crack density or heterogeneity in the material; alternatively pore pressure can increase by interaction of magma and groundwater, decreasing the effective stress. We conclude that the three proposed mechanisms for an increase in b-value reflect an increase of the viscous component in the deformation of a material with viscoelastic phenomenological behaviour, even if the underlying deformation mechanism is of a brittle nature, by crack nucleation and growth.

In other cases one of the mechanisms should be predominant; when there are spatial variations of b-values in areas with no significant temperature gradients we can exclude the last factor and consider only the other two. An example is the mapping of asperities in strike-slip faults of California, such as San Andreas and Calaveras (Wiemer and Wyss 1997): high stress patches along active faults have lower b values of $b \approx 0.5$, whereas segments with creep and post-seismic strain have higher values between 1.1 and 1.3. Therefore, asperities are locked during stress buildup and most of the energy is released during a seismic event, whereas the enclosing segments are weakly coupled, possibly due to high pore pressure and are passively dragged during the same seismic event. This is additional evidence for the concept of a characteristic earthquake for a specific segment. In fact, the variation of the b-value between segments of the same fault demonstrates heterogeneities of active fault at the scale of the segment; if we consider various segments in the same magnitude-frequency plot, we mix different b-values for each segment and we obtain a mean value of $b \approx 1$; the estimation of a characteristic earthquake magnitude from palaeoseismology and historical seismicity plots above the extrapolation of a maximum credible earthquake for each fault from instrumental seismicity underestimate this important parameter in earthquake prediction; a better estimation is obtained if we consider only the b-value inside asperities (Wiemer and Wyss 1997), affording better agreement with palaeoseismic and historical maximum earthquakes and confirming that the characteristic earthquake model is the rule and not the exception. This opens the possibility of mapping of asperities by mapping spatial variations of the b-value along an active fault zone and assigning a characteristic earthquake magnitude on each segment on the basis of rupture area-magnitude relationship. We will see in the next section (Sect. 3.7) that as an alternative we can individualise each segment by mapping the spatial variation of fractal dimension of active fault traces along the fault zone by tectonic methods (Ribeiro et al. 1991) if there is a simple relationship between seismogenic process and the geometry of active faults. Instrumental seismicity on other faults (Wyss et al. 2000) shows that low b-values occur in segments with complex traces even if fault roughness can oscillate during a single seismic cycle.

We conclude that experiment and observation support the view that the b-value reflects the behaviour of the material where seismicity is generated. The same conclusion is also supported by theoretical models which we will mention next.

We will address the simulation by Cellular Automata of a Burridge-Knopoff spring block model (Olami et al. 1992). This is a two-dimensional dynamic system of blocks interconnected by springs and can be used as an analogue of stick-slip processes generating earthquakes. A moving rigid plate is connected by a set of springs to the array of blocks resting on a fixed rigid plate. When the force exerted in one block exceeds a threshold value – the maximal static friction – one block slips; this will redefine forces on its nearest neighbours inducing further slips and eventually resulting in a chain reaction. The elastic constant of springs is α, independent of direction in the isotropic case. As each block is connected to the four nearest neighbours the maximum value for α is one fourth and the model is *conservative*; when α is less then one fourth, the model is *non-conservative* because there is a dissipation of energy in the dynamic system. If α=0 there is no interaction and no movement of blocks. If the average distance between blocks is fixed a characteristic length is introduced into the model, simulating the presence of stable barriers (Bak and Chen 1995).

The mapping of the analogue model by this continuous non-conservative cellular automaton exhibited self-organised criticality (SOC) even if was not tuned to be conservative, demonstrating the generality of SOC. (Bak 1996). The results of the simulation showed a negative linear correlation between the power-law exponent β with the elastic parameter α. As α varied from 0.25 to 0.05, b varied from 0 to 2.5. The physical significance of this simulation is that the viscous component of rheology in a viscoelastic lithosphere increases as the b-value increases for each seismotectonic domain.

Further evidence that a connection exists between classes of Cellular Automata (CA) and rheology behaviour has been presented for stick-slip phenomena (Markus et al. 1996); it was shown by simulation that the CA class changes if we vary the parameter energy dissipation in non-conservative systems; class II is the exception, class III is the rule and class IV become more frequent as we increase energy dissipation; however, there are erratic windows of class IV inside the domain of class III for unknown reasons. This is a possible explanation for the fact that very few earthquakes have been predicted by empirical observation of anomalous precursors; short-term earthquake prediction is not impossible, but undecidable (Goltz 1997; Markus et al. 1999). It is also remarkable that a sand pile is simultaneously solid and liquid, according to the continuum micropolar theory(see Sect. 2.6). Hence, the self-organised criticality state of the lithosphere is on the border of solid automata of class II and fluid automata of class III, supporting Bak's views (1996).

The same conclusions were independently obtained (Steacy et al. 1996) by a two-dimensional cellular automaton with strength heterogeneity, but homogeneous tectonic loading. The b-value increases with increasing range of heterogeneity and decreases with increasing fractal dimension of strength, or roughness, of the active fault zone. Similar models (Steacy and McCloskey 1999) have shown that rough faults produce power-law earthquake distributions, and smooth faults are better described by the characteristic earthquake model.

In a recent study (Kumagai et al. 1999), a self-organised model of earthquakes with constant stress drop and b-value of 1.0 has been proposed. This result is obtained in an anisotropic spring-block model with the direction of slip restricted to the loading direction. The parameter that controls the behaviour of the model is the ratio of particle velocity characterising the initial stress to the one characterising the velocity weakening process. An increase in this parameter brings the system to a minimum stress state, but away from the more homogeneous stress state; decreasing the parameter, the stress state becomes more homogeneous, but away from the minimum. In the first case, the behaviour becomes *subcritical* with a low b-value. In the second case, the behaviour becomes *supercritical* with a high b-value and large earthquakes more frequent than would be expected from the power-law trend of smaller ones. Only fine-tuning of the controlling parameter produces strict critical behaviour with a b-value of 1.0 and constant stress drop. These theoretical arguments are compatible with the relationship suggested above, between b-value and geodynamic setting; if the medium has a viscous component, stress is relaxed and the system is driven to a supercritical state; if the medium is rigid, the system is in a constant subcritical state. Plate boundary zones will be in a critical state, forming a global network with constant seismic activity; they will enclose cratonic areas in a subcritical state, where all the deformation is seismic and oceanic interiors in a supercritical state, with large aseismic deformation and minor earthquake activity. The subcritical state of cratonic areas expresses the fact that they are far from a critical state, but can be subjected to propagation of new plate boundaries across them; the supercritical state of oceanic interiors expresses the fact that they were in a previous critical state when they formed at divergent plate boundaries.

It is satisfying to find out that the conclusions from fractal dimensions in seismicity converge with estimations from the elastic thickness of the lithosphere obtained through quantitative models of elastic flexure under variation loads (Turcotte and Schubert 1982; Middleton and Wilcock 1994; Ranalli 1995) in different geodynamic environments, as referred to in Section 3.4. A survey of results shows that the elastic thickness of the continental lithosphere is much thicker than that of the oceanic lithosphere, except in rifts, passive continental margin and young orogenic belts (Lowrie 1997). This convergence is even more remarkable if we notice that the elastic thickness of the lithosphere is obtained through calculations based on continuum mechanics, and strength of the lithosphere with b-values is obtained through considerations of the mechanics of fractured media. In fact, these two fields of mechanics are largely independent in their theoretical basis and formulation.

3.7 Finite and Incremental Strain in Discontinuous Deformation Using Scaling Laws for Faulting

Every student of geology knows that a picture of a geological object should contain a scale. In fact, many geological objects are invariant to a scale change. They are self-similar or fractals. In tectonics this includes fractures and folds.

The fractal concept covers various aspects of the natural objects. A tectonic object can have *fractal geometry* such as a single fault. *Fractal distribution* of objects can exist, each object having fractal geometry, such as in faults, or not, such as in folds. Fractal distribution can reflect a generation by a *fractal process*, as an expression of a dynamic system. Therefore, the sequence of thrusts is a dynamic process of various types: forward propagation, backward propagation or out-of-sequence which can generate a fractal distribution of thrusts, duplexes, horses or a fractal distribution of different fault types in a linked fault system. Some tectonic concepts can be redefined in terms of dynamic systems, such as fabric attractors (Passchier 1997).

The application of fractal concepts to tectonics and especially to discontinuous deformation has advanced greatly in the last few years (Cowie et al. 1996). We have seen in Section 3.3.5 that generalised three-dimensional fields, either of stress or strain, imply discontinuous ruptures composed of separate segments, at least initially; therefore incremental and finite strain can be estimated using scaling laws for faulting. These were discovered by tectonists after the discovery of scaling laws for earthquakes by seismologists. They realised that these laws were a consequence of theoretical fractal concepts in both cases (Turcotte 1997). Active faults and earthquakes result from the same tectonic process in the brittle schizosphere. Seismicity reflects incremental strain and fault geometry, finite strain. Both are intimately related, but the relation is complex because aseismic slip and deformation are reflected in finite strain, but not in seismicity.

The method of integrating seismic moment underestimates the deformation either, because seismic history is of insufficient length or, a large amount of fault slip is aseismic. It can only be applied to current incremental deformation and cannot be applied to earlier episodes or to estimate finite strain. This stimulated the estimation of finite strain from faulting using slip measurements (Scholz and Cowie 1990); these, in turn, were shown to be related to the length of fault traces by scaling laws, as follows.

In the so-called bounded faults the displacements decrease to zero at the fault tips (Cowie and Scholz 1992a). Then a correlation exists between the maximum amount of displacement in a fault, d, and the length of the fault trace, L, of the type (Cowie 1997).

$$d = \gamma L^D$$

where γ is a constant, usually ~ 0.01 and D is the fractal dimension.

For some authors D≈1, fault growth is self-similar for three orders of magnitude (Scholz and Cowie 1990; Cowie and Scholz 1992a,b; Dawers et al. 1993; Scholz et al. 1993). Other authors claim that D~1.5 (Walsh and Watterson 1988; Marrett and Allmendinger 1991). This disagreement is important because the rheology of faulted media will control the value for D. Therefore we will discuss it in more detail.

Field observations and experiments show that faults nucleate and propagate. If D=1, fault growth is self-similar and the distribution of fault attributes are the same at any scale. Profiles of displacement along faults of variable length must be also self-similar (Cowie 1997), with displacement increasing from zero at the ends of the fault to a maximum near the centre (Muraoka and Kamata 1983; Ribeiro,A; Mateus, A et al. 1991). Assuming an elastic medium cut by a fault and a small inelastic zone at fault tips it was shown (Cowie and Scholz 1992a) that a fault loaded by a uniform remote stress grows in a self-similar way obeying the law

$$d = \frac{C(\sigma_o - \sigma_f)L}{\mu}$$

where σ_0 is the shear strength of the surrounding rock, μ is its shear modulus, σ_f is the fictional shear stress on the fault, and C is a constant that depends on the ratio of the remote stress to the rock shear strength. Therefore, in this model, D=1, but $\gamma = C(\sigma_0 - \sigma_f)\mu{-1}$ will vary according to rock type and tectonic environment. A synthesis of available data (Cowie and Scholz 1992b) shows that the theory which assumes

D=1 is the most plausible one, and that the scatter of data is due to variable γ, because it is the ratio of two material properties, and to the existence of two sets of faults: large faults (L10 km) are two-dimensional being only constrained at their tips, whereas small faults are three-dimensional and are pinned along their entire perimeter; the same phenomenon was observed for earthquakes (see Sect. 3.6.).

As an example of the interpretation of fractal geometry of faults we can compare the fractal dimensions of a large plate boundary transform fault, such as San Andreas (Aviles et al. 1987; Okubo and Aki 1987) and a bounded strike-slip intraplate fault such as Vilariça Fault, NE Portugal (Ribeiro et al. 1991; Mateus and Gonçalves 1993; Mateus 1995).

The San Andreas Fault has a fractal dimension of 1.0008–1.0191 for the main fault trace and 1.1–1.4 by covering a set technique for the entire width of the fault zone. The first parameter measures the irregularities of the main fault and the second one measures the complexity of the entire fault zone.

The Vilariça Fault (Ribeiro 1974) is a Late-Variscan sinistral strike-slip fault reactivated in Meso-Cenozoic times and it is presently active (Cabral 1989). We possess data on the variation of fractal dimension along the slip profile of the fault (Fig. 3.28). The tectonic data suggest that the fault nucleated in the centre, near the brittle-ductile transition. It propagated bilaterally in a few million years. The slip profile is different in the middle, where the failure was outside the elastic field, and follows different laws in the northern and southern sides. These are lithologically very different, with higher-grade metamorphic rocks dominating the northern side and low-grade metamorphic rocks and granitoids dominating the southern side; the difference in material properties can explain those facts. The variations in fractal dimension show a clear inverse correlation with displacement; this means that displacement increases as the fault plane is smoothed by fault propagation.

These results suggest that D could vary along the main trace and will lead to a D slightly above 1. Therefore σ_0 will not vary systematically with fault size for which there is no physical basis (Cowie and Scholz 1992b), but σ_f will decrease systematically with size by smoothing out irregularities at all scales as the fault propagates. This irregularity extends from map scale, with a wavelength of 10^3–10^5 m to outcrop scale with a wavelength of 1–20 cm and to hand specimen scale with a wavelength of 0.5 mm–1 cm, both measured by power spectrum techniques (Brown and Scholz 1985; Scholz and Aviles 1986).

Another relationship is a fractal distribution of displacement or length, x; the member of faults above this value $N \geq x$ is

$$N \geq x = ax^{-D}$$

where D is a fractal dimension.

A similar relationship exist for fault patterns

$$N_{(r)} \sim r^D$$

here $N_{(r)}$ is the number of grid or box elements of size r that are occupied by faults and D is the fractal dimension. If faults are uniformly distributed we must find D=2, if faulting is concentrated in a narrow zone, D=1. Very detailed studies (Barton and Larsen 1985; Barton et al. 1986; Korvin 1992) showed *bifractal distributions*: the lower value of 1–1.2 at lower scales corresponds to the irregular shape of the individual

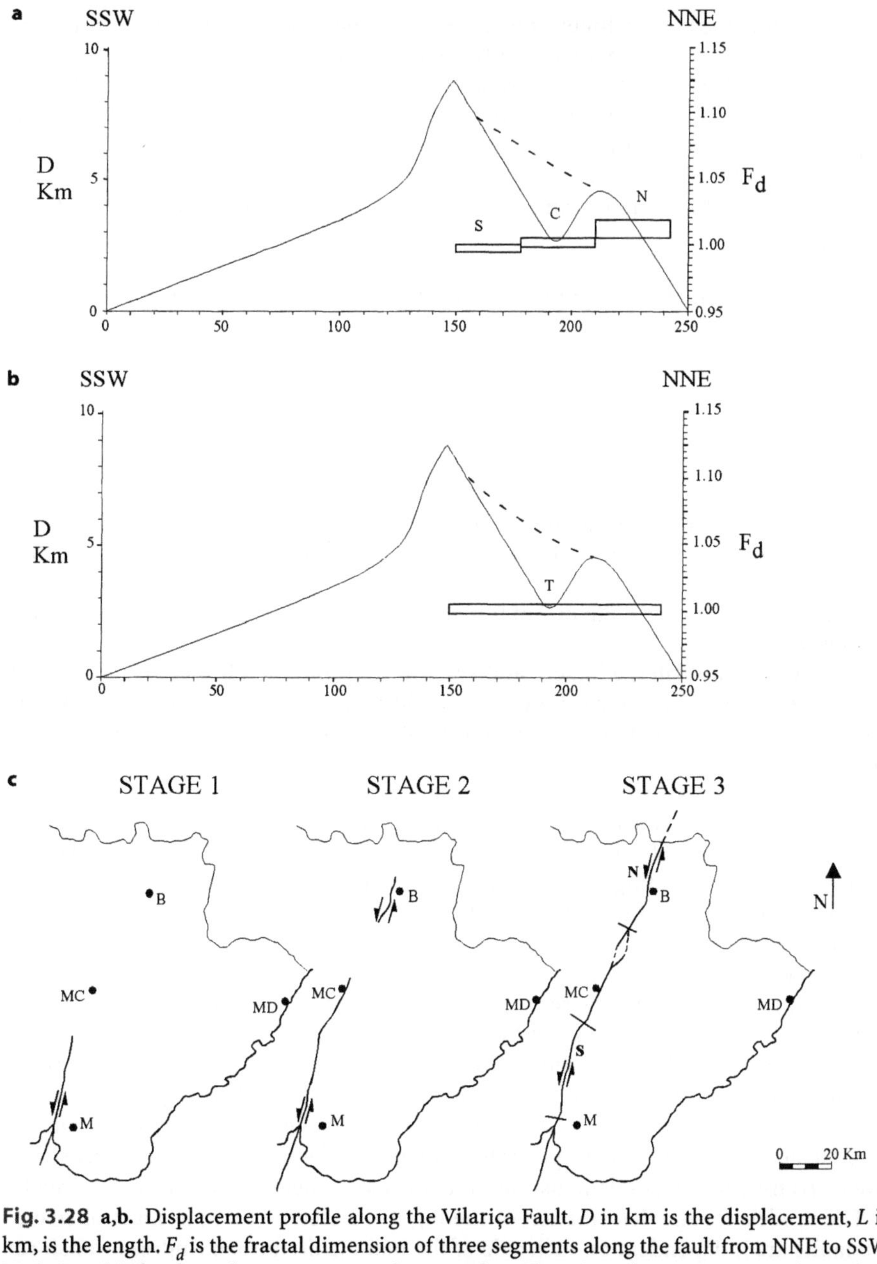

Fig. 3.28 a,b. Displacement profile along the Vilariça Fault. *D* in km is the displacement, *L* in km, is the length. F_d is the fractal dimension of three segments along the fault from NNE to SSW: *N, C, S*, and *T* for these three segments taken together. There is a clear inverse correlation between displacement and fractal dimension (after Ribeiro et al. 1991; Ribeiro and Keil, unpubl.). **c** Main stages in the nucleation and growth of the Vilariça Fault (after Ribeiro et al. 1991). *B* Bragança; *M* Moncorvo; *MC* Macedo de Cavaleiros; *MD* Miranda do Douro. The Douro River is the *full line* and the boundary of Portugal and Spain is shown by *points*. Segments where the fractal dimension was measured: *N* northern, *C* central, *S* southern

fractures, while the higher dimension of 1.5–1.9 is due to self-similarity in mutual arrangement of fractures. This was confirmed by other studies (Barton et al. 1986; Okubo and Aki 1987; Chilès 1988; Hirata 1989).

All the scaling laws referred to above must be related by a model of fault initiation, propagating and coalescence or linkage. This process will generate a linked fault system (Davison 1994) where linked faults branch rather than cross-cut each other.

Some aspects of this model can be derived from the scaling laws. For instance from

$$D=\gamma L$$

when D is displacement, L length and γ a constant we can derive, differentiating in order to time, t

$$\frac{dD}{dt}=\gamma\frac{dL}{dt}$$

but dD/dt is the slip rate, S. Hence,

$$S=\gamma\dot{L}$$

denoting $\dot{L}$ as the propagation rate of the fault.

If S is known independently and γ is assumed for a fault population we can estimate $\dot{L}$=dL/dt.

For instance, a fault with a slip rate of 0.1 cm/year will propagate at a rate of 10 cm/year, approximately.

As strain accumulates, a linked fault system can propagate. If there is only a fault system there are no geometrical problems in fault propagation. However, if two conjugate systems are generated, where the fault intersects they will become locked (McKenzie and Jackson 1989), according to accepted views on kinematics of rigid blocks. Let us examine a zone of cross faulting in this perspective.

If two active faults cross, kinematic compatibility requires that the slip vector of one lies in the plane of both faults (Jackson and McKenzie 1983). One example is a transform fault connecting two ridge segments; along the vertical transform there is opposed movement of conjugate normal faults, even if they are listric. The same applies to transfer zones linking grabens in extensional systems or lateral ramps in thrust systems.

If we apply these kinematic restrictions to a system of conjugate strike-slip faults, we must infer that the crossing fault must deactivate the crossed one. An example of this relationship is the Big Bend region of California, where the NW-SSE dextral transform San Andreas cuts the NE-SW sinistral strike-slip faults, such as the Garlock Fault. Both systems are active, so the kinematics do not obey the compatibility rule referred to above, because it only applies to rigid blocks; the presence of the flower structure of Transverse Ranges demonstrates that the blocks between faults are deformed.

The models for the deformation of the Big Bend assume that the sinistral strike-slip faults are independent on each side of the main fault, the San Andreas transform (Bohanon and Howell 1982), or that the San Andreas Fault was originally discontinuous, being crossed by some faults of the conjugate left-lateral family, oriented N–S originally. Since the Oligocene, the displacement is larger in the dextral system, with minor counterclockwise rotation and large clockwise rotations in the sinistral strike-slip domains, up to 70° or 80° (Twiss and Moores 1992), with basins in the domains

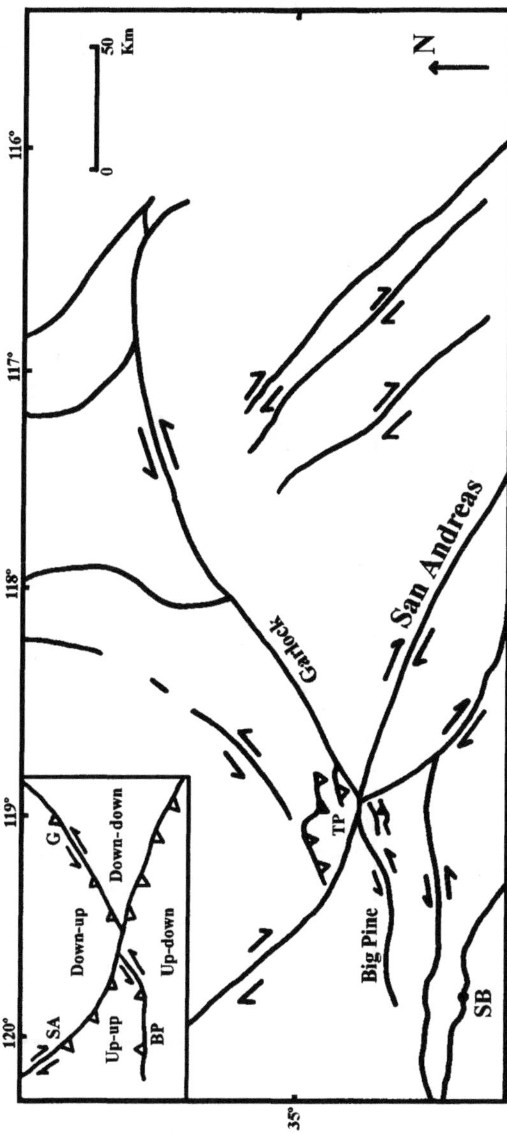

Fig. 3.29. Map of the main active faults of the San Andreas and Garlock-Big Pine fault system. *SB* Santa Barbara, *TP* Tejon Pass (After Bohannon and Howell 1982). *Insert* Reverse-slip components on the active fault system due to listric geometry of strike-slip faults into depth

subjected to extension and folding and thrusting in the domains subjected to compression. However, this model does not account for kinematics at depth, non-rigid behaviour and slip along the Garlock Fault, as stated clearly by its authors.

The geological map of California shows that the blocks displaced by the San Andreas Fault have no tectonic relationship in the present state, with only one exception: the Garlock and Big Pine sinistral faults are only 7.5 km apart in the Tejón Pass. Reversing the slip in the San Andreas Fault shows that they were aligned some time ago (Fig. 3.29).

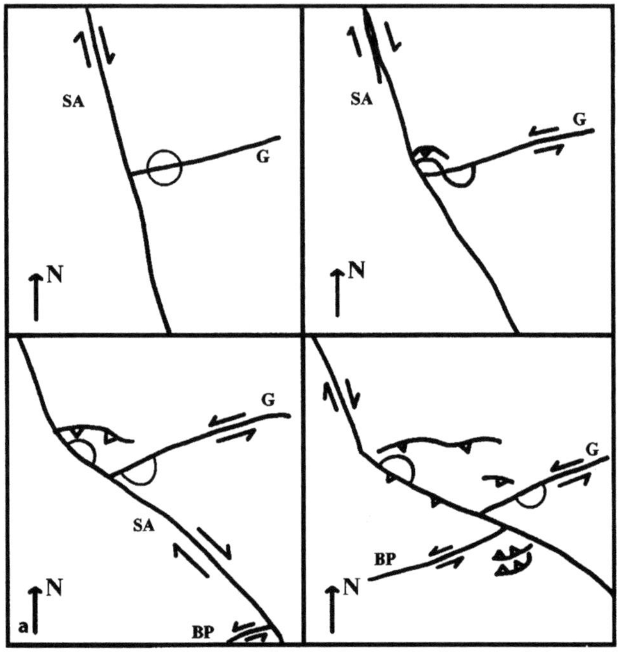
SA
G
N
SA
G
N
G
SA
a
N
BP
G
BP
N

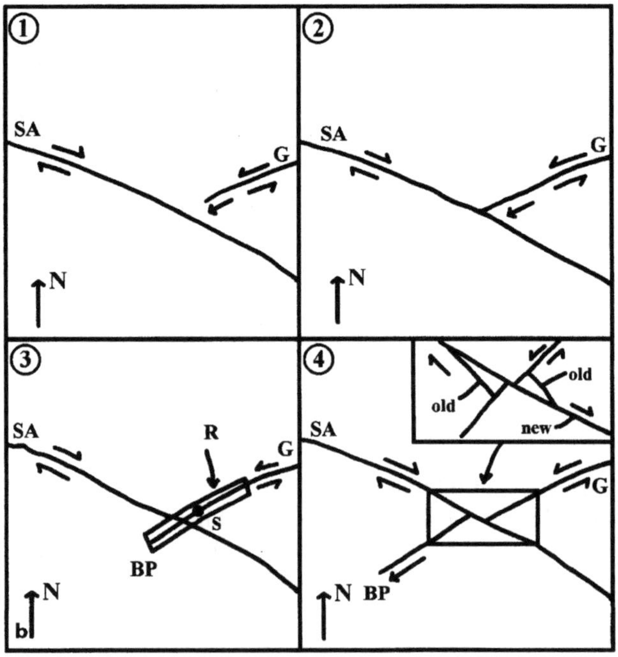
①
SA
G
N
②
SA
G
N
③
SA
R
G
S
BP
b
N
④
old
old
new
SA
G
N BP

We propose an alternative interpretation for these kinematic relationships which is compatible with references on fault propagation based on the scaling laws for displacement-length relationships referred to above (Fig. 3.30).

The physical basis for our interpretations is that during a major seismic event, with surface rupture because all the schizosphere is cut, there is, at the tip of an active fault, a displacement scaled to an increment of lengthening of the active fault, in accordance with the previous scaling law

$$\Delta D = \gamma \Delta L$$

with $\gamma \sim 0.01$ and fractal dimension or exponent of L equal to 1, in accordance with the linear law established above. A displacement of 1 m, corresponding roughly to a magnitude 6.5 earthquake, will cause a lengthening of the fault of ~100 m at the tip of the fault in intact shock or several kilometres if a previous second order fault with favourable orientation is reactivated.

Suppose that the Garlock Fault generates such a seismic event near the junction with San Andreas, during an interseismic period for the last fault. The surface rupture can cross the San Andreas Fault, displacing it just 1 m; stress accumulation will propagate this minor fault with the same sense of lateral slip at the same time as it causes right-lateral slip on the main San Andreas Fault.

We hypothesise that the segment of Garlock Fault to the SW of San Andreas will become the Big Pine Fault, simply because its length, around 97 km suggests it has been activated recently. Given the slip rate of 34±2 mm/year of San Andreas (Sieh and Jahns 1984) in central California, it is inferred that 220,600 years ago the Garlock and Big Pine Faults were in perfect alignment with the same sinistral sense slip. Could the Garlock Fault propagate bilaterally with a tip point in the east and another in the west, at the junction with San Andreas Fault and could the west tip point of Garlock cross the San Andreas at that time? Or, in others words, could Garlock and Big Pine have been a continuous fault at that time?

If this is true then we can estimate the slip rate of the Big Pine segment using the equation

$$S = \gamma \dot{L}$$

with $\gamma \sim 0.01$ and $\dot{L} = \Delta L / \Delta T$ with ΔL=97.5 km and ΔT=220,600 years. We can calculate the slip rate of Garlock to be 4.4 mm/year, which is an order of magnitude below

Fig. 3.30. Kinematic evolution of the San Andreas–Garlock–Big Pine fault system. **a** Interpretation with non-crossing faults. The Big Pine sinistral fault is dragged by the San Andreas dextral fault and becomes fortuitously aligned with the Garlock sinistral fault for a short interval of time. A *circle* is shown as a displacement marker, fixed to the SE block, it is dragged to WSW by sinistral slip in Garlock and afterwards dragged to NNW by dextral slip in San Andreas (after Bohannon and Howell 1982). **b** Interpretation with propagating cross-faults. *Stage 1* San Andreas is active, dextral; Garlock is active, sinistral and is propagating to WSW, as the *arrow* indicates. *Stage 2* The trace of Garlock arrives at San Andreas in Tejón Pass. *Stage 3* A sinistral seismic event nucleates at *S* in Garlock and ruptures a segment, *R*, of Garlock that crosses San Andreas. *Stage 4* San Andreas resumes dextral slip and Big Pines continues propagation to WSW, as the *arrow* indicates. The *insert* shows a detail of old segment of San Andreas, displaced sinistrally by Garlock and the new segment of San Andreas

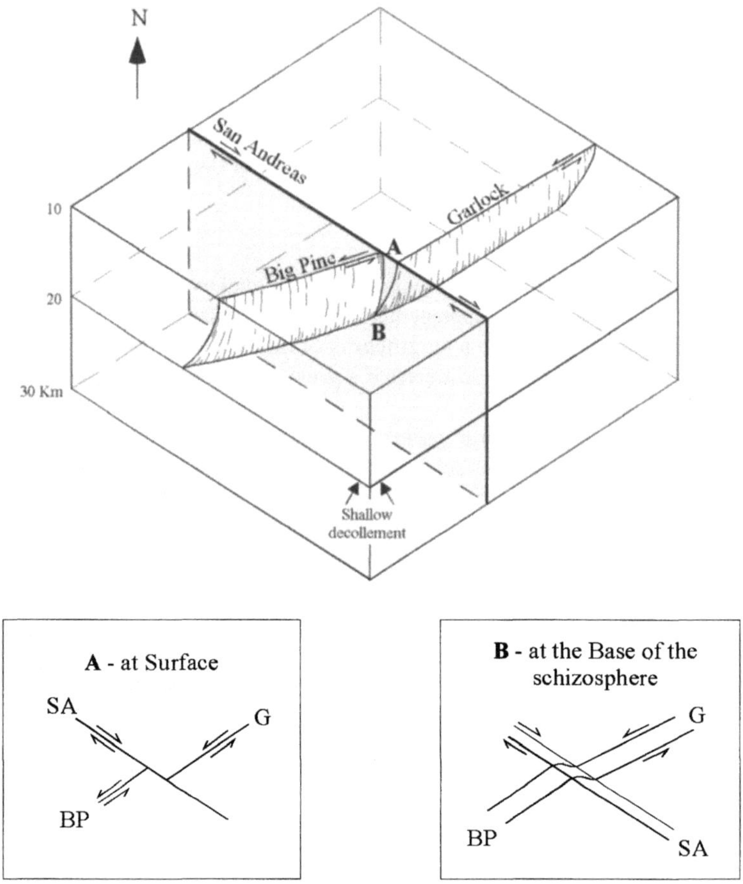

Fig. 3.31. Kinematic evolution of San Andreas–Garlock–Big Pine fault system. The Garlock and Big Pine faults were once continuous and are now being displaced by San Andreas, but the two conjugate families remain active. The Garlock–Big Pine Fault roots in a dêcollement at the base of the schizosphere at approximately 15 km. The San Andreas becomes flat at the base of the lithosphere, much deeper. The *inserts* show: *A* Crossing fault observed relationships at the surface due to brittle behaviour. *B* Crossing shear zone-inferred relationships at the base of the schizosphere, due to brittle-ductile behaviour. The Garlock–Big Pine family is more ductile than the San Andreas because the slip rate is an order of magnitude lower

San Andreas. This result is compatible with the estimated slip rate of 4–9 mm/year for the Garlock Fault (Peterson and Wesnousky 1994).

What are the implications of this model with the Garlock and Big Pine once continuous fault? (Fig. 3.31)

The seismicity associated with the San Andreas deformation zone stops usually at a depth of 15 km. However, the Loma Prieta event was deeper with the hypocentre at 18 km and a component of thrusting higher than usual (Oppenheimer 1990; USGS

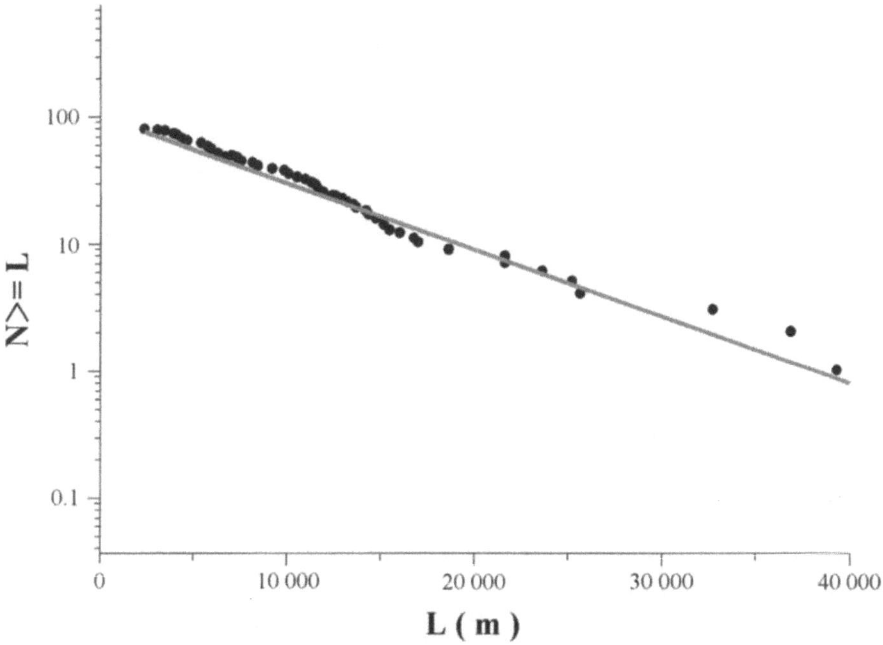

Fig. 3.32. Cumulative distribution of fault length in metres, of the Lucky Strike segment of the Mid-Atlantic Ridge (37°17′N, 32°16′N); *L* is the length and *N≥L* is the number of faults with length above the value in the abscissa. The cumulative frequency is plotted in log scale, but the length scale is linear. The plot is semi-log; so the distribution is exponential. The *full line* is the best-fit solution for the data. (After Lourenço 1997)

1990). This suggests that the secondary faults related to San Andreas are shallow and eventually rooted in a dêcollement at 15 km depth which emerges in the N-S directed active thrusts that limit the Transverse Ranges (Namson and Davis 1988). Therefore, the secondary faults inside the San Andreas deformed zone could be rotated above the Main Fault; even if a second-order fault cuts the main fault during a short period of time it will inevitably be cut by the deeper fault afterwards. In fact, at depth, the dextral shear zone must be later than the sinistral shear, but both can continue to be active (Ramsay 1980). The obtuse angle bisected by the maximum compressive stress is ~120° for the San Andreas-Garlock system as in ductile shears and not acute, ~60°, typical of brittle faults. This suggests that the model of a schizosphere dragged by the plastosphere below it is correct (Molnar and Gipson 1996).

The scaling law for displacement-length must reflect the rheology of faulted medium. If the schizosphere behaves elastically D must be close to 1, for a given population in a single rock type and tectonic province. If the schizosphere is viscoelastic the stresses can dissipate, there is no long-term elastic strength and the stress only reflects the youngest movements (Cowie 1997). The evidence from faults is that both elastic and inelastic deformation occurs at all scales; but the data are not accurate enough to quantify both processes.

The previous scaling laws can be used to estimate finite strain. In theory, this can be done if one of the attributes of fault populations is known, namely length or displacement (Scholz and Cowie 1990). The accuracy and precision increase if we know more attributes, such as fault distribution, which we will consider next.

Continental faults obey fractal distributions as shown before (Ranalli 1980; Cowie 1997; Turcotte 1997). However, oceanic faults are better approached by exponential distributions (Cowie et al. 1993, 1994; Lourenço 1997). This is explained by short- and long-range interaction during fault nucleation and propagation, resulting in strong spatial correlation in the continental domain; in contrast, in the oceanic domain fault populations are generated in sequence by the spreading mechanism and spatial correlation is strongly suppressed (Fig. 3.32).

3.8 Finite Kinematics and Intraplate Strain

In plate kinematics an *instantaneous rotation* can be represented by an *angular velocity vector*, ${}_{A}W_{B}$; it means the rotation of plate B relative to plate A around a specific Euler pole, E; ${}_{B}W_{A}=-{}_{A}W_{B}$ is the rotation of plate A relative to plate B around the antipole E′ (Cox and Hart 1986).

Finite rotations are not vectors as instantaneous rotations, but must be represented by 3×3 matrices (Cox and Hart 1986). This implies that finite rotations are not commutative because the final rotation depends on the order in which the individual rotations are performed, whereas angular velocity vectors are commutative.

The implications are that *angular velocity vector* fields can describe incremental strain, but *finite strain* must be described by integration of incremental strains through tensor fields that vary through time and space. Therefore, to analyse finite strain we have to analyse finite rotations first.

In standard theory (McKenzie and Morgan 1969), an important property of finite rotations in the so-called *three plate problem* has been demonstrated (Cox and Hart 1986). In fact, Euler poles must change position in space in a set of three divergent plates, even if the instantaneous velocity vectors are constant in orientation and magnitude.

This means that a system with three plates is inherently dynamic. In the theory of dynamical systems it has been demonstrated that it is necessary for at least three independent dynamic variables to have sensibility to initial conditions because trajectories in phase space must diverge without intersecting (Lorenz 1963; Ruelle and Takens 1971; Bergé and Dubois 1992, p. 121); this requires at least the three dimensions for space phase. A system with two plates can be deterministic in the classical sense; a system with at least three plates must be chaotic or, in other words, must obey the rules for deterministic chaos.

On the real Earth, with 12 plates in the model NUVEL-1, plate kinematics show that the Euler poles are not fixed through time. In fact, we know that in slow-spreading oceans Euler poles are almost stable in contrast to fast-spreading oceans where the Euler poles spin at fast rates, as mentioned in Section 3.3.2. Between 162 and 119 Ma the Euler pole for the North America–Africa plate pair stays in Iceland; in contrast, between 77 and 32 Ma the Euler pole for the plate pair Pacific-Farallon spins very fast in a non-linear way following a curve which we will describe below.

The reason for this distinct behaviour is known (Cox and Hart 1986). Let us consider a fixed mantle reference frame to which all plate velocities are prescribed. We should note that plates not attached to trenches are very slow and those attached to trenches are rapid; the trenches themselves are approximately stationary, which is due to the fact that if they could undergo rapid lateral motion within the mantle this would require the displacement of large volumes within the asthenosphere. Euler poles of moving plates must move around a fixed Euler pole in the fixed reference frame. Therefore, the Euler poles for a plate pair can remain nearly fixed only if two absolute Euler poles are nearly coincident, or if at least one of the angular velocity vectors is very small. Applying this reasoning to the case of Atlantic-type oceans, we conclude that Euler poles of divergent plates separated by these oceans are almost fixed and in the case of Pacific-type oceans we conclude that they are highly variable.

Even in slow-spreading oceans it is known that Euler poles jump to new positions (Cox and Hart 1986). In fact, the movement between plates is best described by a sequence of finite rotations about slightly different Euler poles. Hence, *stage poles*, fixed for some Ma or for some tens of Ma, were introduced as a better approximation to plate kinematics.

Finite rotations have been modelled by a more elaborate theory than the instantaneous motion along a small circle materialised by a transform fault. The simplest solution is a rotating Euler pole around a specific pole. Observed from another plate, a point in the previous plate describes a cycloid (Cronin 1991) by combining two rotations. An implication of this model is that transform faults cannot maintain pure slip motion during finite rotations; therefore the condition of area preservation in a conservative boundary is violated and some deformation must occur at this type of plate boundary. Fracture zones should be complex curves, not symmetrical across the ridge axis.

Detailed studies of triple junction configuration such as the Afar and Indian Ocean (Patriat and Courtillot 1984) have shown that spreading is episodic between two modes, the effusive magmatic and tectonic modes, corresponding respectively to RRR and to RRF-RFF configurations, typically with a duration of 1 Ma. This allows the triple junction to remain stable under real conditions, less restrictive than the ideal conditions that exclude oblique and asymmetric spreading. This is due to variation in length of transform faults, availability of magma and connectivity of magma chambers and variations in spreading velocities.

In episodic spreading the boundaries of each area of triple junction can overlap in time; the contact between magnetic lineations with different orientations becomes fuzzy instead of a discrete line.

The kinematic evolution of plates depends on the stability of triple junctions (McKenzie and Morgan 1969), represented in velocity space (Cox and Hart 1986), in a combination of geographic and velocity spaces (Patriat and Courtillot 1984), or in ω space (McKenzie and Parker 1974).

If plates deform, the axisymmetric model still applies and can be used to study finite strain along the boundaries between different magnetic lineations for each arm of the triple junction.

In the more frequent case magnetic anomalies are subperpendicular to transform faults. Therefore, in the axisymmetric model the major axis of strain ellipse should be parallel to magnetic anomalies for each arm of the triple junction.

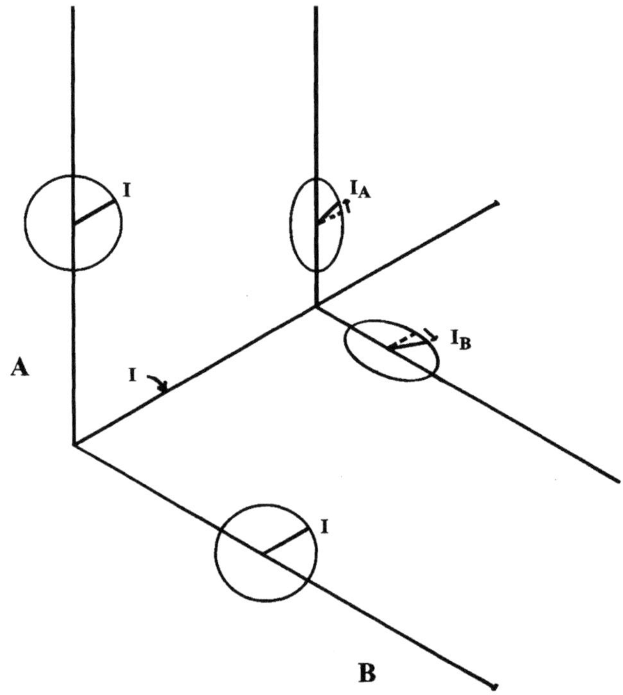

Fig. 3.33. Strain incompatibility at a triple junction. Magnetic lineations are *A* and *B* on each arm of the triple junction and they intersect at *I*. In the domain *A* the intersection is rotated in dextral sense to I_A; in the domain *B* the intersection is rotated in sinistral sense to I_B. There is no compatibility in strain because *I* cannot be rotated simultaneously to major axis of strain ellipse *A* and in *B*

Consider now the boundary between magnetic lineations of the same age, but different orientation because they belong to different arms of the triple junction. As deformation progresses, lines will be rotated toward the major axis of strain ellipse according to the Wettstein equation (Ramsay and Huber 1984), $\tan \phi' = \tan \phi / R$, where ϕ is the original angle between some line and the major axis of strain ellipse and ϕ' the same angle after deformation and R is the relationship between the major axis and minor axis of the strain ellipse; so $R \geq 1$, and $\phi' \leq \phi$. The boundary between magnetic lineations cannot be attracted simultaneously to both arms of the triple junction, so deformation is not compatible in domains with different magnetic lineations (Fig. 3.33).

These internal boundaries must be coherent in the sense that no material lines are ever discontinuous across them (Cobbold et al. 1984). In mathematical terms this requires that the deformation and its time history, the motion, are continuous across a boundary, but their first-order derivatives (gradients), with respect to space or time, need not be so. These coherent strain boundaries may propagate or migrate through a material; if so, the particle velocity field is discontinuous at the boundary (Cobbold et al. 1984: Fig. 3.34).

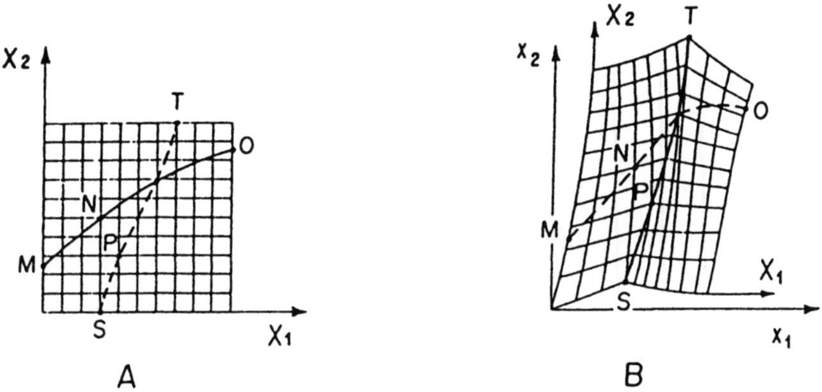

Fig. 3.34. Deformation in the vicinity of a coherent boundary. Position in reference state *A* is relative to Cartesian frame *X*; in current state *B*, to frame *x*. Frame *X* deforms with the material, without loosing continuity; but, slopes of material grid-lines become discontinuous at coherent boundary *SPT*. Boundary propagates from material surface MNO to material surface *SPT*. (After Cobbold et al. 1984; with kind permission from Elsevier Science)

Let us consider a plane tangent to the Earth at a triple junction. If the spreading rate is fixed for each arm and plates are rigid, the motion direction of the triple junction relative to each plate is a straight line (McKenzie and Parker 1974). If the plates deform, the motion direction becomes a curve with convexity towards the slower arm of each ridge pair (Fig. 3.35). This curvature will be attributed, in the standard theory with rigid plates, to an acceleration of the faster spreading ridge in each ridge pair. The problem is: how can we distinguish between an acceleration of spreading-rate with rigid plate from an increasing deformation of a steady-state spreading system?

In accelerating ridges the boundaries are curved, but undeformed. In steady-state ridges the boundaries migrate across the material and we must have a domain where deformation is superposed, with a younger strain ellipse parallel to the slower ridge superposed on an older strain ellipse parallel to the faster ridge. In the domain with superposed deformation, a transition strain state will occur between the domains with no interference and a strain ellipse parallel to each arm of the triple junction; the magnetic lineations parallel to the faster ridge will be rotated to an intermediate position between the two arms (Fig. 3.36). Hence, the original angular relationship between magnetic lineations will be smoothed out; with rigid plates, this angular relationship will be maintained through time. An inspection of a magnetic lineation map of the ocean floor shows that this smoothing effect is present, confirming the soft plate model. A clear example must be referred to in the pattern of magnetic lineations between the Bouvet triple junction and fossil lineation of this junction to the NE, toward the South-African coast. In standard theory, this smoothing effect would be attributed to a decrease in amplitude of magnetic anomalies with the age of the ocean floor related to demagnetisation effects, increase in sediment cover etc. This explanation is invalid because it would affect not only the transition zone between different magnetic trends, but also the whole length of magnetic lineations and this is not the case.

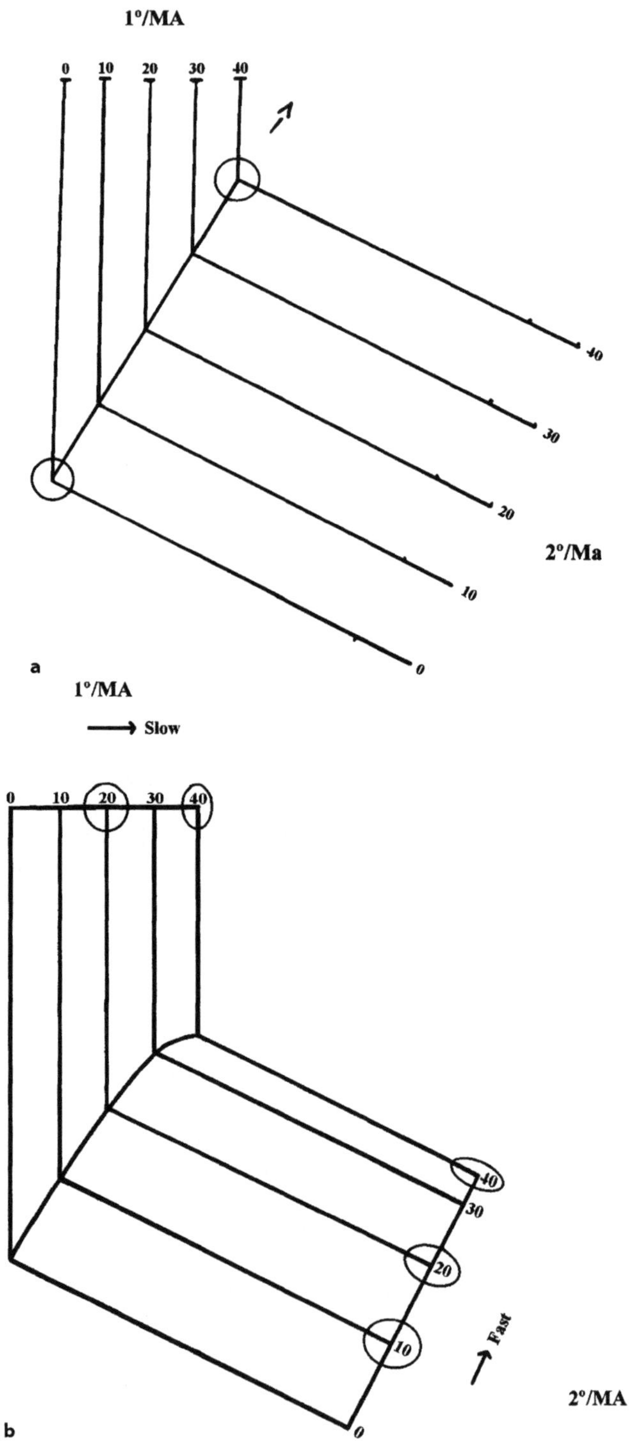
1°/MA
0
10
20
30
40
40
30
20
2°/Ma
10
0
a
1°/MA
Slow
0
10
20
30
40
40
30
20
10
Fast
2°/MA
0
b

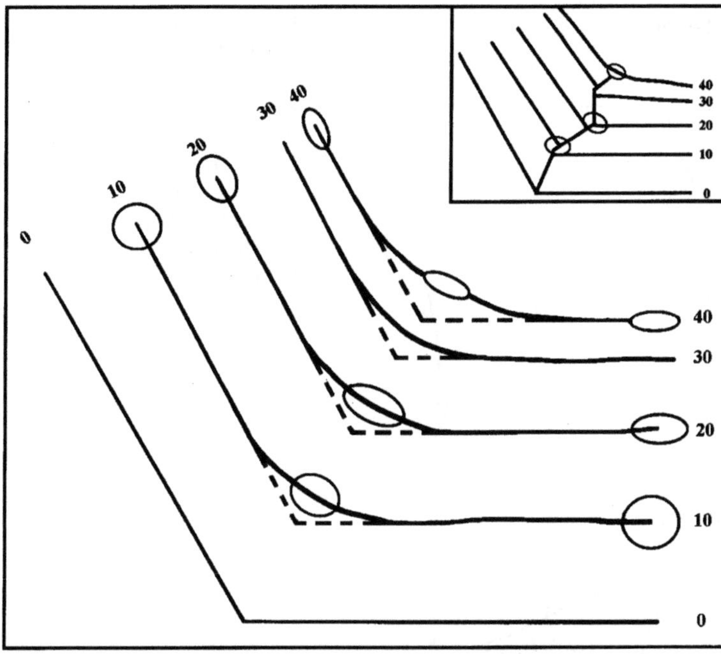

Fig. 3.36. Smoothing at the intersection of magnetic lineation domains. *Insert* Detail of the smoothing mechanism: the straight boundary between the two magnetic lineation domains is, in fact, composed of alternate jumps between these two domains with one trend advancing or releasing relative to the other trend

If there is a superposed deformation in the smoothed area due to migration of the coherent strain boundary it must be reflected in the geoid lineations due to incipient boudinage of the oceanic lithosphere (Fig. 3.37). The pattern of geoid lineations in triple junctions confirms that these lineations are parallel to relative motion of the plates.

In rift triple junctions, RRR, such as the Indian Ocean and Galapagos, the geoid lineations remain parallel to transform faults for each rift. The boundary between domains with different transform directions and magnetic anomalies is clearly visible as triangular zones with the apex in the present position of the triple junction. They

Fig. 3.35. Pattern of magnetic lineations projected at a plane tangent to the spherical surface of the Earth at a triple junction; two arms of the triple junction are mapped; they make an angle of 115°; the fast arm spreads at a constant rate of 20/Ma and the slowest arm spreads at a constant rate of 10/Ma. **a** The plates are rigid; the reference circle remains a circle. The boundary between the two arms is a straight line if the spreading rate is constant. **b** The plates deform according to the axisymmetric model. The spreading rate is constant, but the material line separating the two magnetic lineations domain is curved with convexity towards the slowest arm

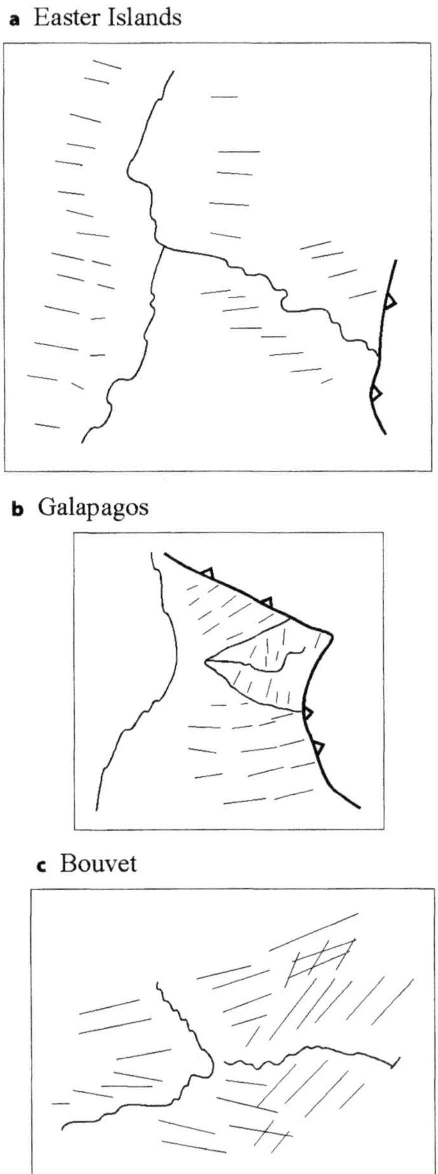

Fig. 3.37. Geoid lineations in triple junctions. **a** Easter Island triple junction: parallel geoid lineations in both arms of the triple junction. **b** Galápagos triple junction: geoid lineations at high angle with distinct boundaries at apex of Galapagos triangle. **c** Bouvet triple junction geoid lineations at high angle with fuzzy boundaries and crossed lineations

coincide with the boundary between different fabrics of the ocean floor, such as the rough-smooth boundary in the Galapagos triple junction.

In the rift-rift-transform fault triple junctions, RRF, such as Easter Island the same rules apply, however, here there is only one direction for geoid lineations because there is only one direction of transform faults.

The angular relationships between different magnetic lineations near present triple junctions of the RRR type seem to be smoothed out as we move to older parts of the ocean floor. In fact, in these smoothed areas of magnetic lineations we find a difference with the linear fabric from the younger part of the ocean floor, given by geoid lineations; visual inspection suggests the presence of crossed geoid lineations; this would imply double incipient boudinage at the superposition zone between two orientations of simple boudinage. This suggests that as deformation proceeds in the older oceanic lithosphere the incipient pinch-and-swell extensional structure becomes a fully developed boudinage; or, in other words, incipient failure occurs at the thinnest domains. At the boundary zone between domains with oblique magnetic lineations, oblique boudinage in two directions must take place, perpendicular to each magnetic lineation. This mechanism is known in extensional mesoscopic structures because in already fractured rocks boudins need not be perpendicular to the principal elongations (Ramsay 1967, p. 113). The present resolution in geoid lineation maps is unfortunately not sufficient enough to confirm this qualitative impression by a quantitative analysis of the signal and confirm the order of superposition between two boudinage directions.

3.9 Finite Intraplate Strain: Tests and Paths

3.9.1 A Paradox in the Pacific

To prove that oceanic plates deform we need strain markers on the ocean floor. These are very difficult to characterise, although it could be possible to find initially circular impact craters that become elliptical by progressive deformation, as could be the case of Tore "Seamount", the object of one of the next sections. In fact, the usual material lines present on the ocean floor are magnetic anomalies, transform faults fracture zones and hotspot traces. A variation in spacing between dated magnetic anomalies can be attributed to the variation of spreading rate or to strain, and in the last case we do not know what the original spacing was, before strain occurred. Therefore, a definite test of oceanic plate deformability is extremely difficult to put into operation because the pattern of magnetic anomalies in a spreading ocean reflects this inherent ambiguity: has the spreading rate changed, or strain accumulated, or both processes occurred simultaneously? Moreover, the angular relationship between transforms and magnetic anomalies will be little altered in the axisymmetric model of deformation, but the original angle is unknown.

This stimulated us to look at a more special situation than a spreading ocean with a single mid-ocean ridge, and to consider a triple junction that must obey some restrictive conditions to be stable (McKenzie and Morgan 1969). The complex evolution of the fast-spreading Pacific ocean could be a good case, specially in the oldest segments of its western side (Larson and Chase 1972; Larson and Pitman 1972) which should have accumulated more strain if the oceanic lithosphere is deformable.

In fact, three trends of magnetic anomalies of the same age (late Mesozoic) have been identified in this area: Japanese, Hawaiian and Phoenix lineations. If we consider

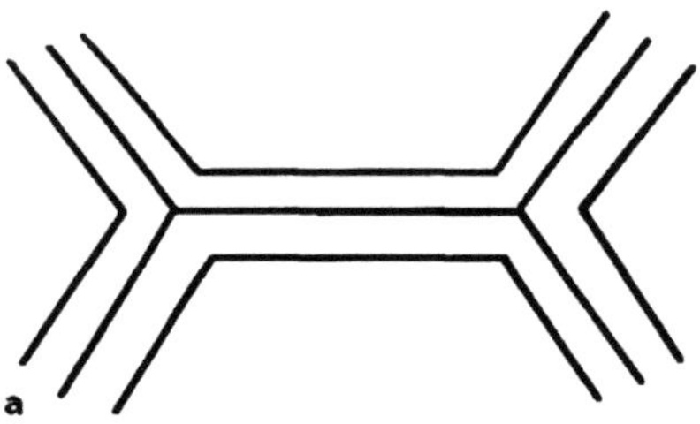

a

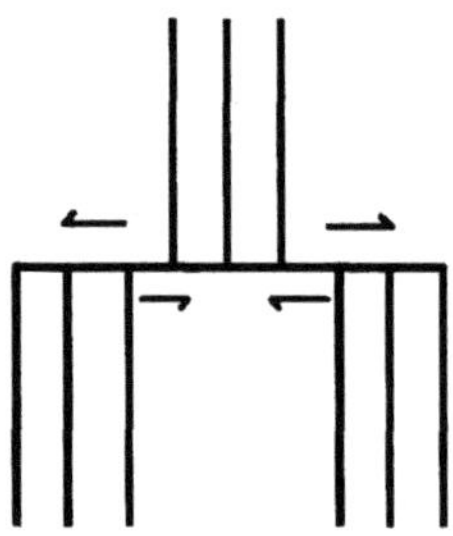

b

Fig. 3.38 **a** Divergent magnetic lineations around two triple RRR junctions.
b Subparallel magnetic lineations in an RRF triple junction

the situation 120 Ma ago, we see that the Phoenix and Japanese magnetic lineations converge as we move away from triple points. This is a paradox which remains to be explained in the rigid plate theory (Uyeda 1978). We pretend that this cannot be the original geometry if the plates are rigid and that this must be due to internal strain of the plates involved, as we will try to prove by the following reasoning.

The stability of triple junctions can be decided by the use of velocity space diagrams, combined with geographic data (McKenzie and Morgan 1969; Patriat and Courtillot 1984; Cox and Hart 1986). Unstable triple junctions can exist during a short period of time and should evolve to a new, more stable, plate configuration. The most stable triple junctions are of the RRR type, because the perpendicular sides of a triangle always intersect in a single point; in unstable triple junctions the three velocity lines do not intersect and the location of the triple junction is not defined in velocity space.

As ridges form a global network connecting all spreading oceans we must expect that they will diverge from triple junctions of the RRR type or have two parallel arms in the RRF type, but must not converge (Fig. 3.38). Ideally, a hexagonal pattern with six triple junctions making angles of 120° maximises the cooling by convection; this is the reason why this pattern forms in some lava flows. On a spherical earth some symmetrical solutions can exist (Chossat and Stewart 1992) with upwelling at the corners and downwelling at the faces of platonic solids.

A particular configuration has been proposed by some authors to explain the pattern of magnetic anomalies in the Pacific. It consists of three triple junctions at the corners of a seed triangular zone. The first problem is the origin of this triangular zone. If this zone is not a microcontinent, but consists of older oceanic lithosphere, this requires a change of plate configuration from an RRR junction to three RRR junctions at a younger stage of evolution. The question is: what is the most stable configuration?

We have seen that to maintain an RRR junction in the rifting mode (Patriat and Courtillot 1984) we should maintain the length of transform faults, the availability of magma and connectivity of magma chambers and spreading velocities. These conditions are not fulfilled all the time in a single RRR junction, causing it to change to the RRF or RFF modes. Therefore, it would be even more difficult to maintain the RRR mode operating at the same time in the three RRR junctions. Eventually, the "triple" triple junction will revert to a single triple junction again (Fig. 3.39).

These interpretations are confirmed by the observation of modern analogues. Along the fast-spreading Mid-Pacific Rise, there are small microplates, such as the Easter microplate (MacDonald 1989); they perturb the tensional stress field of the ridge and induce localised compression in young oceanic crust (Rusby and Searle 1993). These microplates must be short-lived, as demonstrated by the now inactive spreading centres inside the Nazca Plate, 10–25 Ma old. One can argue that these centres were not located at the main ridge, contrary to the Easter Plate.

Microplates located along ridges are due to progressive evolution of propagating rifts and/or overlapping centres (Searle et al. 1995). This induces rapid rotation as "roller bearing" until one of the propagating ridges breaks through the boundary between two driving plates; this will stop microplate rotation and will cause welding of the microplate to one of the major plates. Another possibility is exemplified by the Trinidad microplate, active between chrons M21 at 147 Ma and M14 at 135 Ma at the SE corner of the Pacific Plate (Nakanishi and Winterer 1998); spreading in a fracture zone converts an unstable FFR triple junction into a more stable RRR one. No modern analogue is known for this process.

In a spreading ocean the three arms of a triple junction must diverge, otherwise the ocean will not spread. In the most extreme case, magnetic lineations can be parallel in two of the plates if we have a ridge-ridge-transform fault junction (Fig. 3.38), as happens in the Pacific-Nazca-Antarctic junction. In this case, the width of magnetic anomalies along the transform fault plate boundary will be practically zero.

We conclude that in a spreading ocean, magnetic lineations of the same age cannot converge as we move away from the triple point.

These geometric constraints are valid in general terms and we can define more rigorously the specific conditions that invalidate them and whether or not they apply to the case of the West Pacific where we observe the paradoxical relationship between Japanese, Hawaiian and Phoenix lineations. We will analyse successively three more specialised types of geodynamic evolution (Fig. 3.40).

In a spreading ocean three rigid plates usually meet in one stable triple junction with three divergent arms, but instabilities can develop for short periods of time by creation of microplates with complex geometries. This was probably the case of the West Pacific, where the hypothetical Shatski microplate existed between M21 (137 Ma) and M13 (150 Ma) magnetic isochrons (Sager et al. 1988). Extrapolation to older magnetic anomalies show that these nucleated around a roughly triangular zone in the oldest core of the Pacific Plate (Handschumacher et al. 1988). If adjoining segments of the Pacific Plate are subducted in the West Pacific Bennioff zones, one is

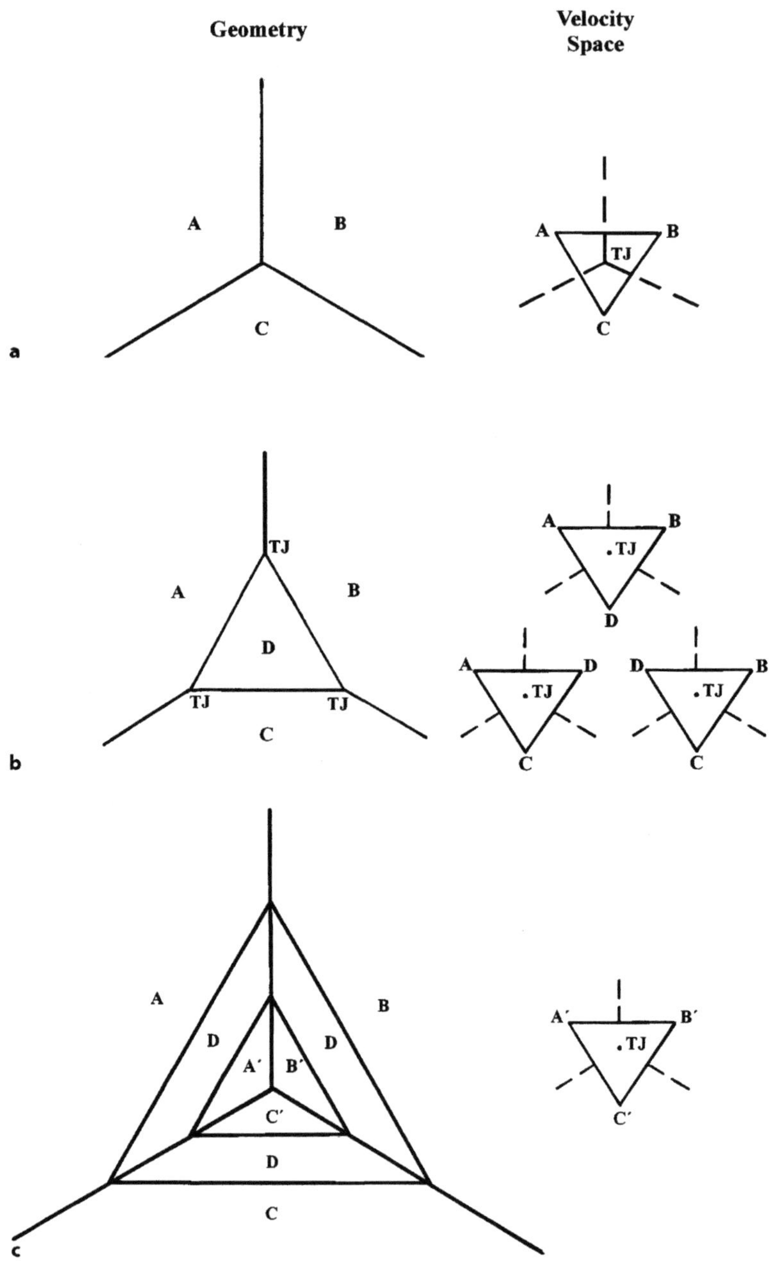

Fig. 3.39. Evolution of microplate at RRR triple junction. **a** Stable triple junction of RRR-type geometry and velocity space. **b** Microplate *D* created at former triple junction; three RRR triple junctions are formed; their stability requires that these three triple junctions are simultaneously stable. **c** Microplate *D* is cleaved and the system reverts to former unique triple junction

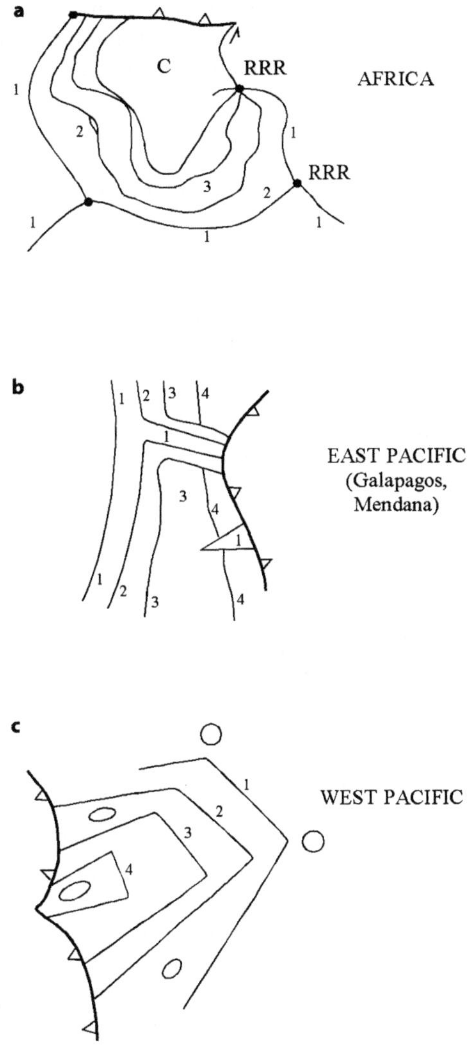

Fig. 3.40. Convergence of magnetic lineations. **a** Around a continent (*C*) or micro-continent, e.g. African Plate. **b** By fracture of plate as it approaches a subduction zone, e.g. Galápagos rift and Mendana Fracture Zone in the East Pacific. **c** By deformation of previously divergent triple junction, e.g. Japan, Hawaii and Phoenix magnetic lineations in the West Pacific. The Pacific Plate, if rigid, seems to nucleate from a seed microplate, inside Magnetic Anomaly 4. Note the difference between the situations depicted in **b** and **c**; in **b** the younger anomalies crosscut the earlier ones; in **c** all the arms of the triple junction have magnetic lineations of the same age

left with an apparent nucleation of a very large Pacific Plate growing from a seed microplate (Fig. 3.40c) because even older and wider oceanic lithosphere has been subducted since the creation of this microplate.

We arrive at the conclusion that unstable microplate geometries which do not obey the previously established law of divergence from the triple point can subsist only for short periods of time; this was not the case for the Pacific-Hawaiian-Phoenix system of magnetic lineations that persisted for at least 34 Ma between M25 (256 Ma) and M1 (122 Ma) and possibly from M38 (171 Ma) until 80 Ma (Larson and Chase 1972).

A second case where magnetic lineations converge as we move away from a triple point is when a continent is surrounded by three or more triple junctions, as is the case with the African Plate. We can exclude the possibility for the West Pacific during the Mesozoic because magnetic anomalies since at least 171 Ma were mapped on this part of the Pacific plate and no remnants of continental lithosphere were given in evidence (Fig. 3.40a).

A third case can be considered, when an oceanic plate has been fragmented by the approach of a subduction zone (Fig. 3.40b); this is the case of the Galapagos Rift and the present propagating Mendana fracture zone is presently breaking the NAZCA Plate by propagating SW from the impingement with the Peru–Chile trench. The space problem resulting from the convergence of two rift zones is solved by internal deformation of subplate between them, with transpression in the Trujillo trough as the Peru–Chile trench is approached.

The reconstitution of the West Pacific 110 Ma ago when the converging Phoenix, Japanese and Hawaiian lineations were formed at the same time precludes this third situation (Larson and Pitman 1972; Fig. 7, p. 3654). In fact, the Pacific Plate was situated in the middle of the Mesozoic Pacific Ocean, thousands of kilometres away from the subduction zones active in the Pacific rim at that time. Hence, this third situation can also be discarded because no fossil subduction zone exists between the Japanese and Phoenix magnetic lineations.

We conclude that the cause of convergence of the Phoenix and Japanese magnetic lineations is the intraplate oceanic deformation accumulated in the Pacific Plate since at least 120 Ma ago.

3.9.2 Paths in Non-Coaxial Deformation and the Role of Anisotropy

In the fast-spreading oceans the variability of the Euler pole position includes non-coaxial deformation; previous anisotropies such as transform faults and magnetic lineations become oblique to incremental strain axes. The relationship between stress and finite strain is, in this case, complex (Biot 1965; Cobbold 1976). Inspection of maps in all oceans shows that, in general, the fracture zones remain approximately perpendicular to magnetic lineations, even in the oldest domains. In coaxial deformation, with compression parallel to transform and fracture zones, the major axis of strain ellipse will stay parallel to magnetic lineations. However, in non-coaxial deformation, as in the Pacific, this condition is not valid. We must propose a strain regime that can explain these facts.

The presence of zones of diffuse seismicity in plate boundary zones in the old oceanic lithosphere requires that deformation must be distributed; the absence of a ductile layer inside the old oceanic lithosphere favours little partitioning, as shown by oblique-slip fault mechanisms of oceanic earthquakes. The consideration of these factors for the old oceanic lithosphere suggests that non-coaxial distributed deformation must be diffuse and with little partitioning.

The study of seismicity in the oceanic lithosphere (Bergmann 1986) shows that in young domains stresses of thermoelastic origin are important as sources for earthquakes; in old domains earthquakes occur in faults favourably oriented and are reliable indicators of the stress field by first-order patterns due to plate tectonics (Zoback 1992). The strong decrease in the seismic moment release rate with age suggests an increase in strength of the cooling and thickening oceanic lithosphere and a low ambient stress field, too low to result in large earthquakes. The recognition of the active structures responsible for intraplate oceanic earthquakes is in many cases difficult because detailed data on bathymetry and local structures are lacking; the oceanic earthquakes seem to be nucleated in zones of local weakness rather than at a stress concentration of local origin (Stein and Okal 1986) because the inferred stress fields agree over very wide areas. Such weakness zones are fracture zones, lines of age discontinuity and areas with fossil tectonic features such as seamounts. However, the main discontinuities are fracture zones which are reactivated by the present stress field, independently of their previous thermal and kinematic history; there is no evidence for internal deformation of block delimited by these fracture zones; hence the magnetic lineations can still preserve the frequent orthogonal relationship with the fracture zone.

An example is a seismic event located in the Austral Fracture Zone of the south-central Pacific (Okal and Talandier 1980) which shows a focal mechanism compatible with the present stress field, given the orientation of the fracture zone opposite to the sense of movement inferred from displacement of magnetic anomalies when it was an active transform zone (Fig. 3.41). However, as seismic events in the intraplate oceanic lithosphere are infrequent most of the slip must be aseismic, a point that will be confirmed later(see Sect. 3.9.4).

One of the few studies that assumes strain in the oceanic lithosphere focuses on the central Pacific Ocean (Goodwillie and Parsons 1992). The authors found a difference in fracture zone separation in the Pacific and Nazca Plates between anomalies 13 and 7; the upper bound of their deformation is 4%. The authors reject the boudinage model for deformation because the fracture zone separation does not increase with age; they attribute the observed deformation to re-orientation of the ancient Pacific-Farallon spreading centre between anomalies 7 and 5. Hence, this is a case of non-coaxial deformation of the oceanic lithosphere. We propose an alternative explanation for the same data. The boudinage model assumes a very small extension perpendicular to plate movement only detected by the geoid data; therefore, it cannot be rejected, on the basis of the presented data that the strain component measured by the authors is only a shear component, although we can have an unknown amount of coaxial stretching associated with it. Hence, 4% is a minimum boundary for finite strain.

The model of distributed deformation is valid for the brittle oceanic schizosphere. It is confirmed by direct observation of basement outcrops in a submersible (Juteau and Maury 1997) and by study of numerous drill cores obtained through the Deep Sea Drilling and Ocean Drilling Programs (Agar 1990; Sect. 3.12). It is also confirmed by observation of obducted ophiolites; in the Oman ophiolite we observed in Wadi Hilti, SW of Sohar, shearing along the margins of the sheeted dyke complex and associated hydrothermal alteration; this requires very early deformation of the ophiolite before its obduction to the present position in Campanian times. In the oceanic crust the most pervasive anisotropy must be due to the presence of the sheeted dyke complex and, in fact, it is reactivated during the oceanic stage. However, the regime of deformation in the ductile oceanic plastosphere must be different from the overlying schi-

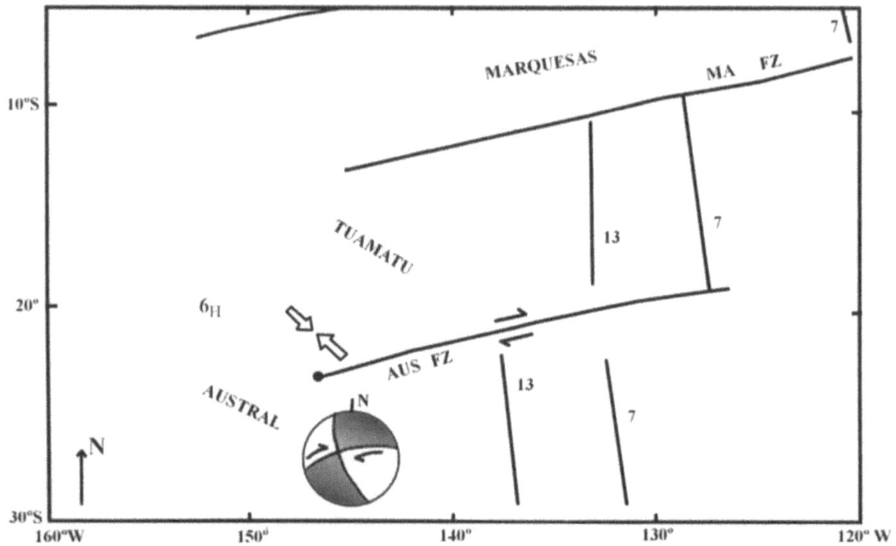

Fig. 3.41. Sense of slip in quasi-aseismic active fracture zones in south-central Pacific. The sense of slip is dextral in the WSW–ESE Austral Fracture Zone, in agreement with present stress field (after Okal and Talandier 1980); but it is opposite to the sense of movement when the Austral Transform was active, sinistral between Magnetic Anomalies 13 and 7. σ_H is horizontal maximum compressive stress; *AUS FZ* Austral Fracture Zone; *MA FZ* Marquesas Fracture Zone

zosphere and what role can be played by anisotropies that may be present in the oceanic plastosphere?

In fact, the propagation of seismic waves has shown that the oceanic upper mantle is anisotropic (Nicolas and Poirier 1976; Anderson 1989). Peridotites show a strong preferred orientation of the dominant minerals, olivine and orthopyroxene. Plastic deformation of these materials shows, in the field and in experiments, that the orientation of slip planes and slip directions tends to coincide with flow plane and flow direction; for instance, in olivine the flow plane is perpendicular to the crystallographic axis b, the slowest compressional velocities and the flow direction is parallel to a, the fastest crystallographic axis. The flow axis is subparallel to the spreading direction in the subhorizontal flow plane, in the present ocean and in ophiolites.

The present azimuthal anisotropy is subparallel to present relative plate motion (Anderson 1989); this means that in the plastosphere there is viscous deformation with small relaxation time that erased the previous anisotropy related to older plate motion; in the schizosphere, on the other hand, older anisotropies are reactivated by the present stress field. This suggests again that the schizosphere is dragged by the movement of the plastosphere.

3.9.3
Anisotropy and Finite Strain Estimation in the Pacific

Accepting the hypothesis of deformability in the western segment of the Pacific Plate, we must examine its implications.

If the Phoenix and Japanese lineations were initially divergent and became convergent by straining, we can have some idea of the minimal finite strain involved and its distribution (Fig. 3.42).

The mean orientation of the major X-axis of the average strain ellipse must be nearly perpendicular to Hawaii lineations, and subparallel to the Phoenix and Japanese lineations, in order to minimise the shear strain deduced from the angle between transform fault orientation and magnetic lineation for the three sets. Therefore, X must be oriented 69±3°E, bisecting the angle between the Japan and Phoenix lineations at M20–M15 of 149–140 Ma age. The approximate parallelism of the major axis of strain ellipse with the magnetic lineations suggests control of strain by previous magnetic fabric in an anisotropic plate. Assuming that these lineations were initially subparallel, compatibility of strain imposes a minimal value for the average finite strain of approximately 1.2 for Rs, the axial ratio of strain ellipse at the nearly horizontal surface of the ocean floor. This is an example that strain gradients can exist even if both trajectories have no curvature (Cobbold and Barbotin 1988), transferred from the plane to the sphere. We note also that the degree of convergence increases for older magnetic lineations reaching 26±2° for M25 (153 Ma) and M38 (165 Ma).

This suggests that the seed microplate with an approximately triangular shape, from which the Pacific Plate apparently nucleated, results in fact from progressive strain modification of some polygonal form that initially was opening to WSW (Fig. 3.43). This interpretation is also favoured by other data. Compression between the Japanese and Phoenix magnetic lineation sets must induce tension in the hinge zone between both sets and the Hawaiian magnetic lineation sets; this effectively happened because the Shastky Rise, a possible hotspot trace is located in the hinge zone between the Japanese and Hawaian lineation sets; there is also evidence of fanning of magnetic lineations in the Magellan Microplate between chron M15 at 136 Ma and M9 at 128.5 Ma (Nakanishi and Winterer 1998) located at the hinge zone between the Hawaian and Phoenix lineation sets; this is a good example of how intraplate deformation, intraplate hotspot activity and changing kinematics at plate boundaries are intimately connected.

We noted that the present maximum compressive stress N45°E is nearly perpendicular to the incremental strain axis, N120°E, but at a high angle to the N70°E finite strain axis. This suggests that anisotropy controls the mechanical behaviour of the Pacific Plate in a non-coaxial regime (Fig. 3.42). We know that the movement of the Pacific Plate changed between chron M7at 25 Ma and chron M5at 7 Ma; most of the finite strain observed in the West Pacific accumulated in the period prior to the change of kinematics; in fact, the major finite strain ellipse axis is subparallel to the relative plate velocity vector at that time. The most deformed domain faces the inflexion point between two convex arcs delineated by West Pacific subduction zones. This suggests a three-dimensional constrained flow regime in the down-going slab (Fig. 3.44). The deformation since this kinematic change is much less intense (Fig. 3.43).

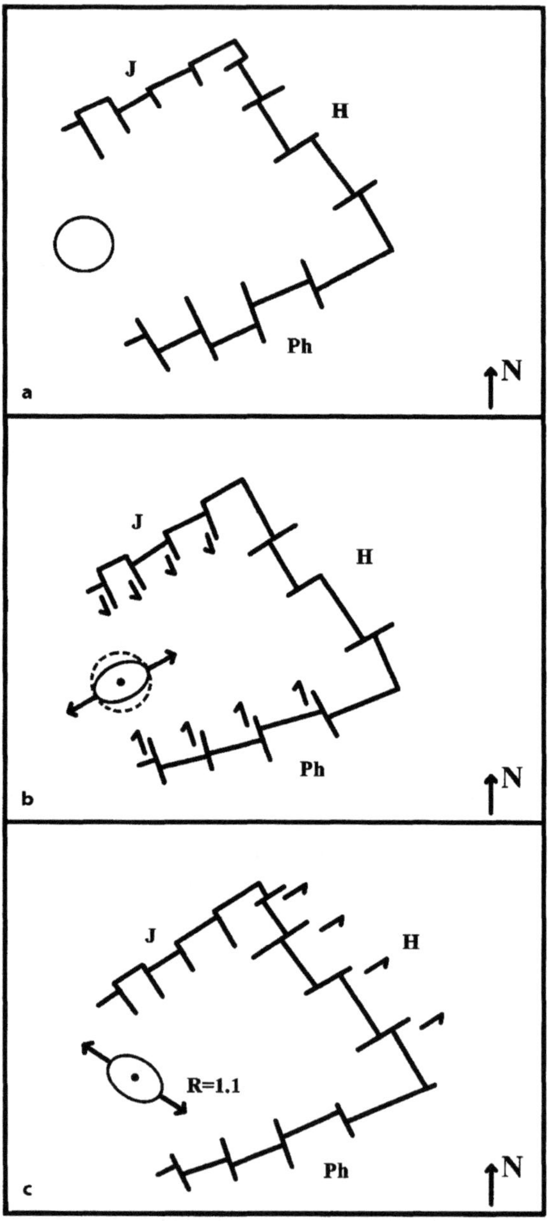

Fig. 3.42. Schematic diagram for strain in West Pacific-deformation of a triplet of magnetic lineations by domino faulting, with imposed sense of shear. **a** Initial stage, around 120 Ma. **b** Situation before the change in kinematics between 26 and 10 Ma. Sense of slip is sinistral across Japan's magnetic lineations and dextral across Phoenix magnetic lineation. The imposed bulk strain is given by a strain ellipse with axial ratio Rs=1.2. **c** Situation since 7 Ma. Sense of slip is sinistral across Hawaiian magnetic lineations. The imposed bulk strain is given by a strain ellipse with axial ration R_s=1.1. Magnetic lineations: H - Hawaiian; J - Japanese; Ph - Phoenix

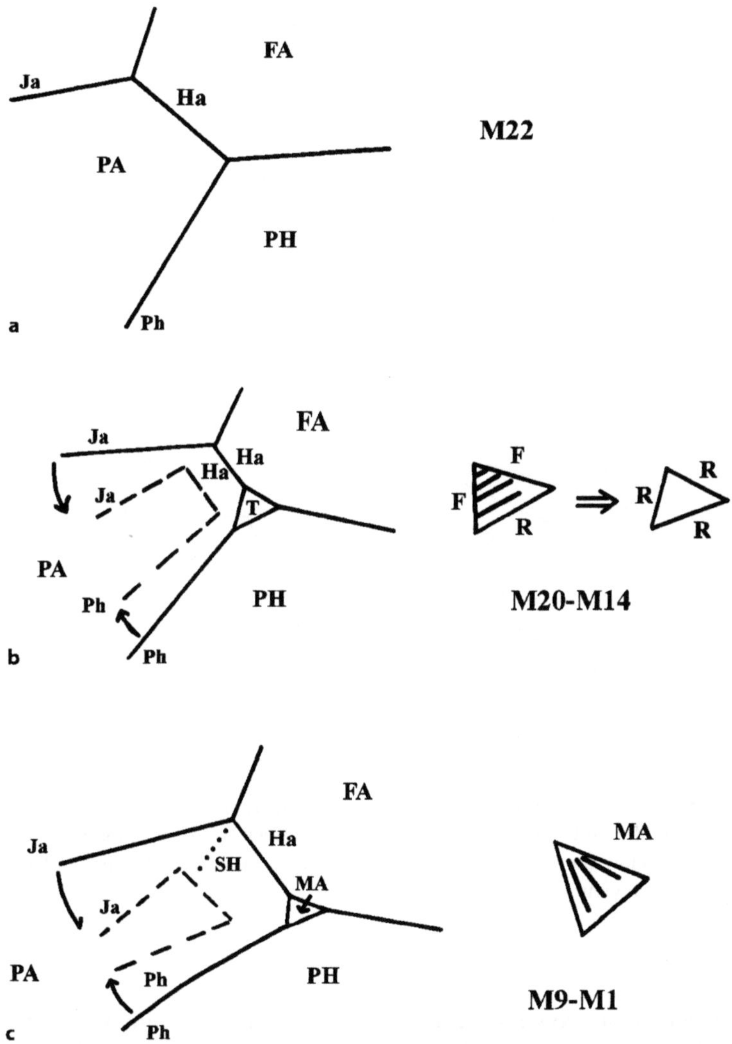

Fig. 3.43. Soft plate evolution for the West Pacific (modified from Nakanishi and Winterer 1998). **a** Chron M22. Synchronous Japanese, Hawaiian and Phoenix magnetic lineations form a triplet opening to SW. **b** Chrons M20–M14. Japanese and Phoenix magnetic lineations start to converge; the Trinidad microplate nucleates and changes from FFR type to RRR type. **c** Chrons M9–M1. Japanese and Phoenix magnetic lineations become convergent; the Magellan microplate is activated with fan magnetic lineations at the hinge between Phoenix and Hawaiian magnetic lineations; the Shatsky Rise forms at the hinge between Japanese and Hawaiian magnetic lineations. Plates: *FA* Farallon; *PA* Pacific; *PH* Phoenix. Microplates: *MA* Magellan; *T* Trinidad. Rises: *Sh* Shatsky Rise. Magnetic Lineations: *Ha*-Hawaiian; *Ja*-Japanese; *Ph*-Phoenix

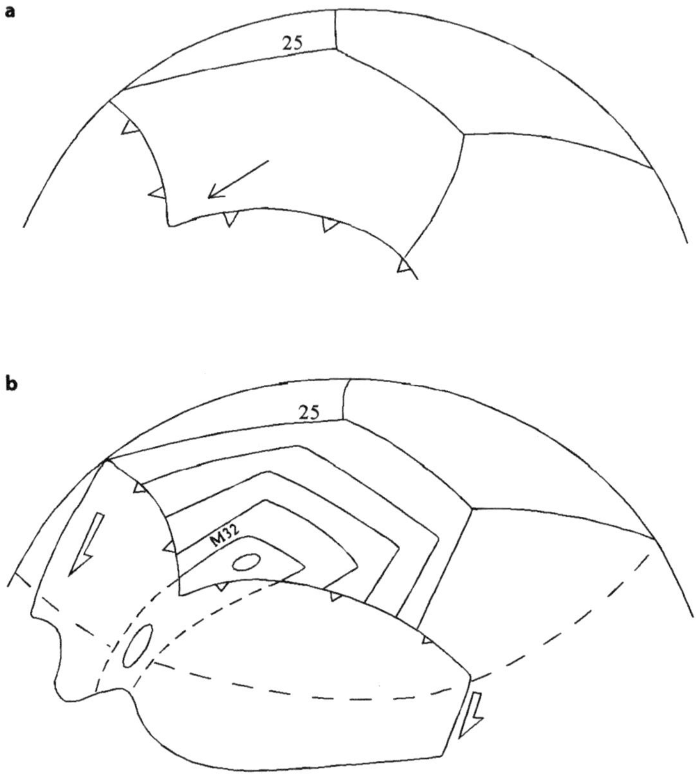

Fig. 3.44. Three-dimensional strain regime in West Pacific is prior to the kinematic change between chron 7 (26 Ma) and 5 (10 Ma). Plate movement is given by *arrow*. Constrained flow in the downgoing slab below the inflexion point at the crossing of two convex arcs. **a** Surface view; **b** three-dimensional view

We conclude that two strain regimes exist, controlled by finite kinematics of the plates involved. In Atlantic-type slow-spreading oceans Euler poles of divergent plates are stable, and the deformation is coaxial; anisotropy is enhanced by parallel shortening, and magnetic anomalies stay orthogonal to fracture zones or become orthogonal if they are oblique to fracture zones (Fig. 3.45). In Pacific-type fast-spreading oceans Euler poles of divergent plates spin significantly and the deformation becomes non-coaxial, the presence of subduction can induce permutation of the incremental strain axes, the final result is that these axes can make any angle with the previous anisotropy, inherited from previous periods of coaxial deformation; fracture zones are reactivated with a minor internal deformation of each domain so magnetic anomalies stay orthogonal to fracture zones that can rotate in a domino-style of deformation. This deformation regime is typical of slip-system domains (Cobbold and Gapais 1986); in this case, only one slip set is present, represented by fracture zones. The kinematic constraints are such that slip in a given direction is easy, but stretching in the

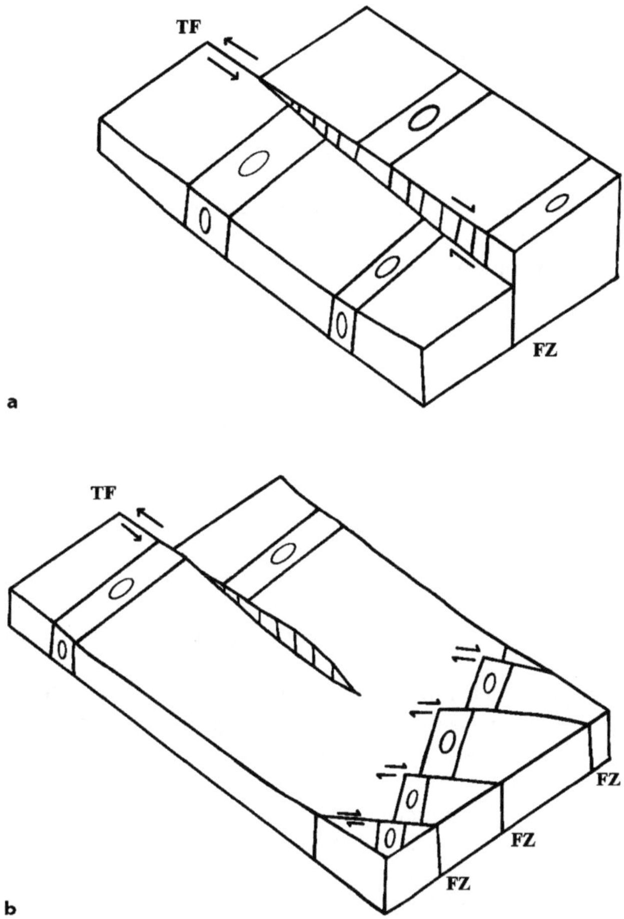

Fig. 3.45. Deformation of oceanic lithosphere. **a** Coaxial regime in slow-spreading, Atlantic-type ocean. **b** Non-coaxial regime in fast-spreading, Pacific-type ocean. *FZ* Fracture Zone; *TF* Transform Fault

same direction is inhibited; this expresses mechanical anisotropy; the motion becomes almost restricted to shearing parallel to slip lines or fibres until a new set of cross-cutting faults deactivates the previous one.

Internal deformation of oceanic plates is confirmed by geodetic methods and underestimated by seismic moment release, suggesting that aseismic deformation predominates over seismic deformation. An indirect proof is the fact that sinuosity of the Mid-Pacific Rise decreases with age.

3.9.4
Tectonics of Abyssal Plains

Abyssal plains are one of the quietest seismic settings on the Earth, together with stable cratonic interiors. According to the standard theory, they should be almost perfect rigid. Therefore, they become a favourite target for disposal of high-level radioactive waste into the seabed (Shephard et al. 1988) because very low seismicity would require very low tectonic activity.

In order to test this rationale, very high resolution reflection seismic profiles were conducted in a number of abyssal plains. It was a surprise to find that sediments of the abyssal plains were shown to be cut by a dense array of faults that will be described in the following.

It has been refuted that seismic discontinuities could be artefacts (Shephard et al.1988, p. 79). In fact, faults are characterised by some features such as: they form linear subparallel arrays parallel to acoustic basement topography; they are subvertical, and can change the sense of dip and vertical separation, normal in deep levels and reverse at higher levels; they cut through all the sediment pile, affecting the sea bottom and giving a gentle slope.

Several possible mechanisms have been considered as causes of faulting (Williams 1987). One is simple differential compaction of sediments over rigid basement topography; they should be normal growth faults dipping away from basement highs, which is not the case.

A more elaborate model is the permeable fault model where faults act as conduits for the vertical escape of pore fluids migrating along bedding. If faults become closer, the compaction is faster and the movement can change from normal at depth to reverse closer to the surface.

The third model is the impermeable fault model; pore fluids migrate up-dip through bedding by an excess lateral fluid-pressure gradient. Therefore, antithetic faults stay normal, but synthetic faults can become reverse.

The second and third models are not incompatible, but the latter explains the observed geometry and kinematics better in most cases. However, it has been clearly stated "alternative explanations of sediment faulting which invoke basement movements merit further attention" (Williams 1987, p. 104).

In fact, the seismic profiling conducted for the high-level radioactive waste proposal has high resolution because it is directed to detect structures in the sediments; if we want to study structures in the basement, we also need deep penetration in order to cross the thick cover of sediments in abyssal plains. Therefore, to study relationships between faulting in the cover and in the basement we need both high resolution and deep penetration, which is impossible to achieve without some kind of compromise. Besides, no studies were conducted from this point of view.

We must test thick-skinned tectonic models involving basement as opposed to thin-skinned tectonic models restricted to the sediment cover. Such a thick-skinned model has in fact been proposed (Fyfe 1992).

We concluded that cooling by conduction explains the correlation between depth of the young ocean floor and its age; for the old ocean floor, the data are less clear and some authors claim that we need additional heat transfer, which remains unexplained and for which we must research an origin, as seen in Section 3.3.4. Hydrothermal seafloor metamorphism predicts that water-cooling convection should decay after 20 Ma. However, convective circulation may still continue under the less permeable sediment cover (Fyfe 1992). In fact, recent measurements of heat flow have revealed anomalies (Shephard et al. 1988) and suggest that convective cooling of the oceanic crust may occur in all ocean domains, because even for very low gradients fluid convection is possible.

If convection is ubiquitous in the ocean domain serpentinisation is the final product according to the reaction:

$$\underset{\text{water}}{2H_2O} + \underset{\text{pyroxene}}{MgSiO_3} + \underset{\text{olivine}}{Mg_2SiO_4} \rightarrow \underset{\text{serpentine}}{Mg_3Si_2O_5(OH)_4}$$

This reaction is highly exothermic. Hence, a low rate of chemical heat production is possible and is a function of the rate of water penetration in the system.

A consequence of hydration will be layer volume expansion with uplift of the sea bottom above the theoretical pure conductive cooling curve, explaining the observed topography. As serpentinite expands it will either seal the system or generate new fractures, and it was shown that permeability generation dominates over sealing (MacDonald and Fyfe 1985).

In the thick-skinned model the geometry and kinematics of the active faults are explained by tectonic inversion, in the compressive stress environment of abyssal plains, of previous normal synsedimentary faults on flanks of basement highs. High-level layers must show net extension separated by a null point of net zero displacement (Biddle and Rudolph 1988; Coward 1994).

All the processes must occur at very low rates, including the decay of chemical heat production, sedimentation and strain inversion rate; a low strain rate will favour stable sliding over stick-slip. Therefore, chemically induced deformation must be aseismic, in both the sediment cover and in the hydrothermally altered oceanic crust and serpentinised upper mantle.

3.10
Conclusion: The Rheology of Oceanic and Continental Lithosphere

Standard plate tectonics and soft plate tectonics differ on the relative strength of the oceanic versus the continental lithosphere. This topic remains very controversial (Allen and Allen 1990, p. 8): the standard theory assumes a stronger oceanic lithosphere, but oceanic plates are thinner and hotter and therefore should bend and shorten more easily under an applied compressive force. The b-values and elastic thiknesses of different tectonics environments are systematically ordered, but in different ways according to each theory (see Table 3.3).

We must address the problem of lithosphere rheology in the full scale of time from short periods – seismic waves and free oscillations, to long periods – convection (Ranalli 1995). Models for internal friction in solids are available in this respect (Karato and Spetzler 1990); *anelasticity* in the short time scale explains attenuation of seismic waves by combining a spring with a viscous element arranged in parallel with another spring, the standard linear solid; *viscoelasticity* in the long time scale is mo-

Table 3.3.

Increasing strength⇓	Standard plate tectonics	Soft plate tectonics
	Deformed belts	Deformed belts
	Continental rifts	Continental rifts
	Shields	Oceans
	Oceans	Shields

delled by a spring and viscous element in series on the Maxwell body where strain increases without limit at constant stress. A more realistic model that will explain both anelastic and viscoelastic behaviour is the Burgers body consisting of a spring and viscous element in parallel arranged in series with another spring and another viscous element (Fig. 3.46).

All solids show viscoelasticity at sufficiently high temperatures and/or low frequencies. This is because internal defects in a crystal can be mobilised and reorganised by stresses well below the yield stress (Lowrie 1997). Anelastic behaviour results when motion of the defects under stress stops for a sufficiently long time. Viscoelasticity results when the defect motion continues indefinitely.

The rheology of materials depends on the following environmental parameters: lithostatic pressure and differential stress, temperature, strain rate and chemical environment. In the lithosphere, strain rate is imposed by tectonic setting; lithostatic stress and temperature are controlled by geothermal and lithostatic gradients, such that, near the surface pressure and temperature must be low. If the lithosphere deforms outside the elastic field near the surface, the only factor that can cause this deformation is chemical environment. Increasing confining pressure, temperature and decreasing strain rate increase the ductility of geological materials by enlarging the viscous field at the expense of elastic and failure fields (Fig. 3.47). In other words, rocks shift from weak and brittle to stronger and more ductile (Heard 1968; Jaeger and Cook 1979; Park 1997). Increasing the chemical potential of some specific phases must have the same effect, in general, and increase the confining pressure (Twiss and Moores 1992, pp. 378–379), because the solubility of water in silicate lattices increases with increasing pressure. This has been only quantified in some cases, such as the effect of partial pressure of oxygen on the creep strength of quartz and olivine by hydrolithic weakening (review in Ghosh 1993; Kohldstedt et al. 1995).

Hence, the most obvious deformation mechanism at low temperatures will be *solution creep* (Twiss and Moores 1992), a class of *diffusion creep* that results from diffusion of dissolved components in a fluid along the grain boundaries; deformation takes place in diffusion creep by transfer of material from areas of high compressive stress to areas of low compressive stress. The rheology that results from solution creep will probably be Newtonian viscosity. In other terms, a constitutive law with a buffer strain, but no yield strength could explain viscoelasticity and transient creep (Lliboutry 1992).

In fact, we can consider that the ductile regime pervades the entire lithosphere (for a recent synthesis, see Engelder 1993). In the upper levels of the schizosphere, low-

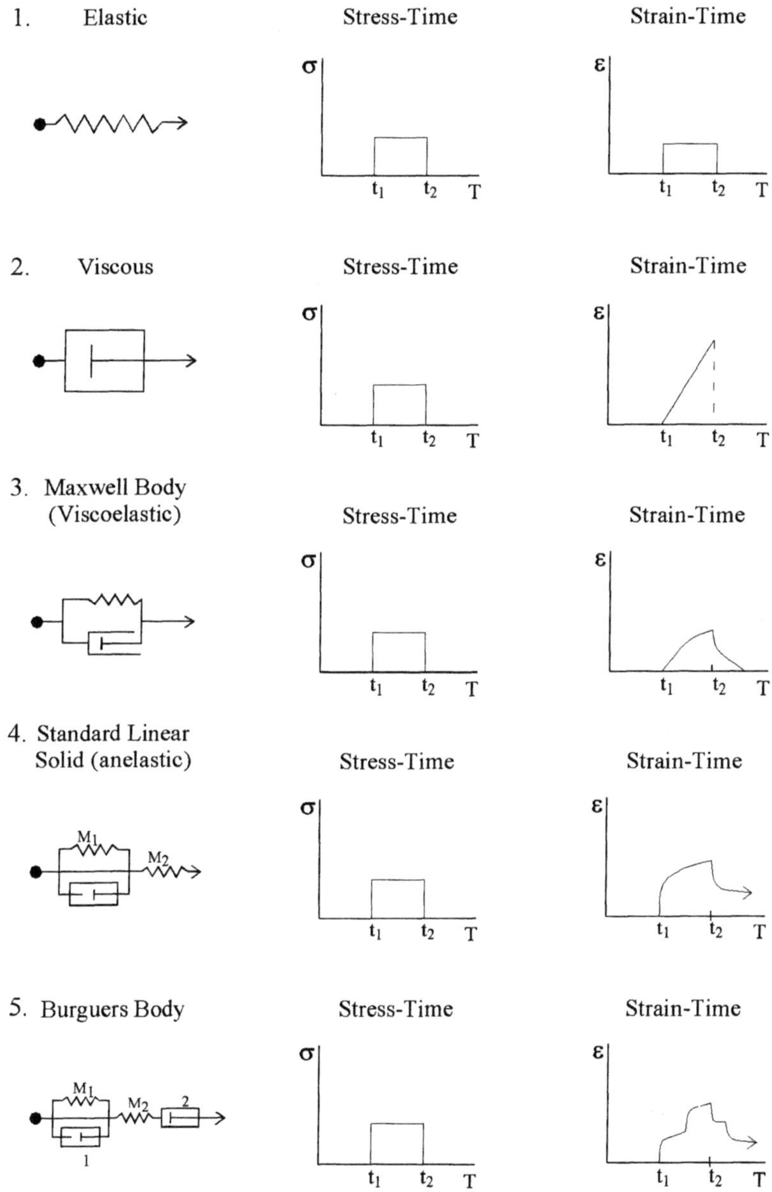

Fig. 3.46. Analogue models of material behaviour; the stress and strain evolution with time are shown by sudden application of a force at time t_1 and sudden removal at time t_2. *1* Elastic with analogue spring. *2* Viscous with analogue dashpot (fluid-filled piston and cylinder). *3* Maxwell body or viscoelastic with spring and dashpot arranged in parallel. *4* Standard linear solid or anelastic with a Maxwell body arranged in series with a spring; M_1 and M_2 are elastic constant for the springs. *5* Burgers body: a Maxwell body arranged in series with a spring and a dashpot; M_1 and M_2 are elastic constants for the two springs and η_1 and η_2 are viscosities of the fluids in the two dashpots. (After Suppe 1983; Karato and Spetzler 1990)

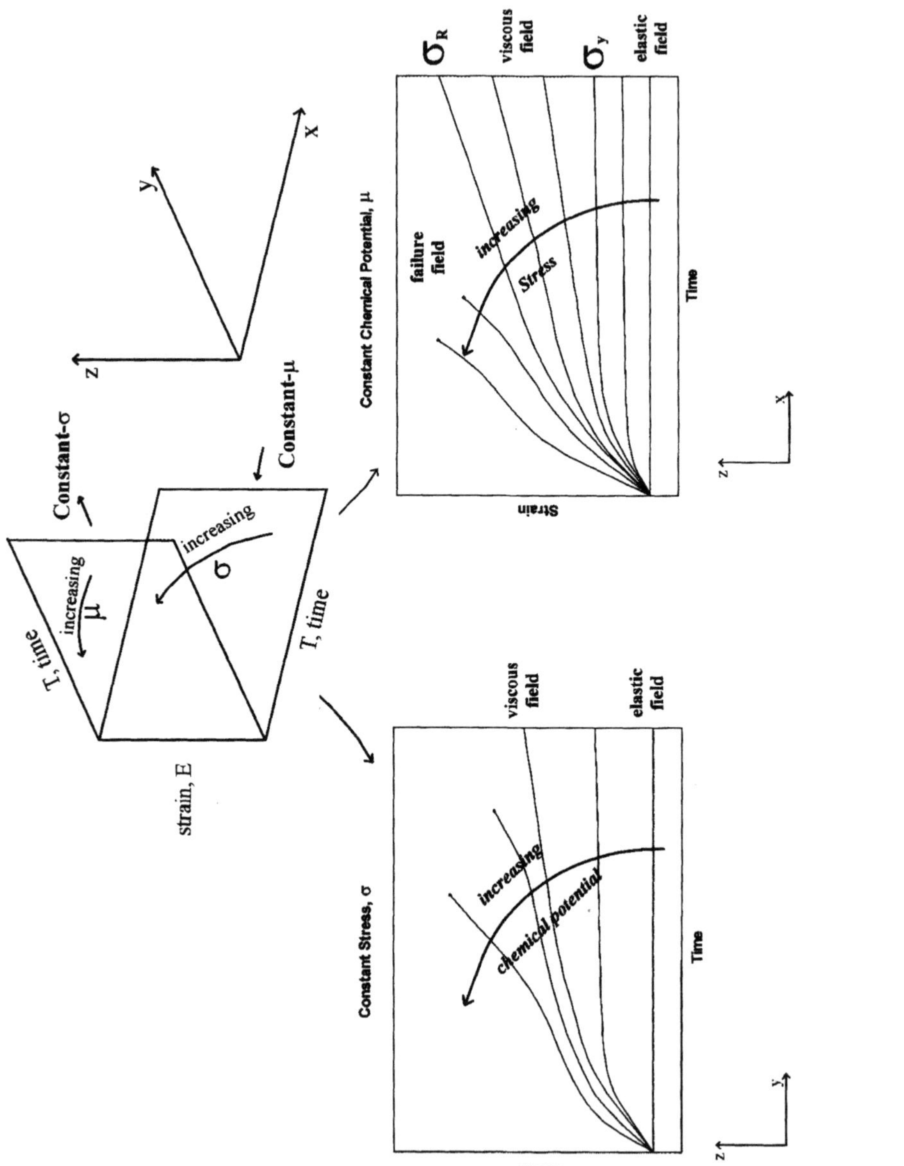

Fig. 3.47. Rheology as a function of parameters in a various three-dimensional space. *XZ Face* Strain, *t* as a function of time, *T* under constant chemical potential (μ) and variable stress (σ). Stress increase leads to enlargement of the viscous field at expense of elastic and failure field. σ_Y Yield stress, between elastic and viscous field; σ_R failure stress between viscous and failure field. *XY face* Strain, ε as a function of time, *T* under constant stress, σ and variable chemical potential (μ). Chemical potential increase leads to enlargement of viscous field. (Adapted from Heard 1968; Park 1997)

temperature ductile deformations include diffusion-mass transfer such as solution creep and work hardening with rate-controlling mechanisms restricted to dislocation glide by twinning and slip. In the brittle-ductile transition at the boundary between schizosphere and plastosphere, dislocation and diffusion creep and other thermally activated processes operate in conjunction with brittle mechanisms to contribute to creep strain; in the plastosphere ductile flow is the dominant deformation mechanism.

The concept of ductile deformation below the frictional-slip regime can explain tectonic and seismologic observations. Many faults on forelands of orogenes show slickenlines due to local dissolution and mineral growth in fibres which are not wear grooves (Ramsay 1980); this expresses aseismic deformation that must be responsible for seismic coupling coefficients below one and viscoelastic rebound. A compilation of differential stress necessary to promote deformation in the upper segment of the ductile-flow regime in the schizosphere (Engelder 1993) shows that the estimated values with paleopiezometers plot below the boundary of the frictional slip regime. This would explain why many thrust wedges in foreland fold-thrust belts will deform in the upper ductile regime with subordinate seismic events; dêcollements of weak rokks such as salt and shale will favour aseismic sliding at low strain-rates.

If we apply this analysis to the problem in question, we will note that the oceanic schizosphere must be viscoplastic. Aseismic viscous deformation can occur, assisted by fluid, in areas affected by ocean floor hydrothermal metamorphism; when the yield stress is attained an earthquake will occur at plate boundaries and, more rarely, in mid-plate environments. In the continental schizosphere, the behaviour is more variable; in cratons, the rheology must be elastoplastic; stresses at the boundaries are transmitted efficiently in the cold and chemically inactive environment; if yield stress is attained a very rare mid-cratonic earthquake will occur. In deformed belts with a higher geothermal gradient than in cratons, some aseismic deformation can occur by active folding in the cover or by aseismic sliding mass transfer (Gratier and Gamond 1990; Gratier et al. 1999). Solution creep is proven by dissolution making asperities which prevent sliding along the active fault zone and by crack-seal in extension structures (Ramsay 1980). When yield stress is attained a seismic event occurs, generating cataclasis along active faults (Fig. 3.48). The rheology becomes intermediate between elastoplastic in cratons and viscoplastic in oceans. These considerations explain the ordered pattern of strengths that we observe according to soft plate tectonics. In this context, the contradiction between higher levels of stress magnitudes, inferred by some authors, for the oceanic lithosphere when compared to the continental lithosphere (Govers et al. 1992) and higher aseismic strains in oceanic intraplate environments that we have estimated so far is more apparent than real. A viscoplastic body can suffer a high strain even if its yield strength is above the one of an elastoplastic body that can only accommodate a minor elastic strain because its ductility was increased by enlarging the viscous field at the expense of elastic and failure fields.

Strain and velocity gradients in the schizosphere are organised on a large horizontal length scale that exceeds its thickness (England and Molnar 1991; Zobak 1992). This suggests that the schizosphere is weak relative to the plastosphere and its deformation passively follows the deeper one, or, in other words, the schizosphere can be treated as a thin viscous sheet whose strength is governed primarily by power law creep in the upper mantle and not by friction in the crust (Molnar and Gipson 1996).

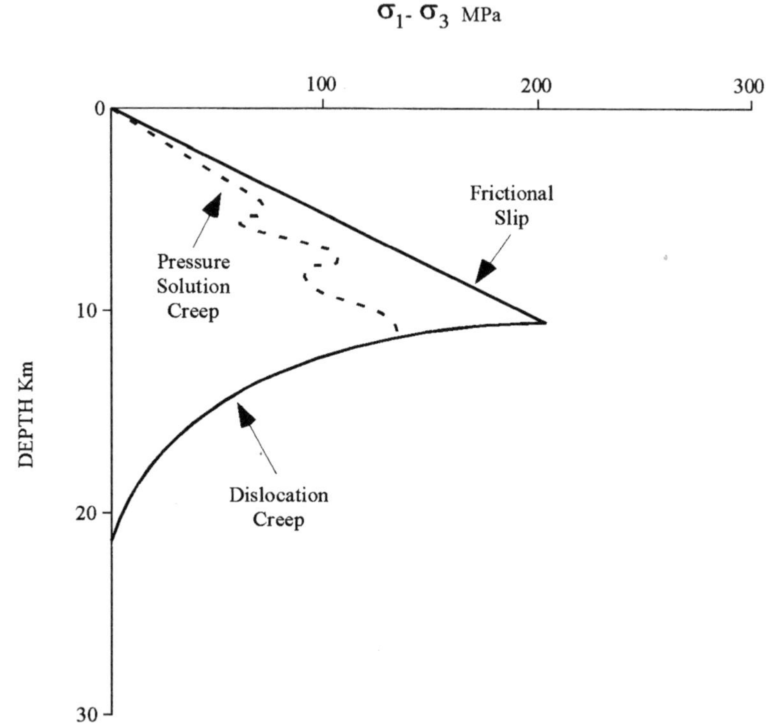

Fig. 3.48. Rheology profile for a real lithosphere composed of a schizosphere with three layers, from *top* to *bottom*: (1) high porous sediments; (2) metamorphic rocks; (3) granitic rocks. The *abscissa* is differential stress $\sigma_1-\sigma_3$, in MPa and logarithmic scale; the ordinate is depth, in km. The strength is much more heterogeneous than in theory-derived profiles. Two envelopes are plotted for the schizosphere: one (*thin line*) corresponds to higher differential stress during earthquakes corresponding to cataclastic deformation of frictional slip-type behaviour; the other shown by *points* corresponds to pressure solution creep; the plastosphere obeys a dislocation creep mechanism of deformation shown by a *thick line*. The strain rate is 10^{-14} s^{-1}, typical for orogenic conditions. (After Gratier and Gamond 1990)

This model is confirmed by considerations of heat flux in actively deforming belts (England and Molnar 1991). This conclusion has implications on the rheology of the lithosphere. As the continental lithosphere is thicker than the oceanic lithosphere their strengths must be a function of strain rate if we keep other controlling factors constant in both cases, namely, heat flow. A continental plate moving at a specified velocity should be stronger than an oceanic plate because the strain rate at the base of the schizosphere will be higher in the thin oceanic plate than in the thick continental one. This is because the coherent moving plate imposes a radial variation of angular velocity as a function of depth.

Although much progress has been achieved in the last few years in the rheology of lithosphere there is persistent disagreement between opposing points of view on two fundamental topics.

Fig. 3.49. Schematic cross section of crust, mantle lithosphere and asthenosphere for different geodynamic settings: ocean, continent and recent orogenes. *As* Asthenosphere, *Cr* crust, *Ecc* extended continental crust, *Ls* lithosphere, *Ma* mantle, *Ncc* normal continental crust, *Tcc* thickened continental crust

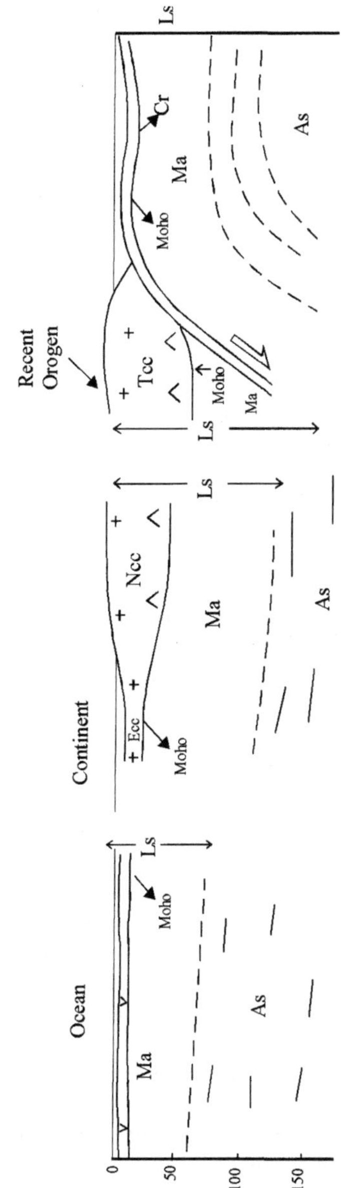

Table 3.4. Rheology of the lithosphere in different tectonic environments

		Shield	Continent collision zones	Oceanic zones	Continental rift zones
Lithosphere	Schizosphere	10–15 km	45–70 km	10–40 km	10 km
	Plastosphere	140–135 km	55–80 km	40–60 km	40 km
		$\eta=10^{-18}$ to 10^{-24} Pa-s	$\eta=10^{-16}$ to 10^{22} Pa-s	$\eta=10^{-16}$ to 10^{22} Pa-s	$\eta=10^{-14}$ to 10^{-16} Pa-s
		$\tau_M>100$ Ma	$\tau_M\approx10$ Ma	$\tau_M\approx10$–20 Ma	
Lithosphere		>150 km	100–150 km	50–100 km	50 km
Crust		40 km	50–70 km	10 km	10–30 km

The first concerns the driving mechanism for schizosphere deformation. In one view which we referred to recently, the schizosphere deformation is driven from below. One alternative possibility is that deformation of the entire lithosphere is driven from the sides by stresses that are comparable in magnitude in the schizosphere and plastosphere (Haines 1998). In both cases, the strength of the whole lithosphere is controlled by the strength of the plastosphere because it is stronger than the schizosphere or, at least has a strength similar to that of the schizosphere.

The second topic, related in some way to the first, concerns the strain regime in the schizosphere and plastosphere. In one view, deformation is distributed on both levels with stick-slip in seismogenic faults above and stable sliding on narrow shear zones below (Thatcher 1995). In the other view, the schizosphere shows distributed deformation along discrete active faults and the plastosphere deforms by heterogeneous shear across the entire width of the active zone (Bourne et al. 1998). Combining this interpretation with the driving forces of the schizosphere by the plastosphere below, the observed slip rates between blocks estimated by geodesy and neotectonics can be derived from slip rates of undeformed blocks outside the active zone.

We can summarise our conclusions on the rheology of the lithosphere in different tectonic environments as follows (Fig. 3.49; Table 3.4).

The rheology is depth-dependent, so the total strength (Ranalli 1995) is given by integrating differential stress as a function of depth in some thickness, H. The base of the lithosphere is defined by a rapid decrease in differential stress which reaches the conditions typical of the sublithospheric upper mantle. This mechanical thickness is above the effective elastic thickness estimated by flexure of the lithosphere (see Sect. 3.4). Plastic layers are characterised by some yield strength. Viscoelastic layers are characterised by the Maxwell relaxation time,

$$\tau = \eta / \mu = \varepsilon / \dot{\varepsilon}$$

where η is viscosity, μ is rigidity, ε is elastic shear strain and $\dot{\varepsilon}$ is viscous shear strain rate. Below τ_M the elastic response predominates and above τ_M the viscous response does. With plastic rheology the viscosity becomes infinite and the Maxwell relaxation time tends also to infinity (Ranalli 1995).

An elastic lithosphere will express stress heterogeneities because it will be a stress guide (Turcotte 1982); a viscous lithosphere will express strain heterogeneities because it will be a finite strain guide. A viscoelastic lithosphere will be a stress guide and incremental strain guide for loading shorter than the Maxwell relaxation time, τ_M, and a finite strain guide for loading longer than τ_M. The study of stress, incremental strain and finite strain fields in the highly heterogeneous conditions of the Pacific Plate suggest that the lithosphere is viscoelastic with τ_M in the range of 10^1–10^2 Ma.

The problem of assigning a realistic rheology to the lithosphere is further complicated by the possibility that the plastosphere has a power-law fluid rheology and not a linear viscous, Newtonian rheology. If this is so, we have to specify the *effective viscosity* as a function of stress and strain rate; in practice, we simplify this problem by attributing an average viscosity to the whole lithosphere (Ranalli 1995, 1997; Bourne et al. 1998).

The global conditions at the lithosphere-asthenosphere transition can be estimated on the basis of phenomenological and microphysical data (Ranalli 1995). The creep strength at geologic strain rates are in the order of 1–10 MPa equivalent to a viscosity of 10^{20}–10^{21} Pa s or a Maxwell relaxation time, τ_M of 10^{-1} Ma. Above is the lithosphere with thickness H, very variable according to tectonic environments (Fig. 3.49; Table 3.4).

In *shields* the schizosphere is thin, 10–15 km, and has a finite yield strength or a Maxwell relation time above 1 Ga, the age of oldest residual stress systems. The rheology of the plastosphere can be approximated by extrapolation of rheology in tectonically active intracontinental domains, such as Tibet (England and Molnar 1997). This strain rate must be in the order of 10^{-18} s^{-1} or lower. Palaeostress estimations for North America (Van der Pluijm et al. 1997) indicate ~20 MPa of differential stress in the interior of the craton (see Sect. 2.5) at the moment of orogenies in its margins. This suggests an effective viscosity above 10^{24} Pa s and a Maxwell relaxation time in the order of 100 Ma. From a tectonic history perspective , shields are composed of an assemblage of cratons of different ages, whereas cratons are characterised by a structure of lithospheric scale, with 300-km-thick *keels*. They remain untouched by tectonic activity for periods of 1 Ga, cooling and becoming very strong, thus they are only rejuvenated by plume-controlled active rifting (Sengör 1999).

In *continental collision zones*, the lithosphere has a thickness of 150 km; the rheological stratification is complex, with a possible weak band in the lower crust. The study of active deformation in Tibet (England and Molnar 1997) shows that the strain rates are in the order of 10^{-16} s^{-1}, corresponding to an average effective viscosity of 10^{22} Pa s, corresponding to an average Maxwell relaxation time of 10 Ma. Pressure solution creep in the upper crust leads to viscosities in the order of 10^{21} Pa s (Gratier et al. 1999). In other *convergent mobile zones*, such as subduction zones, there are no published data but, a priori, the values of average viscosity should be slightly below those of Tibet.

In the *oceanic lithosphere* the thickness is controlled by its age, between 50 and 100 km. The data available (Martinod and Molnar 1995) for mature oceanic lithosphere 60 Ma old in the central Indian Ocean are in the order of 10^{12}–10^{13} Nm^{-1} equivalent to 35–140 MPa of differential stress; the schizosphere is 40 km thick and the plastosphere has an average viscosity of 10^{21}–10^{22} Pa s for a strain rate of 1.5×10^{-16} s^{-1}, equivalent to a Maxwell relaxation time of 10 Ma.

In *continental rifts* the lithosphere is 50 km thick; the schizosphere is only 10 km thick. Modelling (Kuznir and Park 1986) indicates that for high heat flow above 65 mW m^{-2} a tensile stress of 75 MPa can explain whole lithosphere failure. A mean strain rate of 10^{-15} s^{-1} suggests a mean effective viscosity of 7.5×10^{20} Pa s.

We must stress that the main difference between rigid and soft plate tectonics concerns the viscosity of cratonic interiors with a thick continental lithosphere, when compared to the viscosity of the thinner oceanic lithosphere. The contrast between the two versions of plate tectonics can be exemplified by the eastern Mediterranean region; the deformed zones along plate boundaries such as Africa, Eurasia, Anatolia and Arabia curve around and between the oceanic lithosphere of the Black and Caspian Seas (McClusky et al. 2000). This is considered to be in agreement with rigid plate theory. However, we have noted that the North-Anatolian Plate Boundary Zone curves around the continental and more rigid Anatolian Plate with the convexity on the side of the oceanic Black Sea. We infer that continental Anatolia is more viscous than the oceanic Black Sea, because more rigid objects indent the less rigid ones. This interpretation favours the soft plate version versus the rigid plate one. The low seismicity of the Black Sea oceanic lithosphere further suggests that it deforms mostly by aseismic mechanisms, in contrast to the preponderance of seismic deformation in the continental lithosphere. We must note that the North Anatolian Fault follows, at least partially, a minor circle around the Euler pole of rotation of Anatolia relative to Eurasia located by GPS measurements (McClusky et al. 2000). This contrast with the Dead Sea transform, which spirals around the Euler pole of rotation of Arabia relative to Africa (Wdowinski 1998; see Sect. 2.7). Could this distinct behaviour be explained by higher rigidity of the Anatolia/Eurasia plate boundary fault versus lower rigidity of the Arabia/Africa plate boundary fault?

We conclude that the critical parameter that differentiates the standard plate tectonics theory from the soft plate version is the viscosity of the oceanic lithosphere or the equivalent Maxwell relaxation time. In the standard theory it will be in the order of Ga, as in cratons; in the soft plate version it is in the order of 10 Ma, well below its value in cratons. This parameter will control the behaviour of the whole lithosphere, as we noted earlier.

We must review the evidence for this value. Viscoelastic models for flexure of the oceanic lithosphere suggest a viscosity of 10^{22} Pa s for the plastosphere (Martinod and Molnar 1995) or a Maxwell relaxation time of ~10 Ma. We must check whether there is independent evidence for this value.

States of stress in the oceanic lithosphere indicate a change in the stress regime at 10–20 Ma, from extensional and strike-slip near the ridge to compressive and strike-slip for lithosphere older than 10–20 Ma. Let us assume a viscoelastic rheology for this lithosphere and look at the driving forces for the plates (Cox and Hart 1986; Ranalli 1995). Ridge push is proportional to the sine of slope of the ridge flank; it will extend the plate. Mantle drag traction will oppose the movement of the plate; it will compress the plate. Ridge push operates from the surface where the plate is predominantly elastic; mantle drag traction operates deeper because it is due to viscous coupling between the plate and the asthenosphere. The change in the tectonic regime from extensional to compressional at 10–20 Ma is then due to the fact that the viscous response of the plate to mantle drag overcomes the elastic extensional response to ridge push. The Maxwell time of the young oceanic lithosphere must be in the order of 10–20 Ma and a little longer in the old oceanic lithosphere. This result will be important in the soft plate tectonics version of the Wilson cycle.

Note that at almost the same time, 20 Ma, the oceanic lithosphere changes from positively buoyant to negatively buoyant (Davies 1992). Hence, both the mechanical strength and the thermal state of the oceanic lithosphere suggest an important geodynamic change at this moment.

The value that we infer for Maxwell time in the oceanic lithosphere of 10–20 Ma has implications on the interpretation of some of the results that we obtained previously. Plate kinematics is a dynamic system because it is in permanent change. We compared, on the one hand, results from NUVEL-1A that mix data from 3 Ma ago, such as magnetic anomalies and trace of transforms, with present data such as focal mechanism of earthquakes; and, on the other hand, kinematics from instantaneous movements such as those derived from satellite geodetic data. Hence, we face an inescapable problem: with viscoelastic rheology the present state of the system has both an elastic component, which is a quasi-instantaneous elastic response, and a viscous component, which is a delayed response of 0–20 Ma to a changing state of stress input. This problem is minimised in Atlantic-type oceans because they evolve slower than Pacific-type oceans. However, the previous results in the Pacific must be considered rough estimations from which it is difficult to extract more than semi-quantitative conclusions.

In the present state of the art, two opposed schools of thought emerge (Gordon 1995).

One considers that there are significant differences between present kinematics inferred from geodetic satellite data and kinematics models based on anomaly 2A, with approximately 3 Ma, like NUVEL-1 and NUVEL-1A. The present Euler pole inferred from geodetic satellite data, for instance, has shifted significantly from the NUVEL-1A Euler poles (Larson et al. 1997) in both the NNR and relative reference frames. It is approximately 11°05′ west of the NNR/NUVEL-1A pole. Therefore, the kinematics of the Pacific Plate over the last 5 years does not agree with this motion over the last 3 Ma. According to this view, the pole movement is unsteady. For example, the plate motion between Nazca and South America has shown an average of 3.1 ± 1.0 mm/yea per Ma since 25 Ma (Norabuena et al. 1999) due to the increase in subduction resistance by growth of the Andes in the upper plate. The Nazca-Pacific motion is 11 mm/year slower than in the NUVEL-1A model, but this 6% slower rate is close to the one in the NUVEL-1 model.

The second school revised the present-day plate motion model relative to the ITRF 96 reference frame (Qiang et al. 1999); a new model has been introduced, ITRF 96 VEL, because ITRF 96 is dextrally rotated relative to NNR-NUVEL-1A. The revised total angular moment of all tectonic plates shows that ITRF 96 is not an NNR reference frame. A new NNR reference frame, called NNR-ITRF 96 VEL, has been established and it has been shown that it is quite consistent with NNR-NUVEL-1A.

This demonstrates that tectonic plate motions were steady over the past millions of years. Another study (De Mets and Dixon 1999) shows that Pacific-North America motion has remained steady since 3.16 Ma, but at rates faster than predicted by NUVEL-1A. The acceleration of the spreading rate in the Gulf of California is then absorbed by the decelerating slip rate in intraplate faults in the west side of the Baja California Peninsula.

The interpretation of present data on geodetic satellite data must be made in the light of the theory of viscoelastic deformation of the oceanic lithosphere. If oceanic plates are viscoelastic with a Maxwell time in the order of 20 Ma, we must differentiate between elastic response for loads applied during a period less than the Maxwell time and the viscous response for loads applied during a period more than the Maxwell time. The short-term response, elastic if we neglect the anelastic effects, requires that changes in "elastic stress fields and elastic strain fields are so intimately linked that, if one rotates, so must the other" (Biot 1965; Cobbold 1982, p. T8); the orientation of displacement in some point of the plate interior must reflect the orientation of displacement vector at the boundaries of the plate, divergent and convergent. The long-term viscous response must reflect stress relaxation that induces by intraplate strain a modification of the magnitude of differential displacement vectors inside the plate. Hence, a change in kinematics in the last 3 Ma, below the Maxwell time, is registered by the discrepancy, however small, between NUVEL-1A data and present satellite geodetic data in terms of orientation of displacement vectors. In contrast, the long-term effects in magnitude must propagate at a much lower rate and will be registered by departures from rigidity on different points of the plate which we have registered in the plot angular velocity versus angular distance to ridge. If these departures are confirmed by repeated measurements in the near future and differ in a systematic way from the rigid, but unsteady model they are the best evidence that this model is inadequate and should be replaced by a more rigorous model with deformable plates, but more steady at the 3 Ma time scale.

Therefore, in soft plate tectonics theory continental "lithosphere is a chemical thermal boundary layer through which heat removed by conduction rather than convection"; the lithosphere is thicker and colder under cratons than anywhere else. This is the concept of *continental insulation* (Anderson 1984, p. 354). However, the oceanic lithosphere, although a thermal boundary layer, is not a chemical boundary layer because hydrothermal seafloor metamorphism requires advection through this lithosphere. Chemical activity decreases the rigidity of the oceanic lithosphere when compared with the chemical inert continental lithosphere in cratons.

In the rheology of the lithosphere, a fundamental question is: what is the level that drives the flow? The more viscous layer will react in the shortest time interval to a change in boundary forces. Hence, in the case of a sedimentary basin on top of a crystalline basement, *remnant stresses* (Engelder 1993) can be registered in the upper level, which preserves a memory of an earlier stress field, both in magnitude and direction; the crystalline basement will record only the new superimposed stress field. For infinitesimal deformations the elastic and viscous rheology models are linear, without memory and there is a simple correspondence between both that makes them similar; however, for finite deformations elastic materials possess a memory and viscous materials do not. This is because elastic materials store strain energy and viscous dissipate it (Cobbold 1976). The more viscous level is situated below the less viscous level. Another situation exists in the rheological stratification of the lithosphere on top of the asthenosphere; the more viscous layer is now on top of the less viscous layer. A change of boundary forces will reorient the stress field in the lithosphere as a short-term elastic response. However, the magnitude of stress will decay in the long term by viscous relaxation. This suggests that when we analyse viscoelastic models in terms of simple analogues we should pay attention to the order of superposition in space of viscous and elastic elements when they are associated in parallel. Probably, the scale at which the individual elements are associated is also extremely important in the output. As far as we know, these considerations have not been emphasised in models of lithosphere rheology.

A recent review (Gordon 2000b) of lithosphere deformation emphasises its non-rigidity, reading conclusions similar to our own for buckling and faulting regimes in different tectonic settings, but not recognising homogeneous aseismic shortening as a possible mechanism in oceanic lithosphere. Therefore, plate interiors are considered quasi rigid, both in continental and oceanic domains, with strain rates in the order of -10^{17} to -10^{20} s^{-1}. This contrasts to our view that cratons are more rigid than oceanic plate interiors. An interesting suggestion is that the mechanism of deformation of the lithosphere can change from strain-rate hardening power-law rheology to strain-rate weakening self-lubricating rheology as we move from incipient plate boundary zones to narrow plate boundaries, allowing plate decoupling and mantle flow (Bercovici 1995). Eventually a buffer strain (Lliboutry 1992) must be overcome before transient creep changes to steady-state creep. The self-lubricating mechanisms remain unspecified; the most likely candidates are shear heating or the lubricating effect of water (Bercovici 1995). A self-lubrication mechanism is suggested by Pacific velocity data with a maximum of incremented strain and, presumably, a peak in stress level between two domains with lower incremented strain, as we estimated in Section 3.3.3.

3.11 Conclusion: The Driving Mechanism

The Earth is a dynamic planet. This dynamic character is clearly expressed in the evolution of atmosphere, hydrosphere and biosphere, but also in the evolution of the lithosphere; the evolution is slower in the solid geosphere than in the fluid geospheres, but nevertheless expressed in various forms of activity: earthquakes, volcanism, hydrothermal activity. The surface of the Earth tends to be levelled by external dynamics – erosion, sedimentation – and built by internal dynamics – mountain building, volcanism – that are largely a consequence of the relative motion of plates. Hence, we have to explain what moves the plates.

The presence of a geothermal gradient proves that the Earth is cooling and/or that heat is generated in the interior. In the lithosphere heat is transported mainly by conduction and, in oceans, by advection of hydrothermal fluids; flow in the lithosphere must be extremely slow and the rapid movement of plates must be an expression of some other thermodynamic mechanism below the lithosphere. The Earth must be similar to a heat engine. Isostasy shows that an asthenosphere must be present below the lithosphere and heat flow shows that it must be hotter. The dominant mode of heat transport in the asthenosphere must be convection, caused by density differences or by different compositions. The heat regime of the Earth shows that temperature differences are sufficient to cause density differences and it has been accepted since the 1930's that thermal convection must be the key for the movement of plates. The physics of this process is well known and the Rayleigh number of the mantle is estimated to be around 10^6, well above the critical value of 5×10^3 where convection starts (Turcotte and Schubert 1982; Davies 1999).

If convection is the main thermodynamic process in the Earth, an upper thermal boundary layer must exist at the top of the system because the flow is horizontal and unable to transport heat vertically to the surface; heat must be transported by vertical conduction. This is the lithosphere. This cooler and denser layer will, sooner or later, become reabsorbed by the asthenosphere below, causing the inception of subduction. We conclude that thermal convection can explain many aspects of plate dynamics; this was confirmed by analogue and numerical modelling (Davies and Richards 1992), but there are aspects of real plate behaviour that cannot be simulated by simple physical and numerical experiments that will be treated next (Davies 1999).

The fundamental question in the application of convection to the mobile Earth is the nature of the thermal boundary layer or lithosphere (Davies 1999). Conventional fluids do not possess strength in their thermal boundary layers. The lithosphere is divided into plates with angular and segmented boundaries with a variety of sizes and shapes. How can the lithosphere combine strength expressed by its brittleness and the mobility required by plate kinematics? Hence, the lithosphere has an essential role in the convection driving mechanism for plate tectonics. The dominant tectonic mechanism is the *plate mode*, by which heat is removed from the mantle; the *plume mode* is a secondary mechanism by which heat is transferred from the core to the mantle (Davies 1999). Both modes are connected in the sense that plumes can affect the rheology of the lithospheric plates controlling their shape and even leading, in extreme cases, to the splitting of plates.

In physical analogue models at low Rayleigh number convective Rayleigh-Benard cells tend to be equidimensional, that is with horizontal dimensions near vertical dimensions; this is not the case with larger plates that are much thinner than long. Sinking downwellings in physical models are bilateral; subduction on Earth is one-sided, because it occurs mainly between the buoyant light continental lithosphere and the sinking heavy oceanic lithosphere. The differences between simple physical models and real Earth can become clearer if we consider insights from numerical modelling.

In numerical models with a constant viscosity equidimensional cells are produced, simulating Rayleigh-Berard convection experiments. If a stiffer lithosphere is introduced immobility is allowed by low-viscosity plate boundaries and upwellings and downwellings are controlled by plate boundaries (Davies and Richards 1992). This is a better approximation to the real situation, showing that the lithosphere has a role in the organisation of the convective flow. Hence, the question "do the plates drive the mantle or does the mantle drive the plates?"; there is a suggestion for an answer: the plates drive the mantle and are an integral part of the convective flow. In other words, the plates organise the flow (Davies and Richards 1992). Plate tectonics require a highly viscous lithosphere; this does not means that the lithosphere has infinite viscosity. Soft plate tectonics require lateral viscosity gradients in the lithosphere itself. The viscosity is very high, practically infinite, in continents that have the thickest lithosphere, being carried actively as a propelled raft; viscosity is slightly lower in the oceanic lithosphere that is an active upper thermal boundary layer and so the integral part of convection and plate boundary zones, sometimes very wide, have the lowest viscosity, organising the flow pattern.

Numerical models have also shown that three regimes must exist in the mantle convection of terrestrial planets (Solomatov and Moresi 1997; Moresi and Solomatov 1998), in the cases of both Newtonian and non-Newtonian rheologies. In the *small viscosity contrast* regime, the surface mobility includes the whole thermal boundary layer; this is the case in the analogue experiments with conventional fluids. In the *stagnant lid regime*, the surface thermal boundary layer is immobile and convection only mobilises deeper layers; this regime is observed on the Moon and on present-day terrestrial planets , with the possible exception of isolated and episodic downwelling sites on Venus. In the *transitional regime*, surface mobility occurs, but the cold boundary layer is more extended than in the small viscosity contrast regime; this is the present regime on Earth. According to the thesis expressed in this book, the continental lithosphere is quasi-rigid and the oceanic lithosphere has a small viscosity contrast to the underlying asthenosphere and allows mobility of the entire lithosphere. The viscosity drops significantly when we move from the continental to the oceanic lithosphere due to the weakening effect of fluid advection through hydrothermally altered oceanic lithosphere. We can say that the presence of the hydrosphere is fundamental to the operation of the plate tectonic regime on Earth and its absence on other planets (Fyfe 1988; Mian and Tozer 1990), a point that we will consider further on in Section 4.13. Another effect of high fluid content of the lithosphere is the weakening of interplate and intraplate faults required by numerical models. As mentioned in Section 2.7, this effect was well documented in the weakening of the San Andreas Fault boundary (Hardebeck and Hauksson 1999) and it is probable that this effect must be quite generalised in both the schizosphere and plastosphere.

We must stress that even the more sophisticated numerical models have difficulties in incorporating surface plates. The rheology of the plates must be inserted explicitly into the models assuming rigid rafts, separate rheological layers, high-viscosity surrounded by weak boundaries; and the mantle flow must be also prescribed. Therefore, plates are not generated intrinsically in a self-consistent way. Even in the more recent models, where plates are generated in a self-consistent manner by a viscoplastic rheology (Rompert and Hansen 1998), subduction is episodic and a new stiff layer is regenerated; but this style is more akin to convection on Venus than on present Earth.

A very recent synthesis (Tackley 2000) reaffirms the problems caused by the occurrence of plate tectonics on planets. Only the Earth does not have a rigid lid; contrary to other terrestrial planets that had plate tectonics in the past, Mars, or, intermittently, Venus. On Earth, "convection reaches the surface, and oceanic plates are the upper thermal boundary layer participating in the convective motion" (Tackley 2000, p. 2002). The plates must be broken by faulting at the surface, and by distributed microcracking below; in the plastosphere, strain weakening mechanisms must predominate, such as, dynamic recrystallisation reactions with volatile shear heating and previous weak heterogeneities. Self-consistent plate models can incorporate these weakening mechanisms, but numerical models lead to results that do not rigorously match plate motions on the real Earth.

The form of convection required by soft plate tectonics is compatible with all the evidence from plate dynamics. However, if plate kinematics are well understood, many aspects of plate dynamics are still subjected to intense debate; hence we do not possess a definite test in favour of soft plate rather than rigid plate dynamics, only some suggestions which we will refer to next.

The rigid motion of superficial plates can be split into two terms(Cadek and Ricard 1992): the *poloidal field* has components related to the horizontal divergence field, convergent at convergent boundaries and divergent at divergent boundaries; the *toroidal field* has components related to radial vorticity and is needed to describe the strike-slip components at transforms and the spin of each plate around its centre.

Present-day plate motion minimises the ratio of toroidal to poloidal energies (Gable et al. 1991; Cadek and Ricard 1992). This is due to the fact that if the mechanical structure is purely radial, the poloidal field will be driven by the vertical motion related to convective cooling; a pure toroidal field is not associated with any vertical motion and will not help the Earth to cool. Therefore, the lateral viscosity variations could be responsible for the toroidal component of the field through radial vorticity.

In the lithosphere the flow is extremely slow because viscosity is very high. The only body force is gravity and we can neglect the forces due to Earth rotation (Middleton and Wilcock 1994).

In the atmosphere and hydrosphere the flow velocity is much higher and in geophysical fluid dynamics the acceleration due to the Earth's rotation must be included in the description of flow through a Rossby number below one (Pedlosky 1987).

In fact, from the relation $L/U \geq \Omega^{-1}$ we establish that a non-dimensional parameter, the Rossby number, ε, is:

$$\varepsilon = \frac{U}{2\Omega L}$$

where U is the velocity of the fluid, L is a distance at which the fluid moves and Ω is the period of Earth rotation, $\Omega = 7.3 \times 10^{-5}\ s^{-1}$. For the values of U and L in the Earth ε is one order of magnitude below one.

In an Earth with perfect radial stratification, the lithosphere has no net rotation. However, in a model with lateral viscosity variations, there is a differential rotation between the lithosphere and the mantle (Ricard et al. 1991). Taking into account the observations of 14 hotspot traces using the relative motion model NUVEL-1, a net eastward flow of the mantle relative to the lithosphere has been shown, reaching a maximum value of 1.7 cm/year around a pole differing from the Earth's pole of rotation. It is concluded that the "asthenosphere viscosity below the oceans is at least one order of magnitude lower than underneath the continents" (Ricard et al. 1991, p. 8407).

This is an indication that the lateral viscosity contrasts, related to continent/ocean composition, persist deeper than the lithosphere/asthenosphere boundary with higher viscosity in and under the continental lithosphere. In other words, the cratonic viscous structure is more deeply rooted below continents, forming the so-called tectosphere (which we mentioned in Sect. 2.1), while the asthenosphere approaches the surface beneath oceans and almost reaches it below mid-ocean ridges.

Plate movement is driven by and resisted by various forces. These forces must equilibrate in order to maintain steady motion. Theoretical analysis (Lliboutry 1998) shows that *ridge push* is due to gravitational flow which is due to potential energy of the elevated mid-ocean ridge. *Slab pull* is due to the horizontal component of negative buoyancy of the cold oceanic lithosphere being denser than the hotter asthenosphere where it plunges. However, it is more difficult to model because it is difficult to incorporate other effects such as mineral phase transitions in the plunging slab, unbending and suction if the upper plate recedes from the trench.

The hierarchy of these forces can be established by two methods (Cox and Hart 1986; Ranalli 1995; Lowrie 1997). One method investigates the correlation between plate velocity and area for surface forces or length for edge forces, respectively. The other method investigates the correlation between forces and absolute motions of plates inferred from hotspot tracks and no-net rotation torques. Models of absolute rotation incorporating NUVEL-1 data exist for hotspots, HS2-NUVEL-1 (Gripp and Gordon 1990) and for no-net rotation, NNR-NUVEL-1 (Argus and Gordon 1991). The agreement between both models is high for fast moving plates, but low for slow moving plates; it suggests that the hotspot reference frame is not fixed relative to the rigid mesosphere or that the no-net rotation is incorrect from the dynamic point of view.

Both approaches lead to similar conclusions on the hierarchy of forces. Slab pull and ridge push are the most important forces in plate motion. In fact, the correlation between plate velocity and length of slab is high, and the correlation between plate velocity and effective length not cancelled by an opposing ridge is weaker, but both are much more significant than correlation with other forces. Concerning absolute motions it is noted that the torque exerted by ridge push and slab pull are nearly parallel to each other and to absolute motions. However, disagreement persists over the relative importance of slab pull and ridge push. (Fig. 3.50)

If slab pull is dominant, we can explain why downwellings responsible for subduction are concentrated, whereas upwellings are more dispersed. Plates resist being pulled apart, but once a spreading ridge is established it remains a weak spot that persists, passive rifting with magmatism caused by decompression melting (Brown and Musset 1993; Davies 1999). If the slab resistance increases as the slab penetrates it will equilibrate slab pull making it equivalent or less than ridge push, a view maintained by some.

The drag forces caused by relative motion between the base of the lithosphere and the asthenosphere are virtually unknown, but they can be important; if convective flow is faster than plate velocities, the plates are passively driven and the traction has the same sense as the moving plate. If the plate moves more rapidly, the drag force opposes the motion; mantle drag must be larger under thicker continents.

The hierarchy of forces must also be tested by investigating if it is compatible with the generation of first-order, plate-wide stress fields (Zoback 1992; Engelder 1993). In some instances inferences can be drawn as follows: slab pull must overcome slab resistance because down-dip extension is a general feature in circum-Pacific subduction zones; mantle drag must be small, because in North America seismic activity decreases to the leading edge of the plate on the west side; and uniformity in magnitude of stress is more compatible with the ridge push model for stress generation.

These ideas on forces that drive plates must be confronted with the data on strain in the Pacific Plate. It deforms coherently with no torsion, even if the subduction zones surrounding the plate are highly sinuous. This suggests that ridge push is important in maintaining this coherence (Cox and Hart 1986, p. 354). The shortening and extension parallel to plate movement indicate that the two forces equilibrate when ridge push is integrated along the straight ridge and slab pull is integrated along the

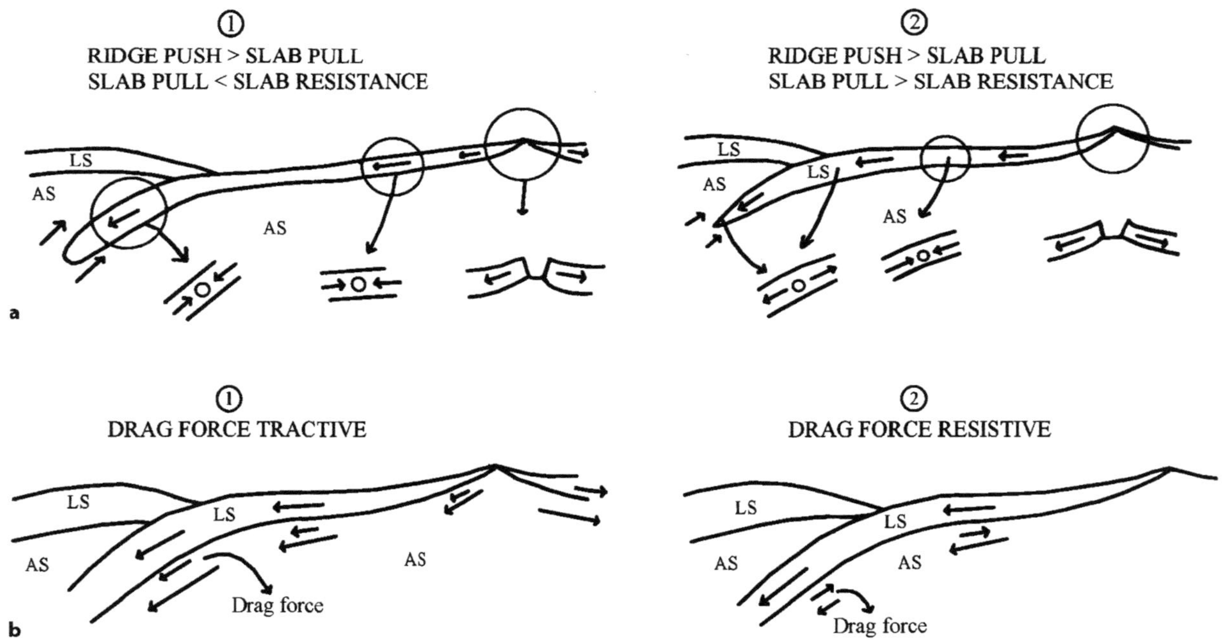

Fig. 3.50. Possibilities for driving mechanism. **a** Ridge push versus slab pull. *1* Ridge push supersedes slab pull. Ridge push is traction near ridge, becomes compressive in older lithosphere and is compressive in downgoing slab because slab drag in the *As* opposes and equilibrates slab pull. *2* Slab pull supersedes ridge push. In young and old lithosphere the same as above, but in the downgoing slab the drag of the *As* does not equilibrate slab pull. **b** Sense of drag force. *1* Drag force between lithosphere and asthenosphere drives motion, with plate velocity in *As* higher than in *Ls*. *2* Drag force between *Ls* and *As* resists motion, with plate velocity in *Ls* higher than in *As*

sinuous subduction zones of the West Pacific. Ridge push is dominant over ≈40° of angular distance to the ridge and slab pull is dominant over ≈60°, suggesting that slab pull is probably more effective than ridge push.

If we compare the stability of subduction zones and ridge, we note that the first are more stable in location than the second. In fact, a migration of the subduction zone will produce displacement of large volumes of the asthenosphere; the dynamic character of subduction is expressed in the variation of azimuth of subduction in response to more instability of the ridge. In the case of the Pacific, the last plate reorganisation occurred between 25 and 15 Ma ago, with only minor changes since then. These minor changes cause shifts of present Euler poles inferred from geodetic satellite data relative to NUVEL-1A Euler poles (Larson et al. 1997), but are considered negligible in the reference frame NNR-1 TRF 96 VEL, quite consistent with NNR-NUVEL-1A (Qiang et al. 1999).

Geodetic methods showed that the Pacific Plate decelerates and accelerates as we move from the ridge to the subduction zones on its western side. This proves internal deformation of slab pull over ridge push. We arrive at the same conclusion if we examine plate kinematics and mass balance at the surface of the Earth; there is no Earth expansion and mass created at ridges must be compensated by mass consumed at trenches (Le Pichon 1968; Le Pichon et al. 1973). Most ridges are oriented N–S; hence, spreading is E–W; the minor N–S component, restricted almost to the Indian Ridge is compensated by convergence in the Alpine-Himalayan continental collision belt. However, E–W consumption around the Pacific must compensate not only E–W spreading in the Pacific itself, but also in the E–W spreading South Atlantic and Indian oceans. As these two are surrounded by passive margins, we are led to conclude that, either, in the Pacific, the subduction rate must supersede the spreading rate, or the Pacific is closing, because slab pull is more effective than ridge push. With ridge plates this would mean that South American and Australian Plates are approaching, but the geodetic methods show that, in addition, there is an internal acceleration of the Pacific Plate because it is being pulled from the convergent boundary on the western side and, possibly, the Nazca and Atlantic Plates are also being pulled from the converging boundaries on their eastern side. Therefore, with rigid plates, the area balance at the surface, to maintain a constant Earth radius as a function of time, implies mass balance at the surface; but with soft plates, the area balance does not imply mass balance: mass excess or deficit at the ridge is compensated for by acceleration or deceleration, respectively, at subduction zones.

We conclude that the soft plate approach reinforces some views on plate dynamics (Davies 1999). The plate mode controls most of the mantle heat budget and the plume mode is a minor component. This means that the mantle is heated from within; in this form of convection upwelling is passive under mid-ocean ridges; the mantle flows upwards between the downwellings, but it is not buoyant relative to the mantle fluid interior, which is well mixed. The plates are active players in convection because they are displaced by the actively sinking cold fluid. Slab pull is more important than ridge push and probably significant at the ridge itself. In fact, the ridge axis has no tensile strength (Stein and Okal 1986), because there is no wide band of intraplate extension in the spreading direction, as we mentioned in Section 2.5. This weakness of the ridge will be enhanced by the presence of a continuous magma chamber under fast-spreading Pacific-type oceans.

The primordial role of slab pull in the driving mechanism of plate tectonics is also demonstrated by the recent solution to an old debate. Thermal convection operates at the scale of the whole mantle, as geochemistry of ocean island basalts suggests or are there independent cells for the upper and lower mantle, because the 660-km discontinuity is not crossed by the sinking slab in subduction zones (Brown and Mussett 1993)? Seismic tomography (Anderson and Dzieworski 1984; Anderson 1989) showed that in some subduction zones the slab is deflected as it approaches the 660-km discontinuity, such as in central Izu Bonin, but in others the fast region crosses the discontinuity (Van der Hilst et al. 1991), such as in central Japan. This suggests that there are long periods of independent convection of the upper and lower mantle and short periods, where avalanches of the upper mantle material sink into the lower mantle, rendering its convection unstable and generating plumes across the entire mantle, from the core-mantle boundary to surface hotspots (Tackley et al. 1993; Fig. 3.51).

These periods were named Wilsonian and Mantle Overturn and Major Orogenies, or MOMO, by some authors (Stein and Hofmann 1994). In the avalanche periods, subduction becomes superfast because the resistance of the 660-km discontinuity to be crossed by the sinking slabs is overcome (Van der Hilst et al. 1991). Superfast subduction will increase the strain rate of the oceanic lithosphere; the strain regimes – homogeneous shortening, buckling, subduction related deformation – will succeed at faster rates leading ultimately to plate collision and orogeny.

Recent studies (Grand 1994; Van der Hilst et al. 1997; Van der Voo et al. 1999) have shown that deep subducted slabs can reach the base of the mantle. They also suggest that the whole mantle overturn would be the dominant regime, both in space and time and the two-cell regime would be local and transient (Davies 1999), and eventually give way to stable one-cell regimes as the Earth cools (Allègre 1999, pp. 287–288). This is a field of geodynamics where we can expect rapid progress in the near future if we continue to intensify the coverage of the whole globe and use improved technological capabilities.

Meanwhile, physical experiments (Davaille 1999) show that even small compositional contrasts can drive convection through density differences. If these are large, the buoyancy is not enough to overcome chemical stratification, except in isolated mantle plumes. If they are small, the buoyancy can generate superswells, like the ones under Africa and the South Pacific which oscillate in cycles of 100 Ma. Magmas are more primitive at hotspots than in superswells, confirming the geochemical evidence for a separate reservoir in the lower and upper mantles. This also suggests that mantle dynamics evolved from the present-day regime from a more stratified initial stage; this agrees with the seismological evidence that the whole mantle overturn will gradually replace the two-cell regime (Allègre 1999).

What will be the role of a deformable lithosphere in soft plate tectonics as opposed to an elastic lithosphere in standard plate tectonics? The organising role of the lithosphere in thermal convection is essential in both cases; the response of the lithosphere to changing dynamics will be delayed in the first case and almost instantaneous in the second case. In standard plate tectonics the energy produced by the heat engine will be dissipated only by mechanical processes at plate boundaries and by cooling both inside and between the plates. In soft plate tectonics the dissipation at plate boundaries and inside the plates will be by both mechanical and thermal processes.

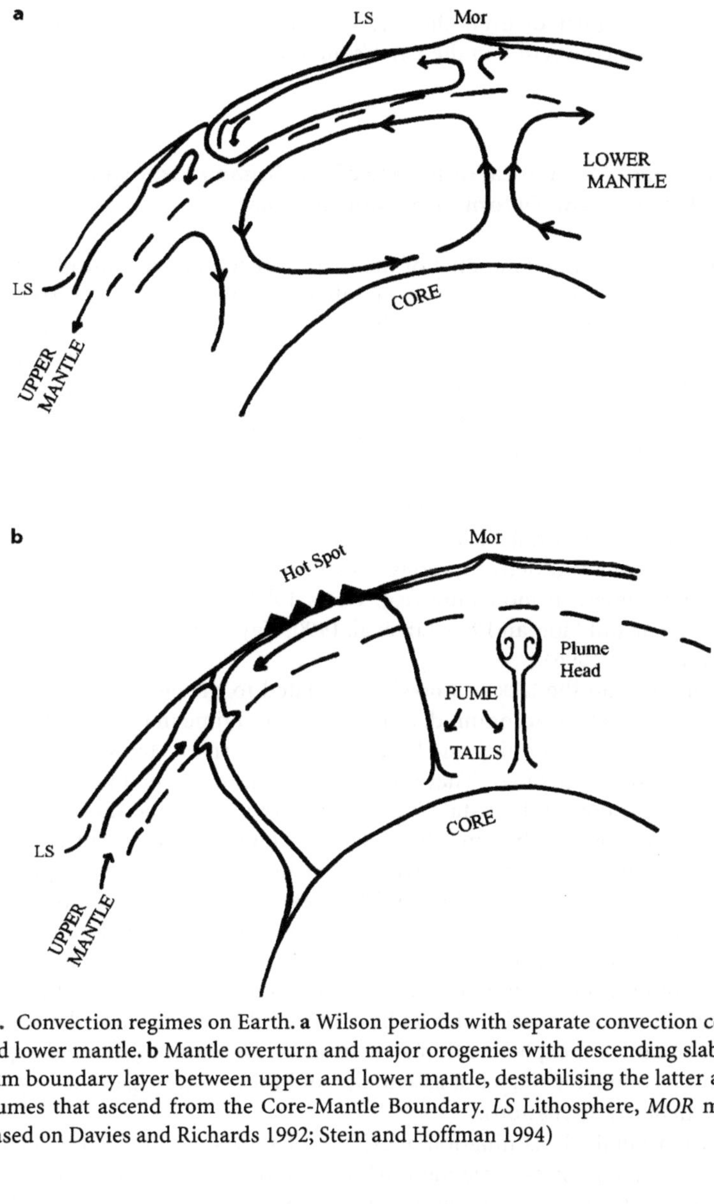

Fig. 3.51. Convection regimes on Earth. **a** Wilson periods with separate convection cells in the upper and lower mantle. **b** Mantle overturn and major orogenies with descending slab crossing the 660-km boundary layer between upper and lower mantle, destabilising the latter and generating plumes that ascend from the Core-Mantle Boundary. *LS* Lithosphere, *MOR* mid-ocean ridge. (Based on Davies and Richards 1992; Stein and Hoffman 1994)

We conclude that soft plate tectonics emphasise the dissipative character of thermodynamic processes inside the lithosphere. The presence of avalanche periods of whole mantle convection has stimulated a discussion about the possible chaotic nature of thermal convection on Earth (Davies and Richards 1992). Convection is surely unsteady at a timescale of 100 Ma, but the lithosphere has a stabilising effect; convective flow is laminar, in part, due to organising role of plates; plume initiation and evolution may be irregular or even chaotic; spherical symmetry-breaking bifurcation will affect thermal convection in the Earth's mantle (Chossat and Stewart 1992).

3.12 Time-Dependent Mechanical Response of the Oceanic Lithosphere; Strain Partitioning and Deformation Mechanisms

Much of our present understanding of the major processes that control the complex evolution of the oceanic lithosphere is mainly the result of scattered direct observations, coupled with inferences based on thermal and mechanical models that consider both geochemical (including petrogenetic) and geophysical data, as well as estimations of physical properties deduced from experiments on monomineralic and/or non-hydrothermally altered rocks (e.g. Agar 1994; Fisher 1998, and references therein). That understanding is therefore chiefly conceptual rather than specific, and many key questions will remain unsolved until additional direct measurements are completed in a variety of ocean crustal ages and settings. This is particularly clear when one examines the available rheological models for the oceanic lithosphere, where major gaps in our knowledge severely constrain the boundary conditions to be imposed on the numerical modelling (e.g. Watts 1982; Bergman and Solomon 1984; Buck 1986; Chen and Morgan 1990; Lin et al. 1990; Hayes and Kane 1991; Tolstoy et al. 1993; Shaw and Lin 1996).

Uncertainties and the lack of knowledge related to the geological processes that strongly influence strain accommodation in space and time in the oceanic lithosphere (including strain partitioning and local variations in stress states) are critical in the understanding of its rheology. Therefore, the evaluation of the integrated effects of magmatism, fluid flow and crustal extension at large scales does not enable an adequate characterisation of the time-dependent mechanical response of the oceanic lithosphere, thus preventing the construction of high-resolution rheological models (especially for the upper oceanic crust). This means that further work is necessary in order to understand the rates and mechanisms by which thermal, fluid flow and chemical variations of the oceanic lithosphere are actually controlled over a continuum of scales. In other words, comprehensive examination of the magmatic, hydrothermal and crustal extension evolution should be considered in any realistic rheological model for the oceanic lithosphere. Microstructural studies, both in situ and experimental, are thus a complementary, but essential, tool in these investigations (e.g. see Agar 1994, for a detailed discussion).

In addition, detailed examination of oriented samples (to provide kinematic interpretations and palaeostress orientations) and more experimental data are needed, particularly on strain accommodation by texturally diverse mineral assemblages similar to those usually found in the heterogeneous suite of altered rocks recovered by dredging of the ocean crust and in their analogues in ophiolite complexes.

Some general comments about strain partitioning and the most probable active deformation mechanisms during oceanic crust evolution can, however, be made, considering the magmatic-hydrothermal crustal extension evolution that is believed to characterise both fast- and slow-spreading systems (for a comprehensive review, e.g. see Agar 1994), thus tentatively responding to the major challenge of the present book.

The key issue is that the oceanic lithosphere deforms by buckling as a consequence of its viscoplastic behaviour acquired after 10–20 Ma, regardless of the magmatic-hydrothermal crustal extension evolution experienced by the crust. As already discussed, buckling of the oceanic crust is preceded by the accommodation of 4–10% of homogeneous shortening, perhaps ascribable to different (brittle) failure mechanisms which converge to similar consequences, i.e. to prevailing aseismic creep. If so, strain partitioning in the oceanic crust as a whole, as well as its stress state variations in space and time, reflects the heterogeneous results of the intricate relationships established between magmatic and hydrothermal processes during crustal extension, which are expected to be notably distinct in fast- and slow-spreading systems.

The architecture of the oceanic crust is, indeed, far more puzzling than commonly assumed in general simple layered models. It reveals strong lateral and vertical lithological and structural heterogeneities, mostly due to multiple magmatic injections subsequently overprinted by polyphase hydrothermalism (e.g. Cann 1974; MacDonald et al. 1988; Karson and Rona 1990; Jacobson 1992; Smith and Cann 1992; Harding et al. 1993; Detrick et al. 1994; Head et al. 1996). These heterogeneities, also well documented in ophiolites (e.g. Moores and Vine 1971; Coleman 1977; Christensen 1978; Christensen and Smewing 1981; Varga and Moores 1985), may therefore be responsible for a significant and variable lithospheric weakening, since they reflect variations of many physico-chemical parameters (such as mineralogy, grain size, permeability, cohesion, etc.) that dramatically alter the time-dependent mechanical response of rocks, and/or changes in local stress configurations (mainly dependent on pore fluid pressure buildups and on the geometry of discontinuities mostly formed at lithological contacts), thus controlling strain accommodation.

According to the available data, narrow, but long-lived magma conduits lead to multiple intrusive events at fast-spreading systems (e.g. Hayes and Kane 1991; Chen 1992; Wilcock et al. 1992; Christeson et al. 1994; Su et al. 1994). A higher background geothermal gradient and a relatively swift translation of the new oceanic lithosphere further from the ridge axis are therefore predicted. In addition, possible deep melt accumulations over a relatively wide region (from 5 to 20 km; Garmany 1989; Barth et al. 1991) will favour aseismic slip during fast spreading, causing strain concentration at the base of the young oceanic crust. In these circumstances, magmatic over-pressuring related to variations in magma supply will tend to restrain normal fault development, thus explaining the low seismicity and relief of these oceanic settings (e.g. Bratt et al. 1985; Parsons and Thompson 1991). However, even in the presence of important tectonic faults, similar to those often found in slow-spreading systems, seismicity also remains relatively low and is usually concentrated at the ridge segment ends, mainly at the inside corners of ridge-offset intersections (e.g. Forsyth 1992; Shaw 1992). This means that the cause of the aseismic behaviour of the fast-spreading ridges must be sought elsewhere.

Episodic magmatic activity is believed to characterise slow-spreading systems, thus enhancing the relative importance of crustal extension by faulting (e.g. Harper 1985; Karson and Rona 1990; Cannat et al. 1991; Hayes and Kane 1991; Chen 1992; Smith and Cann 1993; Su et al. 1994). Changes in fault geometry, fault spacing and displacement (often increasing towards ridge-offset discontinuities) are common and presumably ascribable to variations in lithosphere thickness and fault dips (e.g. Chen and Morgan 1990; Neumann and Forsyth 1993; Shaw and Lin 1993). These intricate structural configurations, strongly influence along-axis valley morphology and often lead to thinner and discontinuous ridge segment ends, where strongly altered gabbros and peridotites outcrop (e.g. Cannat 1993; Cannat et al. 1995; Escartín et al. 1997a,b). Once in a position to be hydrated, peridotites will be serpentinised if the average temperature remains below ca. 400° C, and this concentrates fault slip on the inside corner of ridge-offset intersections, assuming that hydration is first restricted to fault planes (Raleigh and Paterson 1965; Francis 1981; Janecky and Seyfried 1986; Rona et al. 1987, 1992; Escartín et al. 1997a; Moore et al. 1997). However, given the distribution of serpentinites at slow-spreading systems, not only at ridge segment ends, but also at many localities along the axial valleys (e.g. Bonatti and Michael 1989; Tucholke and Lin 1994; Cannat et al. 1995), the chemical softening processes related to non-serpentinising peridotite hydration will clearly influence the rheological evolution of the oceanic crust.

Serpentinisation appears to proceed as a non-equilibrium process that takes place mainly under temperatures ranging from ≈350° to 400° C if the stability field of the most common serpentine polytype in these geological settings (lizardite) is considered (e.g. Janecky and Seyfried 1986 and references therein). Consequently, the maximum depth of serpentinisation will grossly coincide with the position of the 400° C isotherm, which is positioned ca. 6 km below the seafloor at the end of the ridge segments in slow-spreading systems (e.g. Escartín et al. 1997a). This corresponds also to the depth of brittle deformation indicated by seismic activity, strongly suggesting that the maximum depth of seismicity in the oceanic lithosphere coincides with the serpentinisation front (e.g. Bergman and Solomon 1990). However, a general aseismic creep is expected, since the deformation of lizardite aggregates is largely accommodated by shear cracks along cleavage planes, probably suppressing dilatancy (Escartín et al. 1997a,b). As a result of this mechanical behaviour, fluids will be confined within already serpentinised rock domains, enhancing subsequent weakening and aseismic slip of fault zones. This will promote strain concentration as well, contributing to the increase in fault spacing and fault throw at the end of ridge segments (e.g. Escartín et al. 1997b).

Crustal extension along listric faults that detach in low angle ductile shear zones at depth are believed to represent a major structural feature on ridge segmentation at slow-spreading systems (e.g. Mutter and Karson 1992). Thus, accepting that these faults provide suitable paths for deep seawater circulation after significant crustal cooling, hydration should affect deep domains of the oceanic lithosphere, further controlling their mechanical response (e.g. Rowlett 1981; Bergman and Solomon 1990; Kong et al. 1992; Wolfe et al. 1995; Escartín et al. 1997a). Late breccia zones and late chemical transformations assisted by fracturing processes are therefore to be expected in deep plutonic units, overprinting many of the early microstructures either caused by syn-magmatic deformation or ascribable to high-T (>600°C) brittle failure (e.g. Agar 1994), and favouring the stable slip of the fault zones. A serpentinised, weakened upper mantle may as well be developed, assuming that the pathways for fluid circulation are not completely sealed as the crust moves away from the spreading centre. This will provide the means for the accommodation of significant aseismic creep in the cooled upper mantle, preventing also the development of critical detachments between this layer and the lower oceanic crust above. Consequently, the assumption that, far from the ridge, the upper mantle responds as a strong, potentially brittle, layer is not free of controversy.

Although the multiple magmatic events in fast-spreading systems and the presence of abundant serpentinites in slow-spreading systems may explain the relative scarcity of seismicity in near ridge ocean domains, the fact is that other processes must be envisaged in order to justify the crustal homogeneous shortening prior to its buckling. The extensive chemical-softening of primary mineral aggregates during crust alteration by means of hydrothermalism is the major candidate, because:

1. it modifies significantly the mechanical characteristics of the rocks, being often tightly related to stress corrosion mechanisms and to subcritical inter- and intragranular fracturing processes (e.g. E.H. Rutter 1976; Fyfe 1976; White and Knipe 1978; Kranz 1983; Atkinson 1984; Carlson and Herrick 1990); and
2. it is dependent on geological processes that are an inevitable consequence of ocean-spreading magmatism and exert a major influence on the chemical and thermal evolution of seawater and ocean crust (e.g. Wolery and Sleep 1976; Fehn and Cathles 1979, 1986; Sclater et al. 1980; Lowell 1980, 1991; Rona 1988; Cann and Strens 1989; Cathles 1990; Rona et al. 1990; Baker et al. 1991; Travis et al. 1991; Alexander et al. 1993; Johnson et al. 1993; Lowell et al. 1995). Hydrothermal activity starts when the crust is young and will continue for tens of Ma (Staudigel et al. 1981; Gallahan and Duncan 1994). However, the nature of hydrothermal activity in seafloor spreading systems is the result of a complex combination of different parameters. These include the bulk permeability, the chemical composition of fluids and the effect of magma chamber geometry on the distribution of near-critical P–T conditions for fluid flow (which are roughly reached near the transition zone between magma and its crystalline envelope; see e.g. Fyfe et al. 1978; Brikowski and Norton 1989; Mottl and Wheat 1994; Fisher 1998).

The uppermost part of the oceanic lithosphere in fast-spreading systems is mostly composed of pillow basalts, lava flows and breccias, often forming complex interfingerings of variable thickness. Permeability values are thus expected to be high, favouring the expansion of the hydrothermal alteration zones during the earlier stages of off-axis migration. During this high fluid/rock interaction, large and successive, although transient, variations in pore fluid pressure may be developed at distinct, but widespread discontinuities (e.g. microcracks having narrow widths and limited lateral extent, features associated with pillow boundaries, and collapse structures). These enable the nucleation/propagation of distinct fracture arrays at different scales that weaken the crustal strength and promote strain concentration. Discontinuous sealing of fractures would also cause irregular redirection in fluid flow throughout the (altered) rocks. As metasomatic processes evolve, fluid-enhanced deformation increases, and pressure solution creep may play an important role in strain accommodation, alternating with fracturing mechanisms (e.g. Gratier and Gamond 1990; Gratier 1993; Gratier et al. 1999).

At slow-spreading systems, differences in fracture geometry and/or hydrothermal alteration in the uppermost crustal domains are also responsible for strong vertical and lateral heterogeneity. Indeed, fracture configurations related to crustal extension and to serpentinisation can be locally reinforced by means of hydrothermal alteration, and the repeated syn-kinematic fluid flow along pre-existing faults may determine their subsequent reactivation, leading to distributed cataclasis or to distinct brecciation patterns. Contacts between gabbros and serpentinised peridotites may evolve into discrete faults with minor vertical displacements or into distributed fracturing along prominent sets of microcracks formed during fluid-rock interaction. Also, as in fast-spreading systems, a general viscoplastic rheology will be expected after significant fracturing and fluid-enhanced deformation (which may also explain the seismic anisotropy usually identified in the upper oceanic crust; e.g. Stephen 1991; Fisher 1998).

Bulk permeability is, however, expected to decrease in depth, as documented by the progressive shrinkage of the hydrothermal alteration intensity in ophiolite complexes (e.g. Richardson et al. 1987; Gillis and Robinson 1988; Nehlig and Juteau 1988; Nehlig et al. 1994; Gillis and Roberts 1999), and predicted by different conceptual and numerical models (e.g. Lister 1983; Lowell and Germanovich 1995). Therefore, the vertical and along-strike widespread fracturing in the sheeted dyke complex would mainly reflect the orientation of stresses developed during spreading, although other mechanisms might be involved in fracture generation, such as thermal craking, volatile-rich magma expansion and differential thermal expansion of pore fluids (e.g. Knapp and Knight 1977; Nehlig and Juteau 1988; Nehlig 1993, 1994; Fisher 1998; Gillis and Roberts 1999). How can magma chambers influence the distribution of rock permeability at and below the sheeted dyke – high-level gabbros transition?

The contact between the sheeted dyke complex and the high-level gabbros hardly represents a real barrier to deep hydrothermal circulation, as revealed by geological and geochemical evidence from ophiolites (e.g. Gregory and Taylor 1981; Nehlig and Juteau 1988). The fact that hydrothermal circulation may reach the plutonic levels is also supported by fluid inclusion data (Vanko 1988), the geobarometry of hydrothermal vent fluids (Campbell et al. 1988), the characteristics of the seismic reflector present at many spreading systems and thought to represent the top of the magma chamber (e.g. Calvert 1995), and the requirement of a thin boundary layer above the magma chamber in order to supply enough heat to support ridge crest vents (e.g. Lister 1977; Cann and Strens 1982; Converse et al. 1984; Gillis and Roberts 1999). This means that chemically modified seawater is able to penetrate the roof of the magma chamber, its relative abundance being sometimes high enough to promote metasomatism in the underlying layered gabbros (e.g. Nehlig et al. 1994; Nehlig, 1993, 1994). However, (long?) before being reached by hydrothermal fluids, these deep crustal domains may already be highly fractured, due to processes of thermal contraction and subsequent volatile-rich phase expansion, which allow efficient transfer of magmatic heat (e.g. Burnham and Davis 1971, 1974; Baker et al. 1987; Nehlig 1994). One may, consequently, presume that the mechanical behaviour of the sheeted dyke complex high-level gabbros transition may change between brittle and ductile, being extremely sensitive to local thermo-elastic stresses related to dyke injection and magma chamber cooling. According to Alexander et al. (1993), the presence of a non-steady state magma chamber would also be important in controlling late development of extensional faulting, causing as well the tilting of the dykes, and certainly related to retrograde metasomatic processes under higher fluid/rock ratios. These events overprint the early alteration processes which, evolving from diffuse grain scale flow, lead to epidosite formation, an hydrothermal product thought to represent the geological record of the root zone of the hydrothermal circulation in fossil fast-spreading systems (e.g. Nehlig et al. 1994, and references therein).

The size and the geometry of oceanic ridge magma chambers are presumably quite variable (e.g. Brikowski and Norton 1989; Sinton and Detrick 1992; Nicolas et al. 1993). Present-day investigations enable the inference of discrete and relatively narrow magma chambers (<2 km; e.g. Detrick et al. 1986; Langmuir et al. 1986; Shirey et al. 1987), although evidence from some ophiolite complexes strongly suggests the development of much larger chambers and/or their gradual widening in depth (e.g. Nicolas et al. 1988). More or less distinct variations in size and geometry of magma chambers are, however, to be expected and may simply represent distinct stages of their evolution, since a wide chamber will quickly reach a confined morphological configuration as magma cools and differentiates. However, apart from this convergent effect, it should be noted that the influence of the size and geometry of the magma chamber on the time-dependent mechanical response of the oceanic crust is probably not negligible. In fact, these parameters explicitly control the heat flow regime, constraining the transfer of the latent heat of magma crystallisation and, subsequently, the rate of removal of the heat of magma cooling by the hydrothermal circulation of seawater, thus indirectly governing the development of metasomatic processes responsible for the major lateral and vertical heterogeneities usually identified in the uppermost oceanic crustal levels (e.g. Brikowski and Norton 1989; Cathles 1990; Mottl and Wheat 1994; Fisher 1998).

A relatively confined near-critical zone for fluid flow is therefore expected for narrow magma chambers, thus favouring the focusing of hydrothermal fluids into very restricted rock domains. Wider magma chambers will promote the development of a broad near-critical domain for fluid flow, leading to widespread hydrothermal activity that will migrate ridgewards with time as magma liquids shrink, evolving naturally to configurations indistinguishable from those related to narrow chambers. The implications of these two expected evolution trends in terms of heat flow and of the time-dependent mechanical response of the oceanic crust are enormous. More so if we consider calculations by Brikowski and Norton (1989) which show that the hydrothermal activity will remain for nearly twice the lifetime of the magma chamber. One may, therefore, conclude that multiple intrusion events at fast-spreading systems will generally reinforce hydrothermal activity patterns, thus enhancing vertical and lateral heterogeneities of the oceanic crust, especially in the presence of wider magma chambers.

From what was previously said, it seems that the accommodation of significant (<10%) homogeneous shortening by the upper oceanic crust in fast-spreading systems is mainly ascribable to the combined effects of renewed magmatic activity, subsequently complemented by vigorous hydrothermal alteration, which also restrain seismicity and support strain partitioning between seismic and aseismic processes. The time-dependent mechanical response is apparently similar for slow-spreading systems, the decrease in the upper crustal strength being extremely variable and largely controlled by faulting and by the relative abundance of serpentinised peridotites. Serpentinisation processes and subsequent hydrothermal alteration of gabbroic and basaltic rocks, coupled by extensive fracturing at all scales, will thus lead to a mechanically weakened crust, which easily accommodates the homogeneous shortening, delaying buckling onset. Deformation by buckling should, therefore, be earlier and more intense in fast-spreading systems, even after lower rates of homogeneous shortening. Accepting this general interpretation, an evolution from (thermo-)viscoelastic to viscoplastic behaviour is expected for the upper oceanic crustal domains as intense fracturing and hydrothermal alteration proceed. Since the heat flow regime that controls those chemical and physical modifications is mainly active-advective (the heat source being magmatic and crustal cooling mainly controlled by the advection of seawater), a thermo-viscoelastic rheology, locally or intermittently viscoplastic, should characterise the time-dependent mechanical response of deeper intrusive units. This is acceptable for the sheeted dyke complex – high-level gabbros transition, as demonstrated before. However, could we extrapolate that mechanical behaviour for the plutonic units?

As mentioned before, narrow, but long-lived magma conduits should lead to multiple intrusive events at fast-spreading systems, while episodic magmatic activity is believed to characterise slow-spreading systems. Accordingly, the new deep crust will tend to have monotonically decreasing temperatures in fast-spreading systems, and to experience successive re-heating events as it slowly moves off-axis in the other case. From this, wider strength crustal variations at slow-spreading systems may be expected, mostly caused by subsolidus deformation far from steady state conditions that will favour heterogeneous high-T (≥600°C) diffusion creep and intracrystalline slip, subsequently overprinted by annealing (e.g. Karson 1990; Agar 1994). Mineralogical and textural adjustments ascribable to diffusion controlled mechanisms (such as grain boundary sliding) and to the effective competition between work hardening and recovery processes are, conversely, expected in younger deep oceanic crust generated at fast-spreading systems, thus suggesting enhancing of syn-magmatic viscous flow and quasi-plastic flow under near steady state conditions at temperatures above 600°C (e.g. Karson 1990; Agar 1994). We may, therefore, conclude that, when young, the rheology of deep plutonic units at fast- and slow-spreading systems is also expected to evolve from thermo-viscoelastic to viscoplastic.

As a result of the magmatic-hydrothermal-crustal extension evolution that is believed to characterise both fast- and slow-spreading systems, the oceanic crust is thought to behave as viscoplastic some time after its formation (10–20 Ma, according to previous discussions in this volume). At this time, the heat flow regime in the seafloor at fast-spreading systems should become mainly passive-conductive, remaining largely active-advective at slow-spreading systems, grossly agreeing with the age of transition from advective to conductive heat flow estimated for the Juan the Fuca Ridge (<3 Ma), Galapagos Rift (5 Ma), East Pacific Rise (12 Ma), Indian Ocean (40 Ma) and Mid-Atlantic Ridge (70 Ma; see e.g. Mottl and Wheat 1994, and references therein). Nevertheless, as comprehensively reviewed by Fisher (1998), uncertainties regarding the distribution and evolution of permeability within the ocean crust significantly determine the predictions based on the available conceptual models. Consequently, all we can say is that the age of heat flow regime transition from active-advective to passive-conductive differs notably from one ridge segment to another, being mostly dependent on the sedimentation rates and on the roughness of the basement topography (the latter, in turn, a function of the spreading rate, as discussed before). However, despite all these uncertainties, a general tendency exists for a uniform temperature at the basement interface and for homogenised pore-water composition as the heat flow regime become passive-conductive with time. In terms of mechanical behaviour, this means that a gradual evolution from high-T to low-T diffusion creep is expected for both the strongly hydrothermally altered uppermost crustal domains and their deeper counterparts. Within the sediment cover formed in the meantime, progressive dehydration facilitates strain accommodation by means of intense (macro-)microfracturing and low-T diffusion mechanisms, locally assisted by differentiated thermal expansion of pore fluids if non-homogeneous conditions at basement enable the development of heat flow anomalies (e.g. Sclater et al. 1980; Stein and Stein 1994). From this, a prevailing aseismic creep is to be expected at off-axis settings, thus favouring the subsistence of a viscoplastic rheology.

The Wilson Cycle Revisited

4.1
Plate Tectonics as a Dynamic Process

The purpose of this section is to show that plate evolution is a dynamic, non-conservative process. Therefore, plate deformation should occur not only at the boundaries of plates, but also in their interior. We will focus on the aspects of the Wilson cycle (Dewey and Burke 1974; Kearey and Vine 1996) where soft plate tectonics allows a different view from the standard theory. We can envisage that some processes which are known in the geological record can nevertheless be better approached in the context of dynamic systems than in the context of the various branches of linear basic sciences, namely mechanics. We also consider that plate tectonics is mainly a very efficient kinematic description but much of the underlying dynamics are just beginning to be sorted out now. Steady plate tectonic regimes are much easier to explain than the mandatory changes in regime recorded in tectonic history. We will pay special attention to modern analogues of transition situations where the plate tectonic setting is changing either in space or in time. Plate changes in time include (Davies 1999): steady growth or shrinkage of plates, changes in plate velocity, sudden or gradual formation of new plate margins, such as formation of new spreading centres, and finally, their propagation and jumping and formation of new subduction zones. The last case represents a major turning point in the Wilson cycle. All these changes are difficult to deal with in terms of the classical plate tectonics paradigm, and in fact have remained obscure until recently. Therefore, although many of the dynamic processes have been described in the classical literature of plate tectonics, because there is ample evidence that they operate in the geological record, there is still a lack of detailed and quantified studies that must explain these processes. Much work remains undone and we will try to show, in a non-linear approach to plate tectonics, that these processes must be considered inevitable rather than being anomalies that remain unexplained in a linear approach and only assume the steady state of plate kinematics.

Sea-floor spreading is an example of a dynamic process. Suppose that we deal with a small volume of material ascending in the upward branch of a convection cell beneath a mid-ocean ridge. Two points at an infinitesimal distance inside this small volume can be incorporated into the two separate plates that will end in positions thousands of kilometres apart. This is an example of bifurcation, more specifically a pitchfork bifurcation (Thompson and Stewart 1986), in physical space; we borrow the

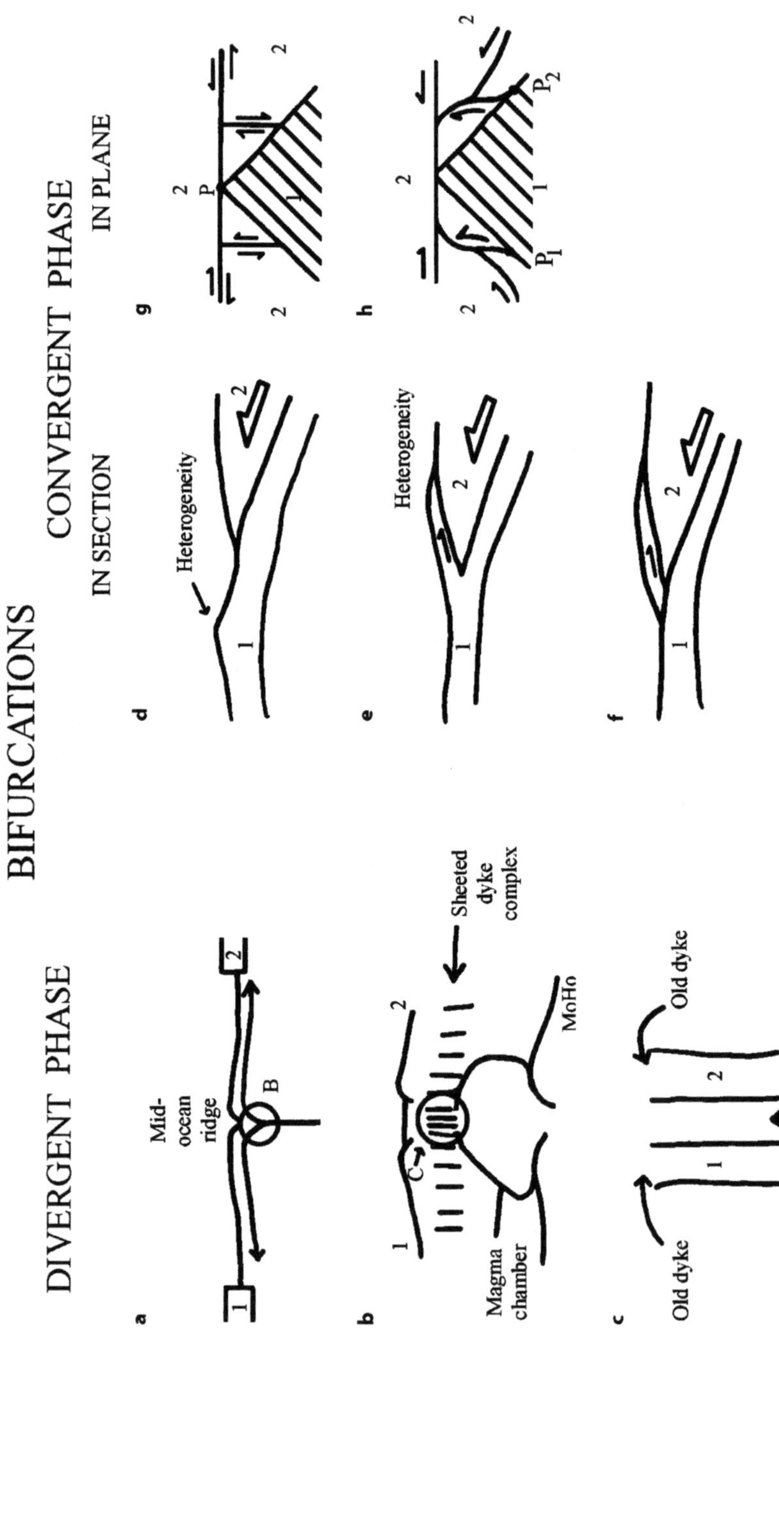
BIFURCATIONS
DIVERGENT PHASE
CONVERGENT PHASE
IN SECTION
IN PLANE
a
Mid-
ocean
ridge
B
1
2
b
C
Magma
chamber
MoHo
Sheeted
dyke
complex
c
Old dyke
New dyke
Old dyke
d
Heterogeneity
e
Heterogeneity
f
g
P
h
P1
P2

nomenclature of non-linear dynamics in phase space to describe a non-linear mechanical process in physical space. This kind of process cannot be adequately described by the usual continuum mechanics approach. In this approach, the functions relating the displacement vector field to original co-ordinates are continuous and by definition, in a strict sense, single valued (Malvern 1969). In non-continuum mechanics these functions are generalised functions or distributions (Sokolnikoff and Redheffer 1966). Some special cases of singularities can be solved applying the theory of complex variables; others can be approximated by numerical methods (Fig. 4.1).

In all phases of the Wilson cycle we find examples of dynamic processes, such as bifurcations. We will see that in some cases of the Wilson cycle a *tectonic flake* due to tectonic wedging will be generated across a convergent boundary. This can be considered as an example of bifurcation in section view; one of the plates is split subhorizontally. The upper part will end upon the top of the other plate, while the lower part will be carried to a depth below the opposing plate. The convergent boundary is a subduction zone that can evolve in different ways. This way the splitting plate is divided into an upper plate, where thrusting is antithetic to subduction, and a lower plate where thrusting is synthetic to subduction (Roeder1973). The upper plate can be entirely oceanic, and we have antithetic obduction (Sect. 4.8). It can be an oceanic plateau, microcontinent or some other heterogeneity inside a closing ocean in the Pacific stage (Sect. 4.9). The effect will be either terrane accretion, or the tectonic flake can occur during continent collision (Sect. 4.10).

Another form of bifurcation in physical space can occur during the Pacific (Sect. 4.9) collision stages (Sect. 4.10). This time the bifurcation is observed in plane view, instead of in section view. A fragment of the overriding plate suffers *lateral expulsion* or *escape tectonics* from a colliding promontory in the other plate (McKenzie 1972a; Molnar and Tapponier 1975). It moves, generally, along large strike-slip faults, towards the re-entrant part of the subduction zone along which convergence takes place.

In all those cases, the process is dynamic in the sense that the block that suffers tectonic flake or tectonic expulsion cannot be decided a priori and probably depends on its boundary conditions, suggesting sensitivity to initial conditions.

Another dynamic process can occur during the Wilson cycle; it is the change of polarity of subduction, or *flip* (McKenzie 1969; Roeder 1973).

In the concept of the Wilson cycle, the idea of periodicity is implicit (Wilson 1966). It was also noted that many times oceans start at approximately the same places whe-

Fig. 4.1a–h. Bifurcation in physical space during the Wilson cycle. In divergent phase. **a** Two close points below the mid-ocean ridge end up at distant places on each plate (*1* and *2*). **b** Detail of **a**, the bifurcation occurs at the sheeted dyke complex. **c** Detail of **b**, a new dyke intrudes, statistically in the middle region of a previous, older dyke. In convergent phase, in section: **d** two converging plates (*1* and *2*). An inherited heterogeneity exists on plate *1*. **e** A flake nucleates at the heterogeneity because it resists movement down the opposed plate. **f** The transient flake generation stops and the previous situation is resumed, with all of plate *1* going down below plate *2*. In convergent phase in section: **g** a rigid plate (*1*) indents a less rigid plate *2*. A point *P* is situated in the tip of the triangular indenter. **h** Indentation proceeds. The domain around point *P* bifurcates and ends up at distant locations, P_1 or P_2

re they opened and closed in the previous cycle. In these cases the notion of inheritance is implied; a mechanism of inheritance could be the presence of root zones with some specific rocks, such as eclogites, whose mechanical properties were favourable to localise the stretching process that initiated the younger cycle (Ryan and Dewey 1997). However, in many other cases, the new ocean does not follow the older ocean and the new rift propagates to a cratonic area that was unaffected, sometimes throughout many millions of years. In those cases there was no inheritance from the previous cycle. Therefore, the dynamic evolution of plate boundaries can follow some periodicity in time, expressed as supercontinent cycles, with a duration of 400±100 Ma (Nance et al. 1988; Condie 1997) but show a larger variability in space.

The dynamic character of plate tectonics is clearly depicted in temporal evolution (Sengör and Dewey 1990). Some opponents of plate tectonics have emphasised the opposition between the episodic character of orogeny and the steady character of plate movements to dispute the importance of the plate tectonics as a driving mechanism for orogeny. Tectonics is an old science, as a branch of geology (Sengör 1979), extending back a least two centuries; plate tectonics is a young branch of the earth sciences, extending back only half a century. So tectonicists of the classical school have resisted some aspects of plate tectonics, insisting on discrepancies between well established facts and theories.

On the one hand, the episodic character of orogeny is not incompatible with plate tectonics, but is rather a consequence of the dynamic character of the theory. Even in a slowly evolving Wilson cycle these are discontinuities superposed on the steady evolution, such as terrane accretion at subduction zones, and collision of buoyant elements such as continents and arcs. These aspects have been integrated in the plate tectonics view of orogeny (Dewey and Bird 1970; Dewey 1972; Oxburgh 1974; Mattauer 1981). On the other hand, accelerations of plates are inevitable because all plates are coupled to each other (and to the asthenosphere?). If plate velocities vary only one order of magnitude between 10 and 1 cm/year, the duration of tectonic episodes is much more variable, between 1 and 200 Ma. Thus, integrated rates can vary at least three orders of magnitude and strain rates for different processes can vary over more than 18 orders of magnitude between impacts and isostatic movements (Price 1975). Even for integrated tectonic rates, there is a broad spectrum between discontinuous short-lived orogeny and steady plate movements.

We conclude that a very rich variety of tectonic regimes must be a natural consequence of continuously changing plate kinematics, and all orogenic events are related even if they are not synchronous in general terms.

4.2 Intraplate Rift Stage

The Wilson cycle starts by breaking a supercontinent. The break can occur along a previous suture or across it. If this break is juvenile and not inherited, a shield must be fractured; we have seen that here the lithosphere is thicker and colder than elsewhere. The heat from shields is removed by conduction alone; it follows that the heat flow has its origin in radioactivity of the lithosphere, with little or no contribution from the deeper mantle. Therefore, shields produce *continental insulation* (Anderson

1984), whereas subducted slabs cool it. Continental insulation requires a long period of thermal softening of the shield by an upward moving mantle plume From geological records it is estimated to be in the order of 100 Ma (Nance et al. 1988). If the break occurs along a previous suture, the effect of continental insulation must be minor.

In continental extension systems (Bosworth et al. 1992) rapid changes in rate of stress are induced. Stress fields are radially disposed around triple junctions but become normally oriented to rift systems to accommodate larger extensional strains. This suggests that these quick rotation rates of stress trajectories, controlled by intraplate stresses produced and propagated from distant plate boundaries, are the dominant forces rather than the drag at the base of the lithosphere.

The process of breaking the supercontinent is still largely unknown in detail. Two mechanisms have been proposed (Sengör and Burke 1978; Park 1988; Condie 1997). They can operate together as a feedback process.

In *active rifting* the lithosphere is thinned by mantle convection, resulting in crustal downing and rifting.

In *passive rifting* the injection of asthenospheric material to the base of the crust induces extensional failure of the continental lithosphere.

Both processes lead to stretching of the continental lithosphere (Wegener 1915; Mckenzie 1978). The continental lithosphere is uniformly and rapidly stretched by a *stretching factor*

$$\beta = \frac{l}{L}$$

where l is the length of a vertical deformed line and L is its undeformed equivalent.

The kinematics of stretching can be achieved by pure shear in the Mckenzie model and by simple shear in the Wernicke model (Wernicke 1985). The rifts superimposed on a previous suture tend to follow the simple shear model and rifts oblique to previous orogenic fabric tend to follow the pure shear model (Ziegler et al. 1998). All the models have to assume some deformation of the stretched lithosphere; the β values can be estimated by the geometry of balanced structures (Roberts and Yielding 1994) or by the nature of stratigraphic fill of sedimentary basins developed above the stretched lithosphere (Allen and Allen 1990).

The β values obtained so far can reach a maximum of 4 (Allen and Allen 1990, p. 82), revealing a considerable amount of deformation in the stretching phase of rifting. This means that the continental crust has been thinned to almost oceanic thickness.

The style of extension depends on its strain rate (Kusznir and Park 1986; Park 1988). Slow rates of extension will lead to strain hardening of the lithosphere because the geotherm may have time to re-equilibrate. Rapid extension will lead to weakening because the temperature rise will balance the effect of crustal thinning that will harden the lithosphere because it will bring the stronger mantle layer towards the surface.

Therefore, for low strain-rates, in the order of 10^{-16} s^{-1}, the strain hardening will transfer deformation laterally, thus widening progressively the zone of active extension, but with a limited value of $\beta \approx 1.5$. In contrast, with high strain-rates in the order of 10^{-14} s^{-1}, strain softening will localise the deformation in a narrow zone, leading to $\beta \geq 2$ and eventually to lithospheric separation.

Divergent plate boundaries propagate across continents (see Sect. 4.3); this requires intraplate deformation at the tip of the propagating rift. These can form complex triple junctions that segment the continent that is breaking up, so deformation occurs along intracratonic rifts, although minor; some of the arms can abort, giving aulacogens (Park and Jaroszewski 1994), sometimes slightly inverted afterwards. Only along the rift arms that became oceans will deformation increase, breaking and splitting the entire lithosphere, and we evolve to the next step of the Wilson cycle, the Red Sea stage.

4.3 Red Sea Stage

In the Red Sea Stage new divergent plate boundaries are born. Mid-Ocean ridges and rises show a clear hierarchy of segmentation (for a synthesis see Juteau and Maury 1997) with at least three or four orders of magnitude: the largest wavelength is of 1000–5000 km and controlled by hotspots rising from the deep mantle. The next is 200–1000 km, reflecting the presence of transform faults and fracture zones. Below is the segmentation of 20–80 km in length introduced by overlapping spreading centres which are produced by propagation of elemental segments of accretion, and less well-defined lower orders that suggest a self-similar structure repeated at different scales of observation. Oceanic rifts are highly dynamic structures: they propagate through the older oceanic lithosphere and they can change spreading direction quite abruptly (Hey et al. 1988).

Major differences exist between fast- and slow-spreading ridges in transverse and longitudinal structure and morphology (Fowler 1990). In slow-spreading ridges there is an axial valley and a rough topography of abyssal hills, whereas fast-spreading ridges (spreading velocities above 3–4 cm/year) are smoother and generally lack an axial valley. These differences are due to the rheology of the ductile plastosphere that thickens as it moves away from the ridge, and to the effect of magma accumulation by underplating. For slow-spreading ridges the depth of the rift valley depends primarily on the strength of the plastosphere and its rate of thickening. For fast-spreading ridges the major factor is the buoyancy of the magma. These correlations confirm the idea that there is continuous tectonic extension alternating with cycles of periodic magmatism of 10,000–50,000 years in slow-spreading ridges and continuous magmatism in fast-spreading ridges. In fact, geophysical imaging suggests that steady state crustal magma chambers are present below fast-spreading ridges but not below slow-spreading ones. This suggests that tectonic extension dominates magmatic accretion in slow-spreading ridges, which is reflected also by the presence of *non-transform discontinuities* (MacDonald et al. 1988), which are more abundant in slow-spreading ridges (Sloan and Patriat 1992), such as the Mid-Atlantic ridge between 28° and 29°N, south-west of the Azores triple junction. Therefore, in all types of plate-boundaries, divergent, convergent but also in transform types, we find examples of diffuse plate-boundary activity in plate boundary zones instead of discrete, and the diffuse character is more well-marked if the rates of differential movement are lower.

Faulting at mid-ocean ridges is probably one of the simplest tectonic processes in Earth, because it affects the neoformed schizosphere. Attempts at modelling have emphasised a deterministic approach in terms of regular spacing and uniform fault displacements. The real structures, however, are distinctly different, with a characteristic random component, well expressed for instance in abyssal hill fabric. Progress has been made recently in the theoretical numerical modelling of the real tectonic structures (Buck and Poliakov 1998); when a realistic failure criterion is met, for instance when stresses generated by plate spreading overcome the internal strength of the oceanic plastosphere (Goff 1998), there is spontaneous development of faults because the model finds its own geometry instead of arbitrarily imposing regular fault spacing. Dynamic models instead of "fixist" models can explain much better the real structure of abyssal hills as a function of rheology, strain rate, cohesion strength and temperature. Physical modelling of slow seafloor spreading (Shemenda and Grocholsky 1994) also supports the results of the dynamic numerical modelling referred to above.

Seafloor spreading requires bifurcations in physical space at the divergent plate boundaries, such as a pitchfork bifurcation if two plates are involved, or a more complex Hopf bifurcation in a triple junction.

The geological observations in obducted ophiolites (see Sect. 4.8) can be used for a detailed description of this bifurcation process, even if its theoretical foundation has not yet been established. These circumstances also illustrate another important methodological aspect – the present is the key to the past, as claimed by uniformitarianism (Lyell 1830, 1832) but the past can also be a key to the present (Schumm 1991).

Let us just follow two neighbour domains in the upward rising column of a mid ocean ridge and follow them until they drift away because they belong to two separate plates.

Beneath the ridge the upwelling material can be arranged in a very complex way: discrete mantle diapirs cause divergence at a high angle to the ridge but also parallel to it (Nicolas 1989). This is why ridges are segmented.

In the magmatic chamber below the ridge, if it exists, the trajectories of each particle can be quite complex in the oceanic gabbros of the ophiolite sequence. From the gabbros we pass upwards, transitionally to the sheeted dyke complex (Gass 1968). The statistics of chilling contacts (Kidd 1977) show us that each domain has a high probability of ending in a specific plate, because dyke intrusion is only a few tens of meters wide. In fact, if there were a unique dyke successively intruded at its middle, all chilled contacts in each plate would be on one side, and this is not the case. This suggests that there is sensitivity to initial conditions inside the intrusion zone and the whole process is deterministic but unpredictable, except in a probabilistic sense.

Mid-ocean rifts must propagate across continents. The situation is occurring now in the East African Rifts in the Afar Triple junction and in the tip of the Mid-Atlantic Ridge, the Artic Ridge (Le Pichon et al. 1973). A propagating rift poses a severe problem to standard plate tectonics, emphasised by Thom (1982), the creator of the Catastrophe Theory, a precursor to the Dynamic Systems Theory. Plate boundaries that end inside a plate "are in contradiction with the hypothesis of plate rigidity (inside a solid the velocity has no discontinuity)" (Thom 1982, p. 3); and "The Theory of Plate Tectonics does not require that the lithosphere be solid (rigid)" (Thom 1982, p. 1).

In the soft plate tectonics theory, this difficulty is removed by intraplate deformation. As seen in Section 3.4, the African Plate is divided into the Nubian and Somalian

Plates. A zone of higher than average intraplate deformation connects the East African Rift to the south.-west Indian Ridge, another left-lateral transform diffuse zone connecting the Artic Ridge to the Kuril-Kamchatka trench through the Verkhoyansk Mountains, conspicuously higher (2389 m) than the Siberian Platform and reactivating a late Jurassic Suture (Voo et al. 1999). Plates are not affected by "rigor mortis" but have their own lives with birth, young, mature, decay and death phases as for the Wilson cycle of oceans. Hence, they sometimes also join and split.

The dynamic evolution of plates has been tentatively described by some authors, using parameter spaces such as the ω space (McKenzie and Parker 1974).

4.4 Atlantic Stage

As seafloor spreading proceeds, the ocean changes from Red Sea width, less than 100 km, to Atlantic Ocean width of thousands of kilometres. In this stage the ocean is surrounded by passive margins.

The present stress fields in Atlantic-type oceans shows that the young oceanic lithosphere is, in a normal fault stress regime, up to 20–30 Ma. The older oceanic lithosphere is in the strike-slip regime along fracture zones or in the thrust fault regime elsewhere (Stein and Okal 1986).

At the Atlantic stage the seismicity is usually concentrated in mid-ocean ridges. Some intraplate oceanic earthquakes can occur, however, rather infrequently, because most oceanic intraplate deformation must be aseismic in the soft plate tectonics model. At passive margins we must have a discontinuity in rheology behaviour, between the viscoplastic rheology in the oceanic domain and the elastoplastic rheology in the continental domain. We should expect a discontinuity in state of stress between both domains and it follows some of the consequences for deformation at passive margins.

Passive margins with stretched continental lithosphere or intermediate lithosphere are in compression. Hence, during their evolution, they changed from the extension regime with tectonic subsidence to the compression regime with thermal subsidence, because cooling continues. Therefore, passive margins must be in a state of tectonic inversion. This is proved by focal mechanisms of earthquakes because thrust and strike-slip events dominate over normal fault mechanisms.

Passive margins are the locus of considerable seismicity (Johnston 1989; Johnston and Kanter 1990). The global seismicity moment release in passive margins and old aborted rifts is significantly above the stable continental crust outside those domains, but some two orders of magnitude below plate boundaries and active regions. The data base (Johnston 1989) for the last 200 years show that in this period of time, eight events with magnitude 7 ($Mo \geq 4\times10^{26}$ dyne-cm) occurred; seven events were on passive margins and one in a failed intracontinental rift. The maximum seismic moments observed are in the order of $Mo=0.7\text{–}2.5\times10^{26}$ dyne-cm corresponding to great earthquakes ($M_{\omega}=7.8\text{–}8.2$) but again their moments are nearly two orders of magnitude below the great plate boundary earthquakes. Some of these earthquakes are located in the projection of fracture zones onto the margin.

Passive margins and failed rifts have been extended at the moment of their formation and should be weaker than the thicker lithosphere around; this view has been challenged by some authors (Coward 1994) who defend that stretching initially weakens the lithosphere but after temperature equilibration, stretched lithosphere become stronger than unstretched lithosphere. This explanation could apply to some young passive margins where thermal equilibration was not achieved but seems contradicted by seismicity in old aborted rifts. Perhaps thermal equilibration requires a longer time than we envisage from the present state of knowledge. In fact, refined rheological models (Ziegler et al. 1998) for the rifted continental lithosphere show that it strongly depends on the kinematics of stretching; particularly in the simple shear model, there is a clear asymmetry in rheology. It is due to accretion of the new mantle lithosphere derived from cooling of the asthenosphere, below the attenuated old infracontinental mantle lithosphere. Therefore, lower plate margins are considerably stronger than upper plate margins, where a softer young mantle lithosphere dominates over a strong old mantle lithosphere. This asymmetry persists throughout the whole Atlantic phase of the Wilson cycle and will be an important controlling factor for the triggering of subduction (see Sect. 4.7).

The source of passive margin stresses and rheology has been investigated by some authors (Stein et al. 1989); they include: flexural stresses due to the removal of ice loads; spreading due to different densities of continental and oceanic lithosphere, compressive intraplate stresses, finally, major sediment loading, with thickness up to 10 km, leads to stress of hundreds of MPa, an order of magnitude above other types of stress. However, passive margin seismicity is not visibly correlated with sediment loading; this paradox can be explained in two ways. In the standard theory, the long-term sediment load contributes to the stress but does not induce seismicity; or this is relaxed in a viscoelastic lithosphere.

4.5 Diffuse Divergent Plate Boundaries: The Azores Triple Junction

4.5.1 Introduction: Plate Kinematics at the Azores Triple Junction

Triple junctions are potential sites where the postulate of rigidity can be tested, as seen in Section 3.8. If the triple junctions are of the convergent type, TTT, a considerable amount of finite compressive strain can be accumulated. In triple junctions of divergent type RRR, the finite strain had no time to accumulate near the triple junction and the test of rigidity is more difficult to apply. However, in an RRF junction, the transform fault arm should cut across the oceanic lithosphere of different ages and this should be reflected in the finite strain regime as a function of age. So let us examine an RRF junction in this perspective, to see if the rigid plate model and the soft plate model lead to results different to those than can be confronted with the real situation.

An example of such an RRF junction is the Azores triple junction, where the Eurasia, Africa and America Plates meet. The boundaries between America on one side, and Eurasia and Africa, on the other side, are the ridge type, the Mid-Atlantic Ridge. However, the boundary of Eurasia and Africa is mainly the transform type, the

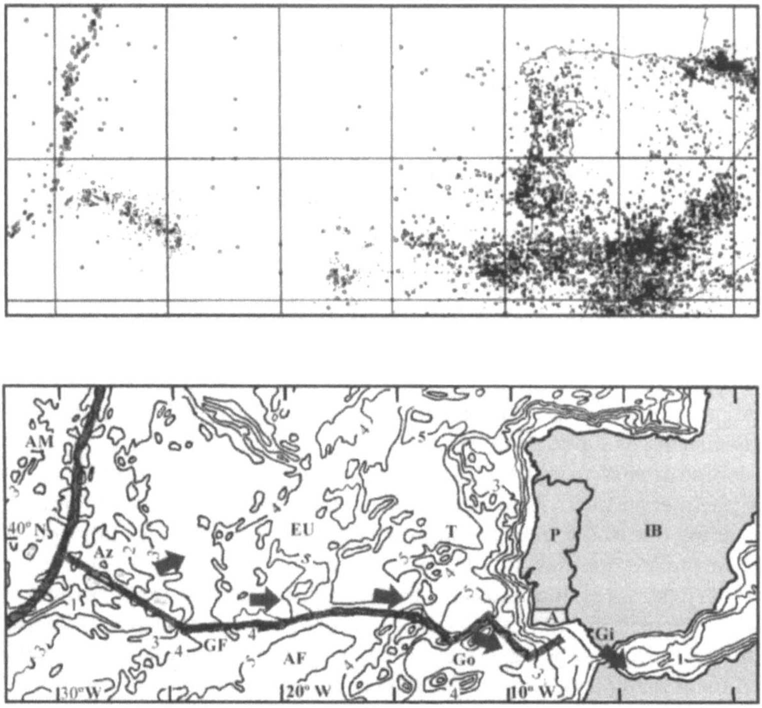

Fig. 4.2. Seismicity of the Eurasia–Africa Plate boundary in the segment Azores-Gibraltar (After Argus et al. 1989 and Cabral 1995). Vectors correspond to plate movement of Eurasia vs. Africa, with a length of 4 mm/year

Azores-Gibraltar Plate boundary zone, connecting the Azores triple junction with the continent collision zone between the Eurasian Plate in the north and African Plate in the south of the West Mediterranean (Fig. 4.2). The present plate movement is transtensional in the Azores (Argus et al. 1989), dextral along the Gloria transform and convergent at the longitude of Gorringe Bank, according to model NUVEL-1. In fact, if we apply all the kinematic information since 3 Ma along this boundary, we obtain a closure of the Africa-Eurasia–North-America Plate motion circuit. Near the Azores, triple junction the boundary between Africa and Eurasia, with an average NW–SE orientation, must have a component of dextral movement, in the order of 0.11 cm/year and an opening component of 0.4 cm/year; so this boundary is a combination of dextral transform and rift. Many authors called it the Terceira Rift (Searle 1980) and others called it the S. Jorge Leaky Transform (Madeira and Ribeiro 1990). In fact, pure transforms are rare because plate movement is a dynamic process and transforms must accommodate components of compression or extension; as discussed in Section 3.8, these are called leaky transforms.

The recent plate movement in the Azores triple junction is well understood. However, the orientation of the present boundary between the triple junction and the western tip of the pure transform Gloria Fault, which is the Terceira Rift or S. Jorge Leaky Transform, remains unexplained.

Plate kinematic reconstructions (Searle 1980) suggest that the movement before 36 Ma was of the RFF type, with pure transform motion in the Gloria Fault, between Eurasia and Africa but also in the East Azores fracture zone, between North America and Africa. At ca. 36 Ma, there was a rapid change in relative motion between Eurasia and Africa, due probably to increasing closure of Tethys. The East Azores fracture zone was deactivated and the Terceira Rift started to open. Since then, there were complex and rapid changes with recent migration north towards the triple point until it reached its present position (Miranda et al. 1991; Luis et al. 1994). The result of plate boundary reconfiguration is that a wedge of European plate was transferred to the African plate.

4.5.2 Rigid Plate Interpretation

Knowing the kinematics of the three plates involved in the Azores triple junction we can predict the tectonics of the Terceira Rift if we know the kinematics of the boundaries between America, on the one side, and Eurasia and Africa, on the other side, and if we postulate plate rigidity. This exercise was made (Searle 1980); knowing the present Euler pole between Africa and Eurasia we infer a transform direction oriented approximately ENE–WSW, parallel to the Gloria Fault, and crossing the Terceira Rift (Fig. 4.3).

Leaky transforms can be modelled by rigid plate kinematics if there is a change in the Euler pole of rotation applied to a pure transform boundary. The slip vector ceases to be parallel to the boundary and becomes oblique to the average direction of the boundary. The oblique movement is resolved in a new transform direction and small orthogonal segments of ridge. A good example is the East Pacific Rise between Baja California and Mexico.

Let us compare this prediction with the real situation in the Azores.

4.5.3 Plate Tectonics at the Azores Triple Junction

The Azores constitutes a plateau with maximum altitude +2300 m above the surrounding sea bottom of –3000 m; the plateau is 200 km wide and oriented WNW–ESE (Fig. 4.4).

The plateau is mainly constructed of alkaline basalt volcanism, with differentiation to a trachytic and even rhyolite pole that are minor constituents. Geochemistry and petrology suggest a hotspot origin for this volcanism (Schilling 1975).

The morphology of the plateau is highly differentiated, due to interaction of magmatic activity and intense tectonic activity. Using recent bathymetric surveys (Lourenço 1997; Lourenço et al 1998) and direct observations in the islands (Madeira and Ribeiro 1990; Madeira1998), we can synthesise the tectonics of the Azores plateau by the occurrence of a complex system of active faults. One family is parallel to the length of the plateau, WNW–ESE; it is composed of two conjugate sets of normal

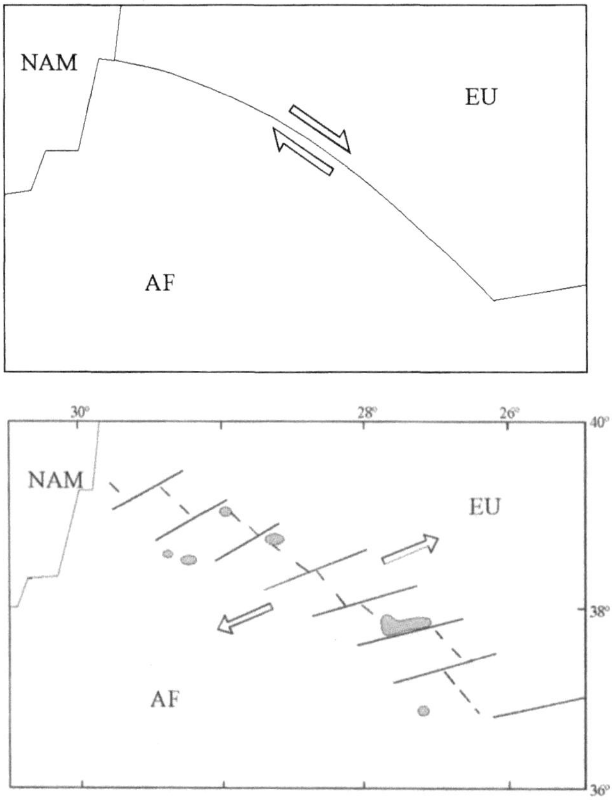

Fig. 4.3. Rigid plate interpretation for the Azores triple junction. A dextral transform between Africa and Eurasia becomes a leaky transform; transforms are *thick lines*, ridges are *dotted*; transforms are *small circles* around the Euler pole Africa-Eurasia. (Modified after Searle 1980)

faults dipping to SSW and to NNE. These faults show a marked component of dextral strike-slip movement. The other family oriented NNW–SSE and is composed of two conjugate sets of normal faults dipping ENE and WSW. These faults show a marked component of sinistral strike-slip movement. Both fault families show alignment of volcanic cones. The tectonic style of the Azores plateau consists of an array of rhomb structures of the horst-graben type, draped by some of the more important volcanic centres. The pattern of faulting can be easily explained by a three-dimensional strain field (Reches 1978, 1983). The corresponding strain regime is transtensional with a dextral oblique normal slip parallel to the WNW–ESE main direction and a sinistral oblique normal slip in the conjugate direction (Fig. 4.5). This strain regime has been confirmed by geodetic methods (Bastos et al. 1998; Pagarete et al. 1998).

The Azores plateau is seismically active (Fig. 4.6). The distribution of instrumental epicentres shows that the present plate boundary is oriented WNW–ESE along the S. Jorge-S. Miguel alignment. This is why we preferred the designation S. Jorge Leaky Transform to the Terceira Rift (Madeira and Ribeiro 1990; Luís 1996). The highest historical magnitude is 7.4 on 9 July 1757; the highest instrumental magnitude is 7.0 on 1 January 1980 (Hirn et al. 1980).

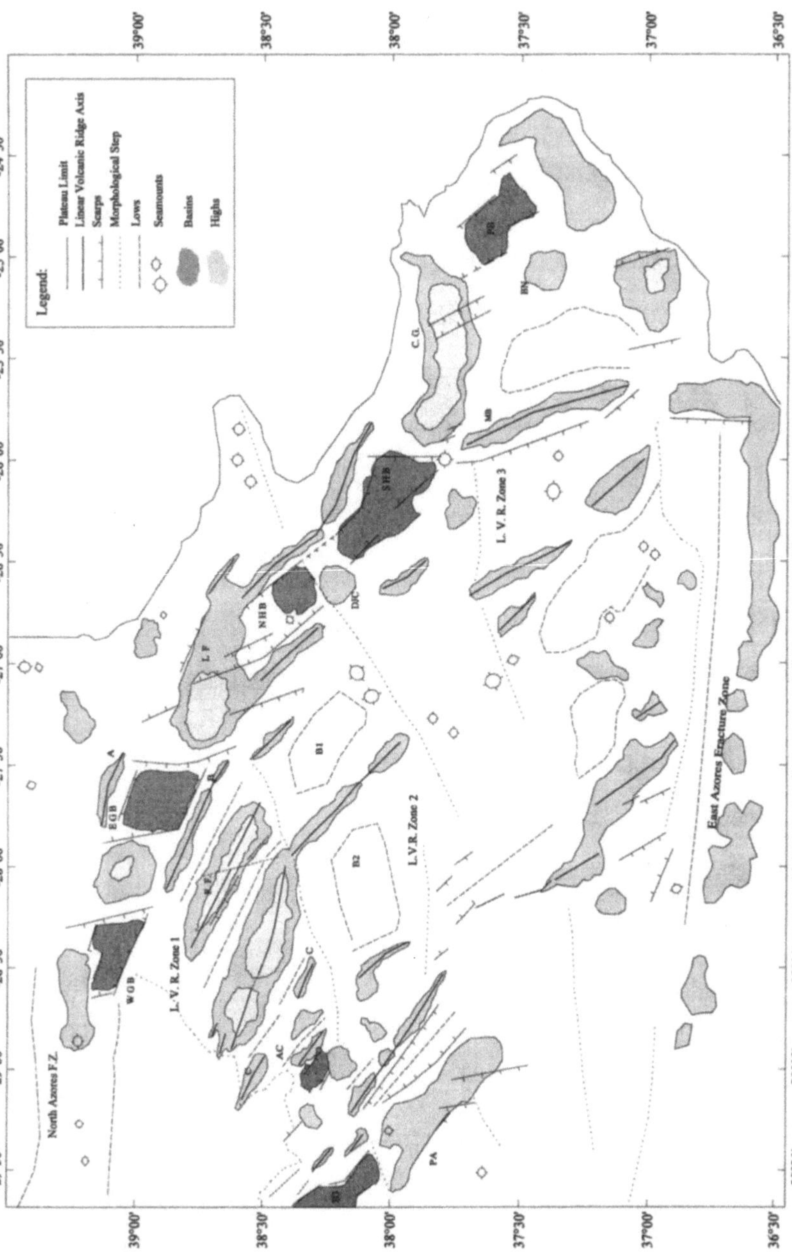

Fig. 4.4. Morpho-structural interpretation of the Azores Plateau bathymetry. *PA* Princesa Alice Bank; *AC* Açor Bank; *BN* Big North; *MB* Monaco Bank; *DJC* Dom João de Castro Bank; *WGB* West Graciosa Basin; *EGB* East Graciosa Basin; *NHB* North Hirondelle Basin; *SHB* South Hirondelle Basin; *PB* Povoação Basin; *RF* Ribeira Seca Fault (Madeira e Ribeiro, 1990); *LF* Lajes Fault; *CG* Congro Graben. A marked clockwise rotation is evident from west to east in the main plateau features. (After Lourenço et al. 1998; with kind permission from Kluwer Academic Publishers)

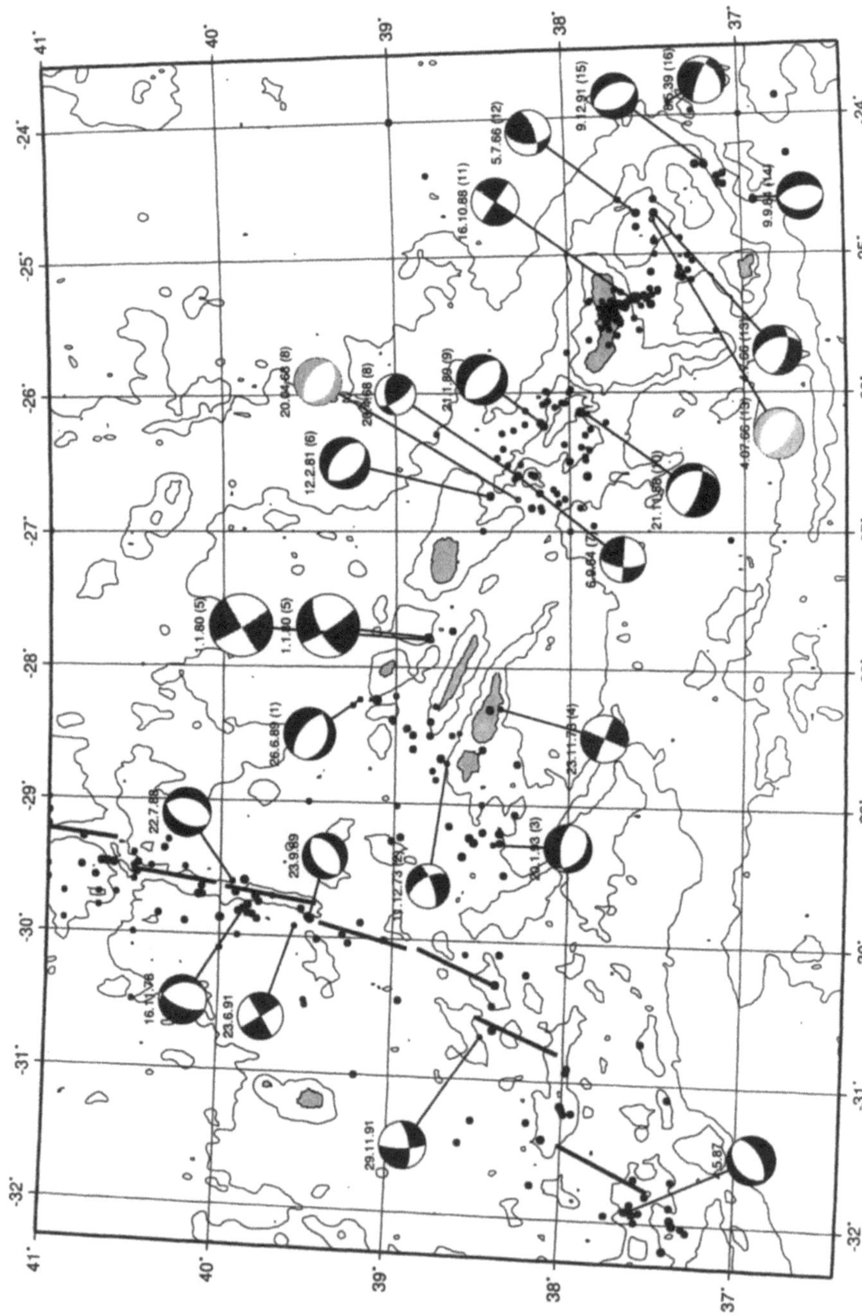

Fig. 4.5. Neotectonic maps from São Jorge, Pico and Faial islands. Note the presence of two main tectonic directions in the three islands. In the Pico and Faial Islands, the intersection of the fault systems corresponds to the locations of both islands' main volcanoes. (After Lourenço et al. 1998; with kind permission from Kluwer Academic Publishers)

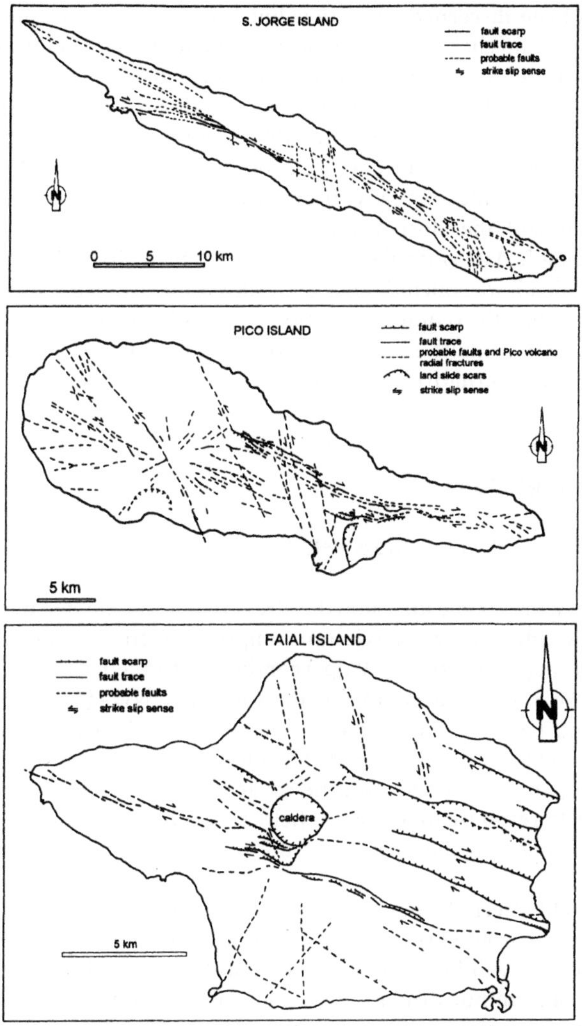

Fig. 4.6. Seismicity maps (M>4) of the Azores Plateau until the present, retrieved from the USGS database. The focal mechanisms are extracted from the same file (*black symbols*), from Buforn et al (1988; *dark grey*) and from Grimson and Chen (1986; *light grey*). (After Lourenço et al. 1998; with kind permission from Kluwer Academic Publishers))

The focal mechanisms of seismic events are as follows: dextral strike-slip with variable components of normal dip-slip in WNW–ESE oriented fault planes; sinistral strike-slip with variable components of normal dip-slip in NNW–SSE oriented fault planes and pure dip-slip normal events in both directions. This suggests decoupling in a transtensional regime; according to the position in time during the seismic cycle we can have pure dip-slip events, pure strike-slip events and oblique slip events. No time pattern has been discerned up to now but very sparse neotectonic observations suggest that strike-slip slickenlines are more frequent on main faults and normal slip slickenlines are more frequent in subsidiary faults (Madeira 1998).

Using all the neotectonic and seismotectonic information we can draw a stress map for the Azores (Madeira and Ribeiro 1990; Lourenço 1997; Lourenço et al. 1998; Madeira 1998). There is a slight curvature on all the features of the plateau, from N75°W at the western intersection with the Mid-Atlantic Ridge to N35°W at the eastern intersection with the Gloria Fault.

Using all available data (morphology, tectonics, seismicity, volcanism, gravimetry, magnetic anomalies), we conclude that the leaky transform rigid model does not fit the data; in particular, the required direction of transform motion across the Terceira Rift with ENE-WSW azimuth is certainly not evident in any of the different data. So we must propose an alternative soft plate model for the Azores triple junction.

4.5.4 Soft Plate Interpretation

Let us take a closer look at the stress field that we reconstructed for the Azores triple junction.

North of the Azores triple junction, the stress field is dominated by the presence of the Mid-Atlantic ridge with σ_3, maximum compressive stress, vertical, σ_2 approximately N–S and σ_1 approximately E–W, perpendicular to the ridge. This stress field is perturbed by the S. Jorge leaky transform, oriented WNW–ESE and with a dextral sense movement. There is a permutation of stress axes, allowing a change from tension regime to strike-slip regime and deviation of stress trajectories: σ_3 becomes NW–SE, making an acute angle with the strike of the transform, σ_2 vertical and σ_1 (see Sect. 3.3.5). In response to this stress field, sinistral strike-slip faults oriented N–S to NNW–SSE develop in the vicinity of the S. Jorge transform as a conjugate set (Fig. 4.7).

This stress field inferred from the active fault pattern is entirely consistent with a transtension that we would expect in a leaky transform. Major displacement and seismic moment release will be accommodated parallel to the dextral leaky transform plate boundary and minor displacement and seismic moment release will occur in the conjugate transtensional sinistral system. The uplift associated with the Azores plateau is probably enough to cause the tensional component that we must add to the pure dextral motion of the transform.

Fig. 4.7. Schematic stress pattern of the Azores Plateau as inferred from the morphological features. *Dots* represent the boundaries between the different Linear Volcanic Ridge domains. *Thicker lines* represent maximum compressive stress orientation (σ_1), *thinner lines* represent the minimum compressive stress orientation (σ_3). The intermediate compressive stress (σ_2) is vertical. *Inset top* T-axis calculated from events displayed in Fig. 4.6, along with calculated standard deviations, the kinematic orientation of the spreading axis, calculated from NUVEL1 Model (DeMets et al. 1990) in the three individualised areas is also shown for reference. *Inset bottom* Proposed schematic regional tectonic model for the Azores domain. (After Lourenço et al. 1998; with kind permission from Kluwer Academic Publishers)

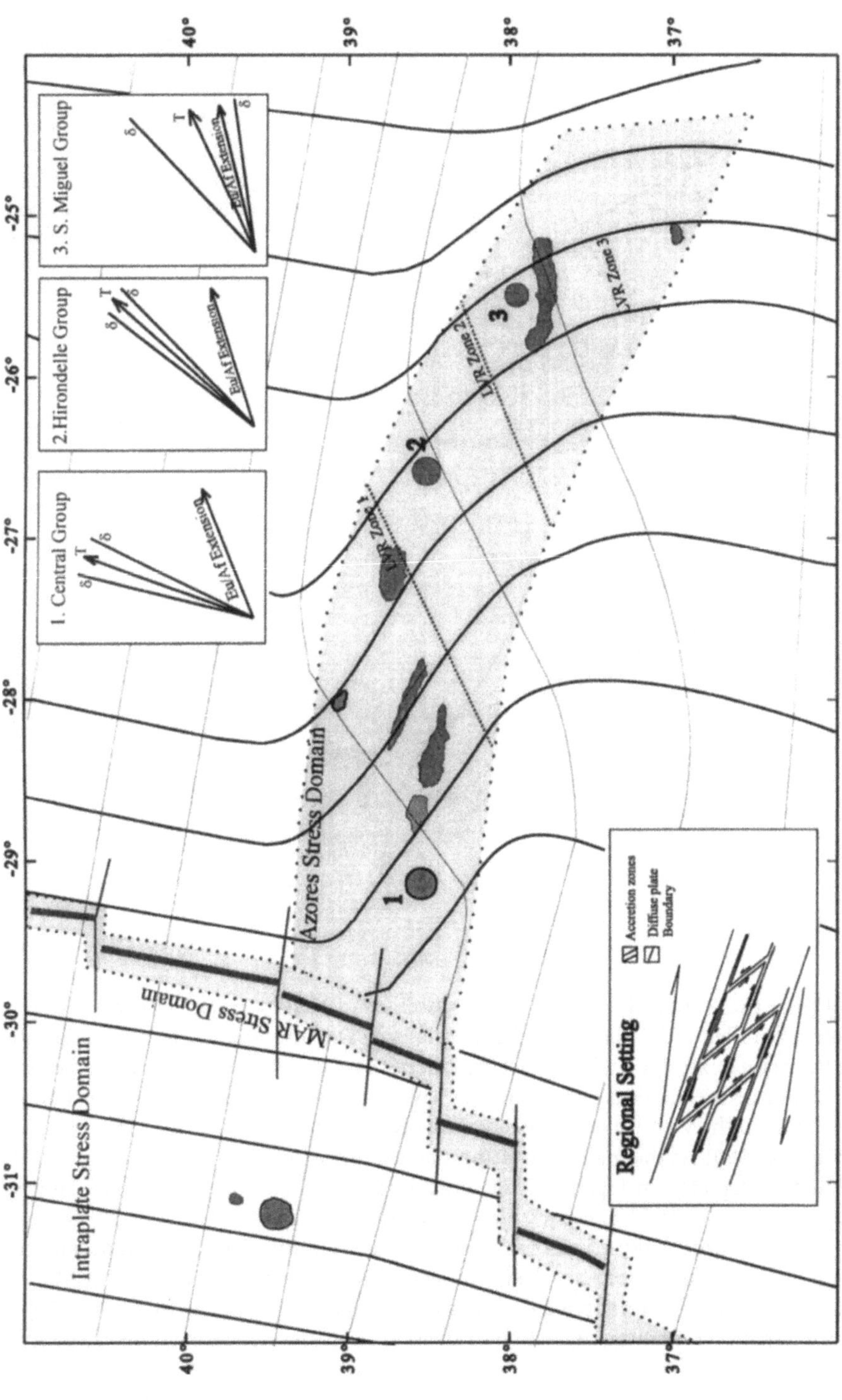
1. Central Group
Eu/Af Extension
2.Hirondelle Group
Eu/Af Extension
3. S. Miguel Group
Eu/Af Extension
Intraplate Stress Domain
MAR Stress Domain
Azores Stress Domain
VR Zone 1
VR Zone 2
VR Zone 3
Regional Setting
Accretion zones
Diffuse plate Boundary

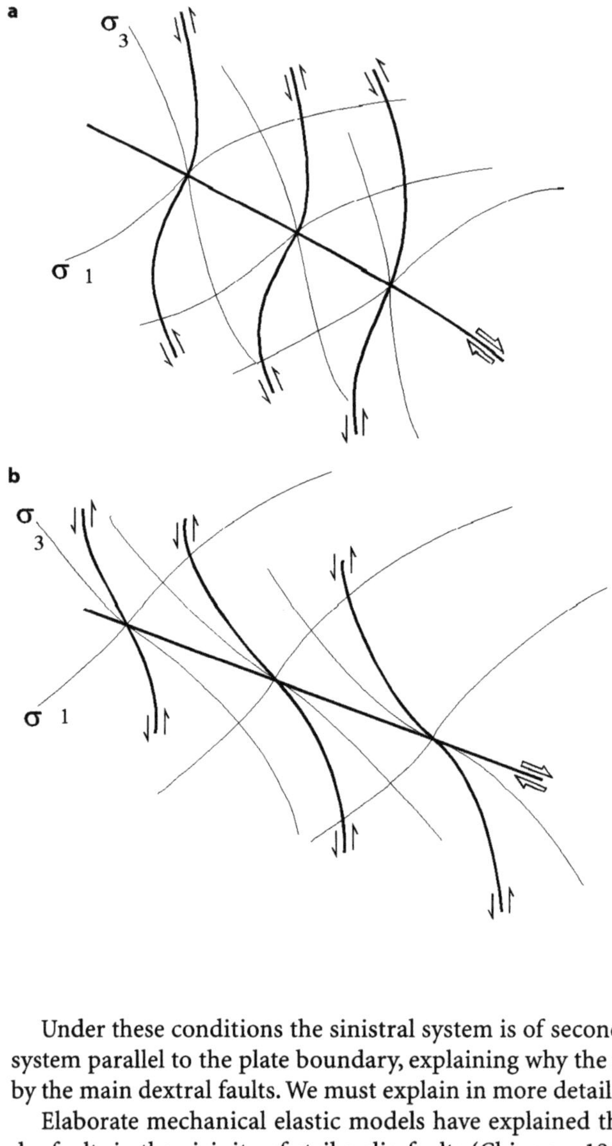

Fig. 4.8. Deflexion of stress trajectories around a major fault (*open arrows*) and curvature of second order faults (*filled arrows*). **a** Elastic rheology; **b** viscous rheology, σ_1, maximum compressive stress, σ_3, minimum compressive stress

Under these conditions the sinistral system is of second order to the main dextral system parallel to the plate boundary, explaining why the sinistral faults are deviated by the main dextral faults. We must explain in more detail this pattern of deviation.

Elaborate mechanical elastic models have explained the formation of second-order faults in the vicinity of strike-slip faults (Chinnery 1966a,b; Price 1968; Price and Cosgrove 1990). As the main fault is approached the principal stress directions are deflected as depicted in Fig. 4.8a. This model can explain successfully some fault systems such as the Najd, Saudi Arabia (Moore and Al Shanti 1973).

In the S. Jorge leaky transform, however, we see a different pattern, because the principal stress directions are dragged toward the main fault system. Hence, the elastic rheology cannot be applied to this case. In fact, experimental work in clay (Reches 1988) has shown that the deflection of stress trajectories depends on the rheology of the faulted material. A viscoelastic material must show a complete spectrum between the elastic solution and the viscous solution, as a function of its relaxation time

(Fig. 4.8b). This proves that the schizosphere in the Azores plateau has viscoelastic behaviour with a short relaxation time. This is confirmed by the fact that the effective elastic thickness of the Azores plateau, inferred from free-air anomalies is only 7 km (Luis et al. 1998).

The fact that the elastic model can be applied to a cratonic area, such as Saudi Arabia when the Najd Fault system formed, and that the viscoelastic model can be applied in the Azores triple junction, confirms our suggestion that the lithosphere effective viscosity must be higher in cratons than in oceanic areas.

The S. Jorge leaky transform is curved, as we referred to earlier. How can we explain this curvature? The plate boundary cuts across the oceanic lithosphere with very different ages from 0 Ma, at the intersection with MAR, to ~55 Ma at the intersection with the Gloria Fault. We know that the stress field in the oceanic lithosphere changes from extensional to compressional around 20 Ma, as seen in Chapter 2.5. If the soft plate model is correct, we must expect extensional deformation in the domain younger than 20 Ma and compression in the domain older than 20 Ma. If we consider the leaky transform as a distributed shear transform zone, the deformation at the walls must change from extended walls to compressed walls as we cross the isochron of ~20 Ma. This will cause a deflection of the shear zone itself in a clockwise tense, in order to maintain strain compatibility (Ramsay and Huber 1984). The strain regime is sub-simple shear (Simpson and De Paor 1993) with narrowing of the shear zone as we walk from the Mid-Atlantic Ridge in the west, to the tip of the active Gloria Fault in the east. This narrowing is expressed by a decrease in the acute angle between conjugate extensional strike-slip faults. This is exactly what happens, confirming the soft plate model (Fig. 4.9).

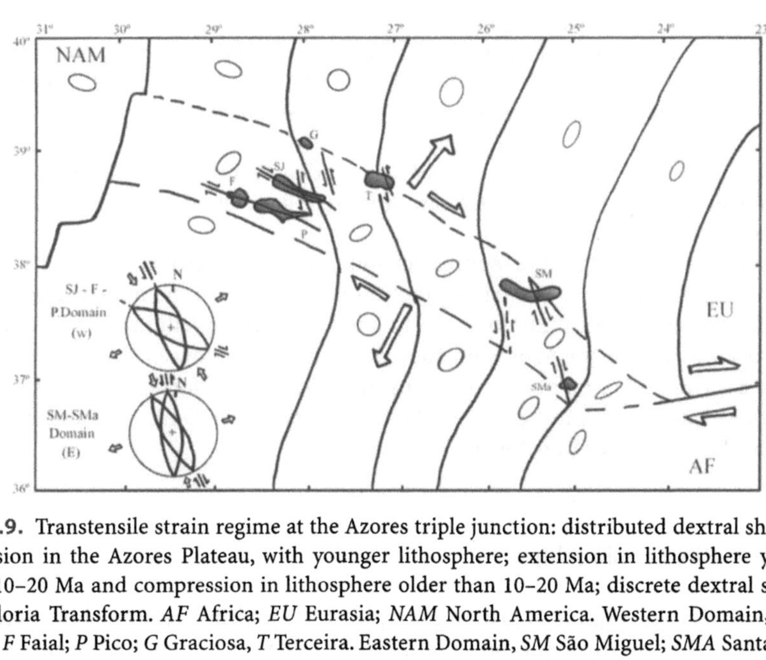

Fig. 4.9. Transtensile strain regime at the Azores triple junction: distributed dextral shear and extension in the Azores Plateau, with younger lithosphere; extension in lithosphere younger than 10–20 Ma and compression in lithosphere older than 10–20 Ma; discrete dextral shear in the Gloria Transform. *AF* Africa; *EU* Eurasia; *NAM* North America. Western Domain, *SJ* São Jorge; *F* Faial; *P* Pico; *G* Graciosa, *T* Terceira. Eastern Domain, *SM* São Miguel; *SMA* Santa Maria

If we understand plate tectonics in the Azores, in general terms, we are still in full ignorance of the dynamics of the interaction of tectonics and volcanism. We can presume that the stress field in the Azores is quite complex and variable, with alternation of strike-slip events on sinistral and dextral faults and pure dip-slip events; in the intermediate periods we could get oblique-slip events. We can anticipate that the pure dip-slip events must be more closely related to volcanism than the strike-slip events. But how does the seismic cycle relate, in detail, with the volcanic cycle? The Azores is a unique natural laboratory for the study of a volcano-seismotectonic integrated approach. It differs from most volcanic areas in that, there is only one volcanic alignment vertical direction normal to σ_1.

4.5.5 Hotspot Activity and Its Significance

Magmatism in the Azores is related to hotspot activity; we can use the hotspot markers to infer something about the interaction between the lithosphere and asthenosphere.

Using absolute motion in the hotspot reference frame, it was demonstrated (Silver et al. 1998) that South Atlantic spreading velocity remained constant over the past 80 Ma; but African Plate absolute motion to NE slowed abruptly 30 Ma ago, due to a collision with Eurasia. This requires that the South America Plate accelerate to the west. Given the minor amount of movement of North America relative to South America, we expect that the behaviour of North America relative to Eurasia is similar to South America relative to Africa, otherwise the Mid-Atlantic ridge would be completely reorganised. This is a clear example of coupling of plate motion. The implication is that the Mid-Atlantic ridge (MAR) moved westward relative to the hotspot reference frame; this movement is well marked in all Atlantic hotspots, including the Azores, and now lies some 200 km east of the MAR (Silver et al. 1998). In fact, the slowly moving hotspot reference frame cannot be fixed to the ridge system, which is more unstable due to plate interactions.

Combining this interpretation with earlier models (Searle 1980) and more recent ones (Luis et al. 1994; Luis,1996), we can propose a general evolution for the Azores triple junction (Madeira 1998).

Around 36–30 Ma ago, the Azores hotspot began its activity, under the Mid-Atlantic ridge at that time. The triple junction migrated from the intersection of the MAR with the East Azores fracture zone to a point further north in the MAR. As the MAR moved west over the Azores hotspot, the movement between the three plates is taken by a series of dextral pure transforms from Pico to the North Azores transform. Where these intersect the hotspot area, the deformation was distributed and the fracture zones became leaky transforms.

4.5.6 Transform-Rift Relationships

The Azores tectonic regime is just an example of the multiple relationships between transform and rifts. These can be systematised (Taylor et al. 1994) according to the angle between faults and extension direction, ϕ, and the angle between the strike of the zone and the extension direction, α. Both decrease in the same way. There is a complete spectrum from oblique spreading centres with $45° < \alpha < 90°$ to extensional transform zones with $15° \leq \alpha \leq 45°$ and transform relay zones with $0° < \alpha < 15°$.

The Azores tectonic regime doesn't fit this scheme because the angle of the Terceira Rift, with the extension direction, is greater than the angle of the main fault system, with the extension direction.

The transtension regime of the Azores is quite distinct from the transtension regime of the Gulf of California, where rift and transforms are orthogonal. Oblique or orthogonal rifting are both observed and must reflect differences in boundary conditions rather than differences in rheology. In the Gulf of California we are looking at a neorift zone of young age, 4 Ma, bounded by the quasi-rigid continental lithosphere. In the Azores we are looking at a rift zone at least 30 Ma old and bounded by the already deformed oceanic lithosphere. Only the obliquity of the transform-rift relationship at the Azores confirms the soft model instead of the rigid model.

4.6 Triggering of Subduction

In the Wilson cycle an Atlantic-type ocean, surrounded by passive margins, must be converted into a Pacific-type ocean, surrounded by active margins. This requires a mechanism for converting passive margins into active margins. This is a fundamental turning point in the Wilson cycle and a difficult problem; in fact, the evidence from old deformed belts is compelling, but modern analogues are not well documented, making it difficult to propose detailed mechanisms for the triggering of subduction.

In the earlier years of the plate tectonics revolution a popular model was advanced for the initiation of subduction (Dewey 1969). The model was based on the concept of the Wilson cycle (Wilson 1966), demanding by itself a conversion of divergent Atlantic-type, passive margins into convergent Pacific-type, active margins. The proposed evolution for past mobile belts in the light of the plate tectonics paradigm, showed that former passive margins were the obvious candidates for the initiation of subduction (Ribeiro 1998b).

With the progress of plate tectonics the need for a detailed mechanism for subduction initiation increased. This mechanism should be based on quantitative modelling taking into account the rheology of the lithosphere involved in the process of subduction. Mainly using extrapolation from laboratory data for minerals, the application of this methodology to the problem of subduction initiation showed that large stresses would be needed at passive margins to trigger the process (Cloething 1982; Cloething et al. 1989), taking into account sedimentary load on thick passive margin basins. Such large stresses seem unrealistic to obtain by simple ageing of the oceanic lithosphere. It was proposed that subduction would be initiated only in weaker zones of transform faults and fracture zones, sometimes very near mature trenches (Mueller and Phillips 1991).

The previous views, based mostly on the experimental approach to the rheology of oceanic lithosphere, were challenged by other, emergent, data. Hence, conflicting views on the relative strength of the oceanic versus the continental lithosphere persist

(Stephenson and Cloething 1991) and they have immediate implications in the problem of initiation of subduction. Some recent data and approaches could eventually resolve this question. Decoupling of a previously locked passive margin, by reactivation of margin faults, could initiate subduction by increasing the deflection of the marginal basin (Erickson 1993; Erickson and Arkani-Hamed 1993). Physical modelling also gave new insights into this problem (Shemenda 1994). Compression of the lithosphere into a passive continental margin produced a buckling instability in the oceanic plate and later, localisations of deformation and finally failure; old inclined faults favourably oriented can also initiate subduction.

Further evidence for subduction initiation was obtained by physical modelling (Faccena et al. 1999). The behaviour of a passive plate margin depends on two parameters: the *Argand number* (Ar) which, as we have seen in Section 2.7, quantifies the ratio between the body force of the continent and its integrated strength and the *buoyancy number*, *F*, that quantifies the ratio between the buoyancy force of the ocean and its ductile resistance:

$$F = L^2 g\Delta_p / \mu\eta$$

where L is the thickness of oceanic lithosphere, η is its viscosity, μ is velocity of convergence and Δ_ρ is the density contrast between the oceanic lithosphere and asthenosphere.

If Ar is low and $F<1$ the ocean deforms by distributed folds; if Ar is high and $F>1$ the continent collapses toward the ocean with subduction and back-arc extension as in the Mediterranean during Neogene. With a low Ar and $F>1$ the passive margins evolve slowly towards trench nucleation; this process is favoured by slow strain rates and is controlled by viscous flow instability in the mantle; a viscosity of 10^{22} Pa s is suggested for the oceanic lithosphere from a scaling of parameters between the laboratory and nature.

Once we accept that oceanic plates can deform internally, a suite of deformation regimes can be established (Fig. 4.22). This pattern is clearer in slow spreading, Atlantic-type oceans (Ribeiro 1996). In this case, the deformation accumulates coaxially, because the Euler pole of divergent plates is stable. The cooling plates respond by homogeneous shortening at the first stage of deformation. Increasing strain leads to the development of instabilities by buckling of the entire lithosphere and diffuse distributed faulting in the schizosphere. The last stage leads to whole lithosphere failure, and in this process an Atlantic-type ocean will be converted into a Pacific-type ocean, the fundamental turning point in the Wilson cycle. It has also been shown (Martinod and Molnar 1995) that the presence of sediments helps the growth of buckling instabilities and, hence, of subduction initiation.

Therefore, to the question "What triggers subduction?", we can answer that increasing deformation in a viscoplastic oceanic lithosphere ultimate leads to its failure in the oldest and heaviest segments near passive continental margins. This process is made easier by the presence of thick wedges of sediments above the transition from continental to oceanic lithosphere. Once the process has been triggered, it is self-sustained by the feedback process of conversion into eclogite faces of the oceanic lithosphere at an incipient trench that would facilitate foundering of this oceanic lithosphere immediately adjacent to the passive margin (Fyfe and Leonardos 1977).

Any model of proposed subduction initiation should not only be tested by using the past geological record, but also by the present plate configuration. In the latter case, the monitoring of plate kinematics by geodetic, seismic and tectonic methods allows us to quantify the dynamic parameters involved in the model. Modern examples of jumps of subduction zones and closing of back-arc basins are Taiwan (Shemenda 1994) and Japan Sea (Ogawa et al. 1989). Modern examples of conversion of passive margin into active margin seem to be reduced to the West Iberian margin.

4.7 From Passive to Active Continental Margin: West Iberia

4.7.1 Introduction

The Azores-Gibraltar Plate Boundary (AGPB) connects the Azores triple junction with the continent collision zone of the West-Mediterranean, separating the Eurasian Plate in the north from the African Plate in the south (Fig. 4.2). As referred to in Section 4.5, plate movement is transtensional in the Azores, dextral along the Gloria transform, and convergent at the longitude of Gorringe Bank. To the east the convergence rate increases.

The AGPB, oriented approximately E–W, crosses Atlantic passive margins, oriented approximately N–S in Iberia and Morocco. The segment of the plate boundary with pure transform motion, the Gloria Fault, determines fairly tightly the relative motion of Africa and Eurasia (Argus et al. 1989), and is a reliable kinematic indicator. Instrumental seismicity, however, shows a distribution that is unusual for a passive margin to the west of Iberia and, to a much lesser extent, in the continental margin of Morocco. In fact, epicentres of seismic events are scattered along a N–S trend in the West Iberia margin, at high angles to the AGPB. Historical seismicity shows a similar pattern, with moderate to high magnitude events along the margin.

A synthesis of seismological and neotectonic data (Cabral 1995; Ribeiro et al. 1996), both onshore and offshore (Rodrigues et al. 1992), led us to conclude that the West Iberia margin is in a state of transition from passive to active (Ribeiro and Cabral 1987; Cabral and Ribeiro 1989; Cabral 1995; Ribeiro 1994a; Ribeiro et al. 1996), and is a possible example of subduction initiation at the former passive margin.

Iberia is a fragment of the Variscan Fold Belt surrounded by extended Meso-Cenozoic margins related to the opening of the Atlantic in the west, the Biscay-Pyrenees in the north, and the Tethys in the south. These margins were inverted mainly during upper Cretaceous to Cenozoic times to give the Pyrenees-Cantabrian Belt in the north and the Betic-Alpine Belt in the south. The interior of Iberia, the so-called Hesperian Massif, was only very mildly inverted. It acts as a more rigid block relative to extended inverted margins. The Iberia Plate was successively connected to the Eurasian and African Plates, with jumps of plate boundaries to the north or south sides of it (Srivastava et al. 1990; Roest and Srivastava 1991), but no significant inversion was supposed to have occurred along the West Iberia margin, which is classically considered a passive margin related to the opening of the Atlantic. The Lusitanian

Basin, sensu lato, was formed since the Late Triassic, related to the opening of the Atlantic (Pinheiro et al. 1996).

4.7.2
Neotectonics

If we look at Iberia from the point of view of active tectonics, we will note that it is a small but massive continental block surrounded by more mobile belts, shown by seismicity, fault activity and topography. The regional tectonic setting changed between the end of Miocene (~5 Ma) and the beginning of Quaternary (~2 Ma), as shown by uplift rates and fault slip rates inferred by the geometry and dating of some reference surfaces (Birot 1964), as follows.

A great part of Iberia is covered by conglomerates deposited by sheet-flood in pediplains and talus, the Rañas. These are due to an increase in tectonic activity and, or climatic change from humid and hot in the Pliocene to hot but drier in upper Pliocene–Quaternary. The age of the Raña deposits is determined by two facts: they are deposited on top of upper Pliocene marine sediments present in some nearshore areas and incised by Quaternary terraces along the main rivers. So their age is between 2.5 and 1.5 Ma (Villafranchian).

A large part of the stable Iberia Massif is dominated by a well-developed erosion surface, called Meseta (Little Table) by Iberian geomorphologists. It is higher, 800 m, in the northern Meseta, around the Douro Basin, than in the southern Meseta, tilted to the west from 900 to 200 m, around the Tagus Basin. It is a polygenic surface traced through the Tertiary with the last retouch before the deposition of the Rañas, so its age is also determined to be in the interval 2.5–1.5 Ma.

The beginning of the Quaternary is, in fact, marked by the incision, stronger in the north than in the south, of the Quaternary main river courses. The northern Meseta has functioned as an endorreic basin up to the deposition of the Rañas, with the Douro capturing, at the time, the endorreic system by headward erosion (Birot 1964). The southern Meseta underwent a different drainage evolution because endorreism, if it existed, should have ended much before, perhaps in the Oligocene.

The nearshore area of West Iberia is characterised by a discontinuous plateau, between 0 and 200 m and varying in width from 0 to 20 km. In some places it cuts marine sediments of the late Pliocene age and is covered by sediments, both marine and continental that abound in the interior against an old sea-cliff, sometimes coincident with the zones of tectonic movement. Taking into account sea-level curves, the age of the old coastal plain can be Early to Late Placencian, depending on the presence of one or two high sea level stands at 1.6 or 2.4 Ma to create the possible polygenic coastal plain. The plateau is incised by lower wave cut terraces in continuity with fluvial terraces, along the main rivers, which are of Quaternary age.

The vertical movements can be studied using the usual methodology: in the nearshore situations the datum is the sea-level curve for Upper Pliocene and Quaternary. A step-by-step approach is followed from the nearshore situation to the interior, using the continuity between erosion surfaces of marine and fluvial origins as we go upstream in the main valleys.

In the coastal region a maximum uplift rate of 0.1–0.2 mm/year is estimated, consisting of a mean value for the last 3.5 or 1.6 Ma respectively. Greater uplift rates occur in NE Portugal, where average rates of 0.1–0.3 mm/year in the last 3 or 2 Ma respectively were inferred. These uplift rates decrease in the NW corner of Iberia, generating the Rias of Galicia, where estuaries were invaded by Holocene transgression. Using all available information a tentative map of vertical movements in Upper Pliocene and Quaternary was prepared (Cabral 1995).

It is concluded that the estimated uplift rates in West Iberia are intermediate between typically passive continental margins and active margins.

The reference surfaces are cut by active faults, allowing its geometry and kinematics to be established.

As in most continental areas the active faults are mostly inherited, in this case, from Late Variscan brittle faults and, more rarely, major ductile shear zones of the Variscan basement or faults related to the opening of the Atlantic and Tethys Oceans, mostly in the Meso-Cenozoic basins.

The vertical separations of reference surface allow us to characterise the main component of faulting in the majority of cases, even if a minor component of oblique slip is not well constrained.

Faults with a main reverse component of movement are oriented NE–SW inland but change to a N–S orientation in the margin. The most important delimitate the central Cordillera, a pop-up structure inherited from Miocene that separates the northern Meseta and Douro Basin in NW from the southern Meseta and Tagus Basin in the SE. The Lower Tagus Fault corresponds to the inverted margin of the Lusitanian Basin. The Portal da Galega Ramp Valley limits the Variscan Iberian Massif north of Porto.

Faults with a main strike-slip component of movement occur inland, oriented NNE–SSW; they develop pull-apart basins in releasing bends and push-ups in restraining bends, showing the sinistral sense of movement.

Faults with a main component of normal movement are oriented NW–SE conjugate, and occur only nearshore.

The West Iberia margin is densely cut by active faults with variable degrees of activity ranging from very low, 0.005 mm/year, to moderately high, 0.5 mm/year (Cluff et al. 1982), but generally below 0.2 mm/year. This tectonic situation contrasts with the intraplate regime predominant in the Eurasian Plate, where slip-rates are generally lower, and approaches the situation of intraplate regime adjacent to plate boundary at the northern part of the Mediterranean or the SW United States.

4.7.3 Seismicity and Its Relationship to Neotectonics

The main features of seismicity are clearly depicted in an epicentre map (Fig. 4.2). Iberia behaves as a block with very low seismicity surrounded by more active areas. In the SW dense seismicity occurs at the eastern end of Gloria Fault and in the Gorringe Bank, with intra-oceanic subduction of Eurasia below Africa (Udias et al. 1976; Udias and Buforn 1994). To the east the seismicity becomes more diffuse in the continental domain of the collision zone. In the north there is moderate seismicity in the Pyrenees and a decrease to the west on the Cantabrian margin. To the west a zone of diffuse moderate to high seismicity occurs along the West Iberia margin.

Historical earthquakes with Richter magnitude above 6 occur on the SW coast, Algarve and along the Tagus Valley; tsunami were reported for oceanic earthquakes.

The released seismic energy in a historical period follows the same trends. Focal mechanisms of earthquakes will be considered in the section on stress fields.

The distribution of b values (Martins and Vitor 1990) shows regional differences between the West Iberia margin and Gorringe area, with b values between 0.61 and 0.66, and the area of southern Spain–northern Morocco, with b values of 0.80, which are significantly above. Another study (MOPT 1992) shows the increase in b value from the stable Hesperic Massif (b ~0.4) to the margins (b between 0.8 and 1.1). This confirms the geodynamic concept of an Iberian Plate, with a cratonised core surrounded by mobile belts.

The quality of data on focal depth of seismic events is low, due to poor coverage by the seismographic network and a lack of precise data on crustal structure. Nevertheless, the available data (Cabral 1995) shows some general tendencies, as follows: in the interior of Iberia focal depths are, at a maximum of 15–20 km. To the south in the continental collision zone, focal depth can reach 50 km; in the west, focal depth can also reach 90±10 km, even in domains with thinned continental crust, showing that activity must reach the upper mantle.

In general terms, we infer that seismicity is well correlated with neotectonics, because historical and instrumental events predominate over the most active and longer faults, if we take into account an error of approximately 10 km in epicentre determination and the geometry in depth of active faults. The main discrepancy is located in the region around Évora, with diffuse seismicity but no major active faults. This could be explained if the Vidigueira Fault has listric geometry; in that case the diffuse seismicity could be generated in a large flat on the thrust, the Vidigueira fault being the surface expression of a frontal ramp (Brum 1990). This hypothesis can be tested by reflection profiling (planned in the SW Iberia project, Europrobe; Ribeiro et al. 1996).

Some remarks must be made about seismicity pattern on the West Iberia margin. Seismicity is more dense and diffuse onshore than offshore. A band, 50 km wide, of more dense epicentres is located along the margin, on the continental side of it. Across this band we find alignments of epicentres coincident with active faults along the main submarine canyons (Nazaré, Tagus, Cabo S. Vicente). These active faults oriented WSW–ENE to E–W correspond to transform directions related to the opening of the Atlantic and the generation of the Lusitanian Basin. Focal mechanisms indicate they are dextral strike-slip faults in the present stress field, as we will see in the next section, with maximum compressive stress oriented WNW–ESE. This suggests that the band with a higher density of epicentres corresponds to a dip-slip N–S structure, parallel to the margin. As the faults with that orientation both offshore and onshore are reverse faults, dipping mostly to the E but with some conjugate dipping W, we must infer that the major N–S structure is a thrust zone uplifting the eastern continental block, relative to the western seaward marine block. Therefore, this thrust zone must be dipping to the east (see section 4.7.5).

4.7.4 Stress Trajectories

The stress field that produces tectonic structures must be considered as a dynamic system (Harper and Szymanski 1991) that evolves in space and time. To understand the present stress field we must consider that it is not static and that, most probably, a component of memory should be present, because non-dissipative systems are the exception and not the rule in nature. So we will compare two states of the system, one corresponding to a local maximum of inversion structures in Iberia due to a paroxysm of continental collision of Eurasia and Africa around 6.5 Ma ago, and the other as close as possible to the present state, because we can use information of other sources than tectonics to characterise it.

During the Miocene, frontal collision between Eurasia and Africa occurs, generating the Betics-Rif deformed belt and inventing structures in the Meso-Cenozoic cover of Iberia and reactivating structures in the Variscan basement (Ribeiro et al. 1979, 1990). The palaeostress field for the upper Miocene can be determined using macrostructures, such as transport direction in a balanced cross section of the thrust system in the Meso-Cenozoic cover of Lusitanian and Algarve Basins, but also in the Variscan basement, with mostly lateral ramps of thrust bringing the Variscan over Miocene sediments. We can also use structures at the mesoscopic scale to constrain palaeostress field trajectories (Letouzey and Tremolières 1980; Lepvrier and Mougenot 1984).

The structures of that age observed in West Iberia are all compatible with a maximum compressive stress oriented N 150±10° and horizontally.

In the Algarve Basin (Terrinha et al. 1990; Terrinha 1998), thrusts are oriented ENE–WSW to E–W, sometimes affecting the basement and predominant sense of movement to S. Dextral strike-slip on NW–SE and sinistral strike-slip on NE–SW faults completes the kinematic picture of simple inversion of the Basin, oriented approximately parallel to the Alpine Front in the Betic Cordillera.

In the Lusitanian Basin the inversion is more complex because the NNE–SSW main orientation of the Basin is highly oblique to the shortening direction. Most thrusts are oriented ENE–WSW, parallel to the Betics, and verge both to NNW and SSE. The thrusts are connected by lateral ramps: most of these are oriented NNE–SSW to N–S and show sinistral movements, and some are transpressional.

The lateral ramps result from reactivation of older extensional faults related to opening of the Basin and, therefore, to the Atlantic. The conjugate system NW–SE dextral is also present inside the pop-up between Arrábida, in the SE, and the Nazaré-Lousã Thrust to the NW. The NE–SW sinistral system rotates clockwise and the dextral NNW–SSE system rotates anticlockwise inside the pop-up (Ribeiro et al. 1992). Joints in the Miocene of the Lower Tagus Foreland Basin related to inversion of the Lusitanian Basin, are generally oriented N 175° to N 145° (Wilson et al. 1989), which is consistent with the other data.

In both basins Alpine inversion structures are formed above dêcollements in the Hetangian evaporite-clastic complex. The Variscan basement was also deformed by ENE–WSW thrust during Miocene generating the Central Cordillera pop-up. The similarity in orientation and style of the basement structures to those on the cover suggest that they also occurred by detachment, but their larger scale indicates that the detachment is deep and involves much of the crust (Ribeiro et al. 1990). The major NE–SW to NNE–SSW faults show evidence of left-lateral strike-slip movement during Miocene in the Vilariça and Verin-Régua Faults.

Present and recent stress fields are inferred for the last 2 Ma, approximately coinciding with the Quaternary. We can use information from neotectonics but also from

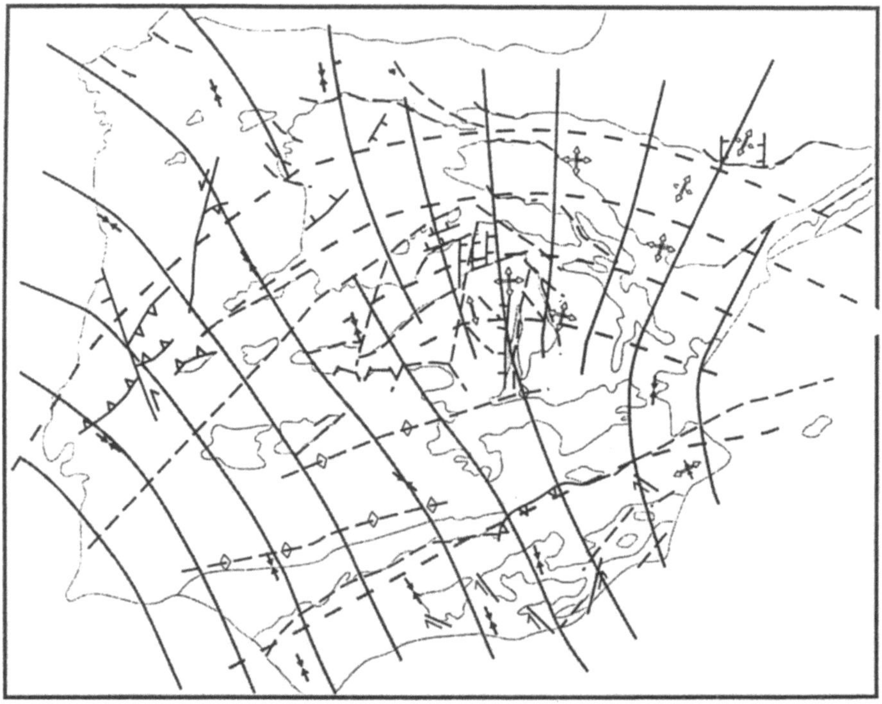

Fig. 4.10. Compilation of stress state data from Quaternary to the present day from fault-slip data, focal mechanism solutions, bore-hole breakout analysis and active faults. *Thick solid grey lines* denote reconstructed S_{hmax} direction, *thick dashed lines* denote S_{hmin} trajectories. (After Andeweg et al. 1999; with kind permission from Elsevier Science)

other, independent, sources such as in situ stress measurements and focal mechanism of earthquakes (Fig. 4.10). Hence, we can enlarge the database (Ribeiro et al. 1996) and we can also dispose of an independent check on fault slip analysis of data. All the data were treated and classified in quality ranking in accordance with the methodology of the World Stress Map (Zoback 1992).

We obtained 17 borehole breakout data, mostly of quality C; ten tectonic stress indicators resulting from inversion of fault slip data using available software (Etchecopar et al. 1981; Etchecopar 1984; Etchecopar and Mattauer 1988), the palaeofocal mechanism solution according to the methodology proposed by Zoback (1992), and fault attitude and primary sense of offset.

Most of the fault-slip data used regard recent reactivation of inherited faults affecting the Variscan basement and Meso-Cenozoic cover. However, we can also find neoformed fractures at various scales. At the mesoscopic scale we can find joints of tension origin (Engelder 1993) and systematic orientation in Upper Pliocene and Quaternary sediments at some specific locations (Setúbal and Dagorda) oriented NW–SE. At the macroscopic scale we must refer to lineaments visible in satellite imagery. In the Lisbon–Leiria area they are at least 50 km long, continuous, subparallel and crosscut active faults without displacement (Ribeiro et al. 1995; Kullberg 1996). We favour a tensile origin (Nur 1982) for these structures, which can explain the fact that they are parallel to neoformed joints. Their length suggests that they represent tension fractures at the lithospheric scale.

An investigation of published solutions of earthquake focal mechanisms and some new data using finite motion catalogue data are synthesised, with most quality B data being due to low earthquake magnitude used.

The data set comprise 32 reliable stress indicators of quality A to C showing a mean σ_{Hmax} azimuth of 135° with a standard deviation of 20°. We must notice that on average the tectonic data are rotated 20° clockwise and the focal mechanism data deviated 15° anticlockwise to the mean value of azimuth, while borehole elongation is consistent with the average trend. This discrepancy can be due to diachronism between tectonic data of Upper Pliocene to Middle Pleistocene age and the geophysical stress indicators referring to the present stress field, or to insufficient coverage of data. We favour the first interpretation because, as in other situations (Tikoff and Teyssier 1994), borehole breakouts document stresses in the cover, which behave viscoelastically, possibly memorising remnant stresses (Engelder 1993) from a previous stress regime, while the earthquake data reflect elastic behaviour of the highly viscous Variscan basement, reflecting more nearly the present stress regime.

These lines of reasoning are compatible with a progressive rotation of S_{Hmax} from NNW–SSE in Upper Pliocene to Middle Pleistocene, shown by tectonic fault slip data, to NW–SE later Upper Pleistocene data on the borehole breakout data of the cover and finally to the WNW–ESE trend in Holocene and present time, as shown by neoformed joints and lineaments and earthquake data.

The stress regime is predominantly thrust faulting in Iberia and Northern Africa and changes to strike-slip faulting westward in the oceanic domain.

If we look at the Quaternary field of the whole of Iberia (Andeweg et al. 1999; Herraiz et al. 2000), we note that there is also a deviation from NNW–SSE S_H trajectories in central Iberia to a N–S to NNE–SSW S_H trajectory in the NE border of Iberia, the Pyrenees. This is symmetrical to the deviation observed in the SW corner, at the southern end of West Iberia margin.

The stress field for Upper Miocene and Quaternary in the West Iberia margin shows different patterns that can be used as a tool for geodynamic interpretation (Figs. 4.11, 12).

If we look first at macrostructure indicators, we must differentiate two domains: the nearshore region and the interior. In the interior the same faults with approximately the same kinematics were active in Upper Miocene and Quaternary, suggesting no large deflection of stress field between these two periods. The situation nearshore is different. In fact, some of the WNW–ESE thrust systems in the Meso-Cenozoic cover of the Lusitanian basin that were active in Upper Miocene, such as the Arrábida chain, because they were nearly perpendicular to maximum shortening direction, became deactivated. Other fault systems, oriented mostly NNE–SSW, moved as left lateral strike-slip in Upper Miocene and continue to be active in Quaternary but with an increase of the thrusting component, oriented now WNW–ESE; this is particularly clear for the salt walls that limit the depocentre of the Lusitanian Basin.

Based on these indicators we suggested (Cabral 1995) a deviation of stress trajectories in Quaternary from maximum compressive stress from NNW–SSE in the inte-

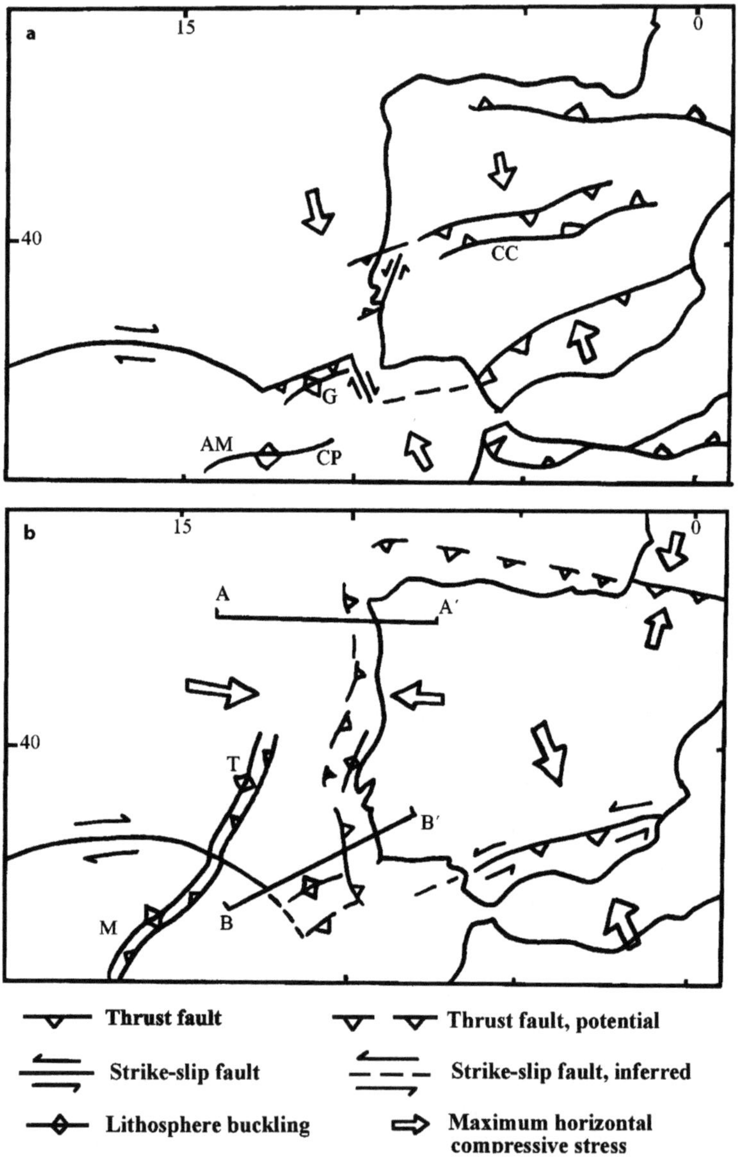

Fig. 4.11. Geodynamic of Iberia. **a** End of Miocene (6.5 Ma); **b** Quaternary (0.2 Ma). *1* Thrust fault; *2* thrust fault, potential; *3* strike-slip fault; *4* strike-slip fault, inferred; *5* lithosphere buckling; *6* maximum horizontal compressive stress. *AM* Ampère Bank; *CC* central Cordillera; *CP* Coral Patch Bank; *G* Gorringe Bank; *M-T* Madeira-Tore Rise; *A–A′*, and *B–B′* location of profiles in Fig. 4.7

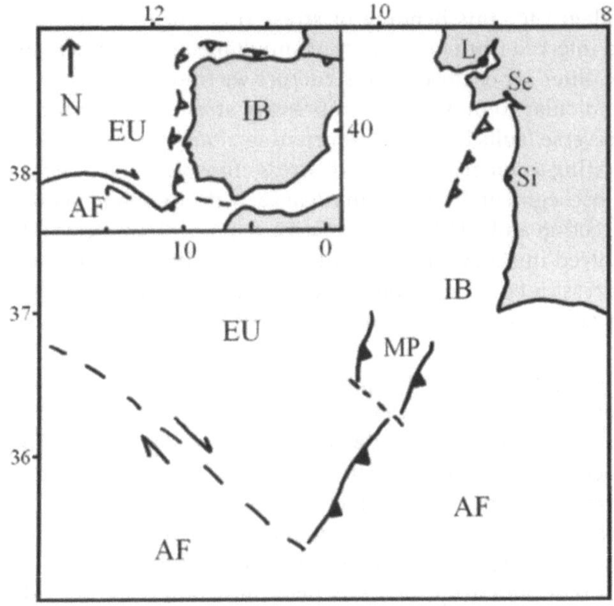

Fig. 4.12. Present active tectonics in SW Iberia. *AF* Africa Plate; *EU* Eurasia Plate; *IB* Iberia Plate. *Symbols* for active structures as in Fig. 4.11. *MP* Marquês de Pombal thrust fault that generated the 1755 earthquake; *Si/Se* thrust fault, possibly blind, that produced the 1858 Setúbal-Sines earthquake

rior to WNW-SSE in the nearshore regions in the last 2 Ma. In fact, if a deviation of stress trajectories existed in Upper Miocene, due for instance to sediment loading in the margin, it would have been parallel to the margin. In the Quaternary it is perpendicular to the margins. This would imply a change in the tectonic regime of the margin, eventually from passive (Upper Miocene) to active (Quaternary), which we will discuss further on. One could argue that the rotation in time of stress trajectories deduced for the present field would not imply a rotation in space. In fact, the fault slip data are the only data existing in the interior, because no well data and earthquake focal mechanisms are available there. Therefore, the earlier field will be preserved in the interior and the younger in the margin. We do not favour this interpretation for two reasons. First, the evidence from macrostructures shows that the reverse faults, oriented ENE–WSW, that were active in Upper Miocene and Quaternary, and the lateral ramps associated with them, imply that the maximum compressive direction is nearly stable between both periods; because of a component of strike-slip faults, as in Vilariça (Cabral 1995) the younger planes of movement show an increase in reverse fault movement, so some rotation of the stress field is required by the data. The second reason can be clearly stated as follows (Harper and Szymanski 1991, p. 19): "Understanding stress genesis in a non-linear system is almost certainly necessary, not only for extrapolation of individual stress determinations in time, but almost certainly also in space".

Further evidence for this bending of stress trajectories comes from the type of stress ellipsoid inferred from kinematics of simultaneous movement in different fault systems. In the outer arc of the curved structure we find normal faults suggesting extension perpendicular to maximum compressive stress. In the inner arc, in the contrast, we find reverse faults indicating shortening along the intermediate stress direction and suggesting a constrictive type of stress ellipsoid. This constriction increases in the inner-most segment of the arc, in Algarve, with simultaneous shortening in the ENE–WSW direction and NNW–SSE at a large geological timescale (Dias et al. 1995), eventually resolved in shorter interseismic periods with permutations of maximum compressive stress in both directions (Reches 1983), leading to a large dispersion of maximum horizontal compressive stress, observed in focal mechanisms (Dias et al. 1997).

A different interpretation for the present stress field in Iberia has been advanced by some authors (Andeweg et al. 1999). The Quaternary stress trajectories are oriented NNW–SSE in central Iberia and deviate symmetrically from that mean direction: in the West Iberia margin they deviate to a more W orientation and in the North Iberia margin they deviate to a more E orientation. This would have an effect around the corner of NW Iberia at the intersection of two passive margins at right angles: the N–S oriented West Iberia and E–W oriented North Iberia margins. It will be explained by the fact that the intraplate stress field must reflect the boundary conditions and the distribution of plate driving forces around Iberia; these will include ridge push at the Eurasia America plate boundary and continental collision and intra-oceanic convergence at the Africa Eurasia plate boundary, E and W of the meridian of the Gorringe Bank respectively; the forces induced by topography must be added. We disagree with this interpretation for the following reasons. First, the boundary between Iberia and Stable Europe is clearly active from the seismic point of view; the stress trajectories deviation reflects the fact that the boundary between the Iberia microplate and the large Eurasia plate is seismologically active. The same argument applies to the boundary between Iberia and the Atlantic domain to the west, which must be an incipient active boundary between both domains. A second reason consists of the fact that it does not explain the difference between the stress field orientation in the West Iberian margin in Miocene times, when this margin was inactive and in Quaternary to present times, when the margin is activated. A third reason derives from the fact that the potential energy provided by the differentiated morphology is dynamically supported. Therefore, topographic effects cannot be introduced ad hoc, but must themselves reflect plate boundary driving forces in a self-consistent way.

4.7.5
Deep Structure of the West Iberia Margin

The deep structure of the Hesperian Massif and its margins is well differentiated, reflecting their divergent tectonic histories. In the Hesperian massif the lithosphere is 80 km thick (Badal et al. 1993). The Moho is at approximately 27 km and the boundary between upper and lower crust is at 8–14 km (Paulssen and Visser 1993), approximately coinciding with the thickness of the schizosphere, around 10 km, from the hypocentre depth location (IGN 1992). These values are consistent with the normal continental crust of the Variscan age (Blundell et al. 1992).

To the west of the Hesperian Massif lies the Galicia Bank; its lithospheric structure can be summarised based on geophysical evidence (Boillot et al. 1992; Whitmarsh et al. 1993); a thinned continental crust, 10–3 km thick overlies a wedge of serpentinised mantle 4 km thick and thinning to 0 km towards the normal continental crust of the Iberian massif. Hence, the normal peridotite mantle was converted during continental thinning into serpentinite, probably because listric normal faulting in the crust provides an easy access to descending fluids coming from the ocean and submarine sediments.

Further to the west, the ocean-continent transition is marked by a serpentinite ridge of diapiric origin. This ridge is well defined in the Iberia Abyssal Plain but less well located in the Tagus Abyssal Plain (Pinheiro et al. 1996). If we compare the distribution of seismicity with deep structure, we note that the epicentres become denser as we cross the boundary between stretched and normal continental crust; the stretched continental crust and the oceanic crust show the same pattern of seismicity with sparse and concentrated epicentres, whereas normal continental crust shows more dense and diffuse epicentres.

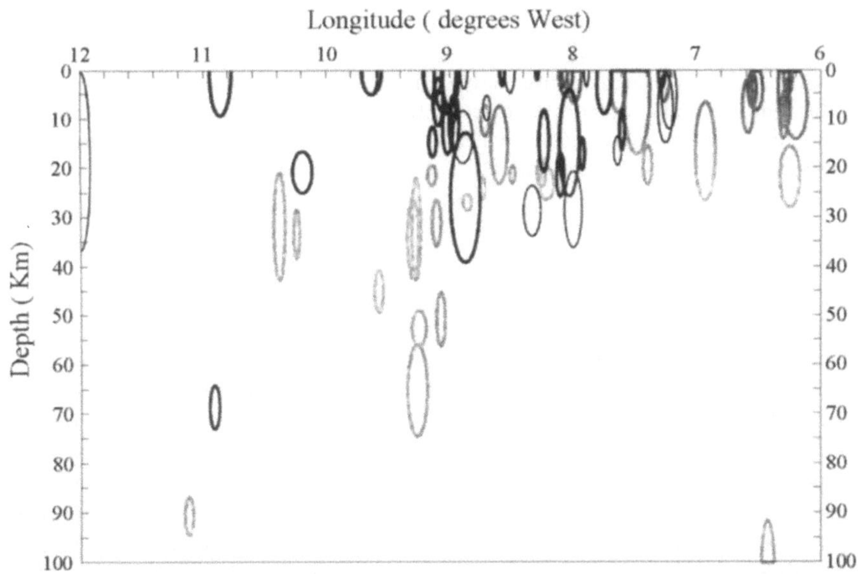

Fig. 4.13. Projection in an E–W plane of hypocentres in the West Iberia margin (area between latitude 36°50′N and 42 N and longitude 6 W and 12 W) excluding events related to the South Iberia margin with seismicity related to the Azores Gibraltar Plate boundary (after L. Matias). For each event the best location was chosen with a number of phases ≥10, with standard deviation of arrival times ≤1.5 s and horizontal and vertical errors ≤20 km and ≤40 km respectively. The "ellipse errors" represent the maximum horizontal and vertical errors of the proposed hypocentre location. Different latitude domains are represented by different shades: *light grey*, between 36°50′ and 38°30′N, *dark grey*, between 38°30′ and 39°30′N, *thin black line*, between 39°30′ and 42 N. (After Cabral 1995)

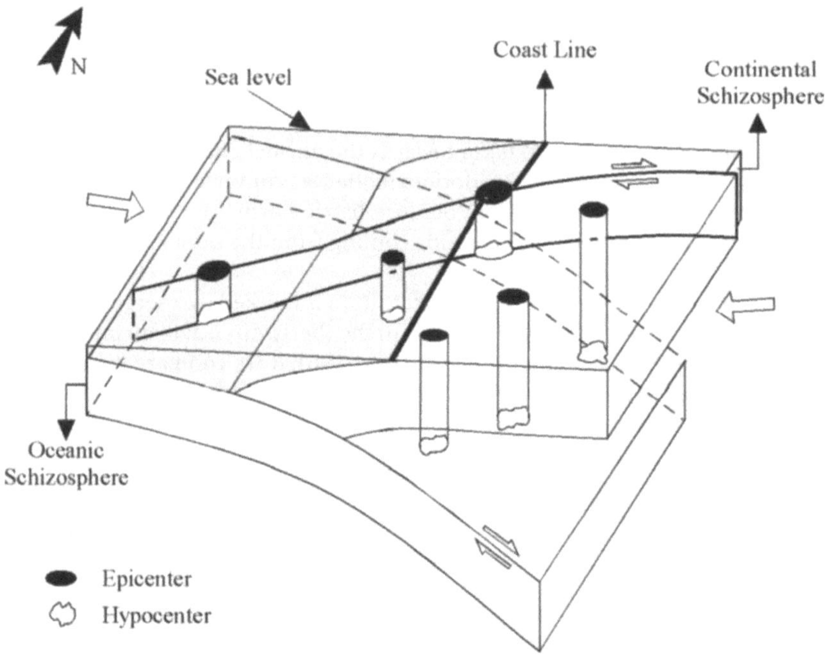

Fig. 4.14. Seismicity at a nascent active margin. The distribution of epicentre and hypocentres reflects the geometry and kinematics of the active structures; along the subduction zone the epicentres are displaced towards the continental side; above the incipient trench there are no seismic events; strike-slip active faults are marked by straight alignment of epicentres

We conclude that the correlation between active tectonics, morphology, seismicity, stress field and deep structure favours the presence of a large active thrust structure between the normal continental crust of the Hesperian Massif and the stretched continental crust of the West Iberia margin, and parallel to it. This hypothesis is confirmed by hypocentre depth along an E–W traverse; the seismicity in shallow on the Hesperian Massif (<10–15 km) but reaches 90±5 km, below the margin (Fig. 4.13). The hypocentre trend, projected in an E–W profile along the margin defines a surface dipping to the east 30±5° (Cabral 1995). This thrust must emerge in the break between the talus and the abyssal plains to the west but at this specific location there are no epicentres other than along the dextral active faults oriented WSW–ENE. This is what should be expected if we know that in the uppermost crust (0–5 km) there are no hypocentres (King and Brewer 1983), because stress corrosion forbids stress concentration in the wet cracks at this level. So the epicentres on the subvertical dextral faults correspond to the vertical projection of hypocentres below 5 km; however, the epicentres along the major E-dipping thrust plane should be displaced towards the continental side of the thrust trace, which is what really happens in the are studied (Fig. 4.14).

Many reflection seismic profiles were shot across and along the West Iberian Margin during the Iberian Atlantic Margins Project (Banda and Torne 1995). They have been interpreted in the passive margin tectonic context; although no through going thrust has been imaged, it is clear that the slope break at the base of the continental margin coincides with increasing seismicity on the landward side. This suggests the presence of an active structure, which for some reason is not in evidence on those profiles. Nevertheless, a major thrust structure (Zittelini et al. 1999) was imaged recently in the SW tip of the margin. It has a length of 60 km and it is oriented NNE–SSW. The hanging wall shows a vertical separation of 1.1 km affecting the most recent sediments. It was named *Marquês de Pombal Thrust* (Zittelini et al. 2000), in honour of the man who rebuilt Lisbon, because it is the most probable source of the destructive Lisbon earthquake and tsunami which took place in 1755, as we will show later in Section 4.7.6. Other active thrust structures were imaged in this sector, on the hanging wall of the Marquês de Pombal Thrust and in the SE limit of the Horseshoe Abyssal Plain (Hayward et al. 1999). These data show that a major crustal feature corresponds to the boundary between the oceanic and continental domains, truncating the Goringe Bank on its eastern end; this structure is a transform that accommodates differential shortening in oceanic and continental domains in late Miocene (Tortella et al. 1997) but becomes part of an active thrust zone to W since then.

4.7.6 Seismotectonic Synthesis of the West Iberia Margin and Initiation of Subduction

All the characteristics of the present regime in the West Iberia margin converge on the conclusion that this margin is in a transition stage from passive to active. Rates of neotectonic activity and rates of seismic activity derived from instrumental and historical seismicity; rates of uplift; convergent stress regime and bending of stress trajectories near the margin; rapid spatial and temporal changes in the stress regime and heterogeneity of stress ellipsoid; all these indicators referred to previously are compatible with a dynamic situation that evolves rapidly at a geological time scale and are incompatible with the stability of a typical passive margin.

Different stages in the transition from passive to active should occur along the margin. We infer that a whole failure has occurred already in the southern sector south of Estremadura Spur; buckling and simultaneous crustal faulting are in process in the central sector between the latitude of Setúbal and Nazaré, and crustal faulting without buckling characterises the northern sector. This variation in space suggests that an incipient subduction zone is propagating from S to N, probably nucleating in the intra-oceanic subduction zone of Gorring Bank, a segment of the Azores–Gibraltar Plate Boundary.

In fact, the three segments are characterised by slightly different tectonic styles and regimes with the central segment being an intermediate between two extreme regimes in S and N (Fig. 4.15).

The southern segment shows minor uplift of the southern Meseta and moderate to large magnitude historical earthquakes concentrated near the continent-ocean boundary. The depth of the hypocentres is compatible with whole failure of the lithosphere along a gentle dipping, incipient subduction zone. Stress will be released in the continental hanging wall and slow convergence will produce only minor uplift.

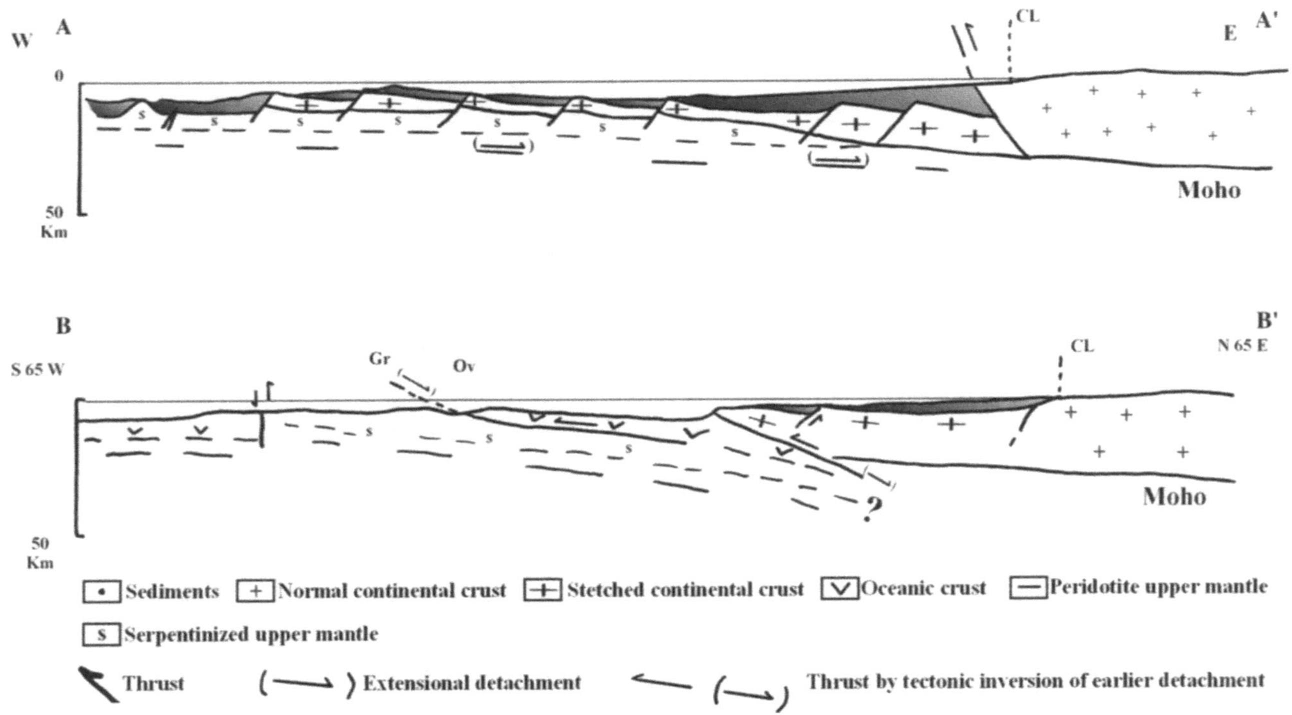

Fig. 4.15. Deep profiles across the West Iberia Margin. Location of profiles in Fig. 4.7. ***A*** Profile across the Galicia Bank; ***B*** profile across the Gorringe Bank. *Cl* Coast line; *Ge* Gettysburgh; *Or* Ormonde

The central segment covers the Estremadura Spur and adjacent onshore domain, with diffuse seismotectonic activity. Onshore we attribute the >50-km-long lineaments, oriented parallel to maximum compressive stress deduced from focal mechanism of earthquakes and neotectonics to lithosphere buckling, with the axis perpendicular to the lineament trend and wavelength in the order of 100 km. The SE limit of the lineaments coincides with the Lower Tagus Fault Zone, showing that it is a major reverse fault in the SE flank of the buckled lithosphere.

The northern sector shows the largest uplift rates in the Quaternary; diffuse seismicity to the E of the boundary between the normal and extended continental crust of the Galicia Bank, with active ramp valley formation at the Pontal da Galega. The convergence across the margin produces the uplift by compression of the continental lithosphere with a very large wavelength (>500 km) through all northern Meseta.

The geodynamic situation that we propose for the Quaternary in the West Iberia margin should be put in the larger context of the plate evolution of Africa and Eurasia in the Late Cenozoic.

In the Late Miocene continental collision of Eurasia with Africa created the Betic-Rif Chain and intraplate pop-up structures (Atlas in Africa and Cordillera Central in Iberia) oriented ENE–WSW, subperpendicular to the plate convergence vector. The same amount of shortening should be produced in the oceanic lithosphere because no transform motion is detected in most of the western passive margins of Iberia and North Africa, and stress trajectories in end-Miocene are not significantly deviated across the West Iberia margin. The shortening in the oceanic lithosphere proceeded by buckling, generating seamounts (Gorringe, Ampere) above intervening abyssal plains (Tagus, Horseshoe), with the same axial trend, ENE–WSW, as the continental structures and only cut by the Mazquês de Pombol Transform.

Recent convergence of Eurasia and Africa continues at a rate of 4 mm/year, according to NUVEL-1. The seismic activity around Iberia suggests that Iberia is an independent microplate with a quasi-rigid core and mobile boundaries. A rough estimation might suggest that the convergence across the West Iberia margin is in the order of 1 mm/year, approximately one quarter of the convergence of Eurasia and Africa. The kinematics of the Iberia microplate must explain the increasing convergence across the West Iberia margin from north to south. Two solutions are, a priori, possible.

The first solution assumes an Euler pole of rotation of Iberia, relative to stable Eurasia, located north of Iberia, in order to explain the maximum of angular displacement in SW Iberia. In this case, Iberia is laterally expulsed to the west because it is caught between two larger plates, Eurasia and Africa.

The second solution assumes an Euler pole inside the Iberian microplate itself; Iberia will spin in a dextral sense induced by the dextral sense of rotation between Eurasia and Africa, the larger plates.

Both solutions can be discriminated using geodetic data. The existing data (Argus and Gordon 1996; Larson et al. 1997; Eloségui et al. 1999) show rates of movement of stations inside Iberia, relative to stable Eurasia, below 1 mm/year. Furthermore, no clean pattern of rotation can be discerned; longer time series are needed to solve this problem. Meanwhile, we can look at indirect evidence.

We favour the spinning of Iberia versus lateral expulsion based on stress trajectories inside Iberia (Andeweg et al. 1999), referred to above. On the northern boundary, across the Cantabrian–Pyrenean range the stress trajectories change from NW-SE in the west to NE–SW in the east. This is compatible with the dextral rotation of Iberia around a pole located somewhere in North Iberia. Dextral expulsion to the west would require a constant sense of dextral transpression in the Cantabrian–Pyrenean range. Moreover, we notice that (Andeweg et al. 1999) the deflection of stress trajectories is less curved in the southern boundary of the Iberian plate than in the northern boundary. We explain this effect by assuming that the amount of the rigid rotation component of deformation is larger in the more viscous northern boundary than in the less viscous southern boundary with more intense and younger heating effect of the Betic orogeny.

The solution of a spinning microplate would perhaps explain the vertical movements of Iberia. The vertical uplift is higher in the northern Meseta, in the order of 0.1–0.2 mm/year than in southern Meseta, which has less than half this value (Cabral 1995). This is not compatible with vertical movement induced only by collision of Eurasia and Africa in the southern boundary of Iberia. The southern Meseta is exorrheic in the Cenozoic and tilted to the west, probably by failure of the lithosphere along the SW Iberia margin. In contrast, the northern Meseta changed from endorrheic to exorrheic only 2 Ma ago, due to rapid uplift of the previous endorrheic Douro Basin (Birot 1964).

The model of a spinning plate will explain the change of stress state in the West Iberia margin, from NNW–SSE compression in Late Miocene to the present WNW–ESE compression. The anisotropy of the previous buckled oceanic lithosphere will resist further compression because shortening is at a small angle to the axial direction of the buckles. The present stress field will mainly produce the N–S active structures that we see in the margin and that we described in the previous section. To the west of the previous buckles that generated the seamounts, further compression will force the oceanic lithosphere to respond by the NNE–SSW present buckling in the Tore-Madeira Rise, accentuating a previous heterogeneity inherited from the accretion stage (Pinheiro et al. 1996; see Sect. 5.3).

The present plate tectonics regime for Iberia suggests that progressive independent movement of the Iberia Plate from Eurasia, relative to the adjacent North Atlantic, will impose increasing convergence in SW Iberia, with minor dextral rotation of Iberia around a pole in the north. As an effect, extension will be applied in NE Iberia in NW–SE direction. This would explain the opening of the Baleares Basin.

The story of Continental Drift, seafloor spreading and plate tectonics shows that we can prove that a phenomenon is occurring, even if we cannot propose an underlying mechanism with reasonable confidence. Therefore, we will propose a mechanism for subduction initiation at the West Iberia margin, knowing, a priori, that the evidence for subduction initiation in this domain is much stronger than the evidence that the proposed mechanism is operating (Ribeiro et al. 1992).

The present reactivation of the continental margin of West Iberia must be controlled by the previous extensional history in Mesozoic and by the derived crustal structure. A conceptual model has been developed recently (Whithmarsh et al. 1993; Pinheiro et al. 1996), locating the ocean-continental transition off West Iberia and assuming a lithospheric detachment dipping to the E as the mechanism that controls the extension of the margin during the Mesozoic. The ocean-continent transition is located nearer to the coastline as we move from north to south: ≈200 km W at Galicia

Bank; ≈150 km W at the Iberia and Tagus Abyssal Plains so the width of the N–S band with thinned continental crust increases from south to north.

If we try to correlate crustal structure and seismicity, we can infer that the epicentre density is low in the oceanic crust and thinned continental crust but increases markedly as we cross the boundary to the normal continental crust, to the east, where the main zone of movement should be located. At the latitude of the Gorringe Bank the oceanic crust transition is even nearer, ≈50 km, to the proposed trace of the normal continental crust.

This pattern cannot be explained without some amount of convergence in the margin. In fact, if it was in isostatic equilibrium, the stretching that caused the thinning of the continental crust should increase to the north, because the width of thinned continental crust increases in that direction if this structure has not been modified since the stretching stage. However, the topography is higher in the north, over the Galicia Bank than in the south, over Tagus Abyssal Plain. We infer that the oceanic domain is not in equilibrium, and it is being increasingly pushed down from the north to the south. We estimate that a band at least 100 km wide of thinned continental lithosphere is missing in the southern sector of the margin simply because it was already subducted in the incipient subduction zone.

We can explain the reactivation of the margin in compression if we reverse in Quaternary the sense of movement in the lithospheric extensional detachment dipping to the east. This is the essence of the mechanism for subduction. We must refer to the fact that the concept of an E dipping detachment explains well the structure of Gorringe Bank (Auzende et al. 1978) with an extensional shear zone dipping E between the upper mantle exposed in Gettysburg and the oceanic crust section exposed in Ormonde (Girardeau et al. 1998). In addition, the proposed detachment coincides with a very important terrane boundary in the Variscan basement (Ribeiro and Silva 1997), a probable suture between the Iberian Variscides and the Late Precambrian Avalonia.

The model presented here is supported by preliminary models of lithosphere strength in inverted passive margins produced by simple shear mechanisms (Ziegler et al. 1998). Lower plate margins are stronger than upper plate margins because they preserve old continental mantle lithosphere instead of accreted asthenosphere below thinned old mantle lithosphere. So, subduction zones may initiate near the boundary between the old and young mantle lithospheres of the upper plate with only minor compression on the conjugate lower plate margin.

4.7.7
Rates of Tectonic Activity at the West Iberia Margin and Implications for Seismogenic Potential

We must think about the implications of the model of subduction initiation in SW-Iberia margin, both at the regional and global level.

We see in Iberia that a small plate is in the course of individualisation; it shows a hard core surrounded by more active areas. The active deformation of Iberia proceeds on two scales.

At the largest scale we have positive undulation or anteclises with a half-wavelength of approximately 500 km and amplitudes of hundred of metres. This amplitude is larger in northern Meseta, 600–800 m and lower in southern Meseta, 100–700 m. The Bouger anomaly is negative and stronger in the northern Meseta than in the southern Meseta (IGN 1992). This suggests thickening (inverse boudinage) of the entire lithosphere, 80 km thick (Badal et al. 1993).

Superposed on the first order deflection we have second order deflections associated with active faults, causing tilting and minor uplifting. These deflections have half-wavelengths of 20±10 km and amplitudes of 100–200 m. This suggests buckling of the upper crust with dêcollement at approximately 10±2 km on the basis of balanced profiles. This agrees with the thickness of the upper crust of 8–14 km (Paulssen and Visser 1993) and the thickness of the schizosphere of 10 km as mentioned in Section 4.7.5.

Rheological stratification of lithosphere and crust (Martinod and Davy 1992; Jin et al. 1994) can explain the hierarchy of two orders just described for the active deformation and topography (Ribeiro and Cabral 1997).

These scale considerations confirm the view that Iberia is behaving coherently, as a plate in the course of individualisation.

An obvious question is: how much time will it take to change from the previous geodynamic setting of the end of the Miocene to the new geodynamic setting of present time? The propagation rate of the new plate boundaries will, of course, depend on their slip rates (see Sect. 3.7). In this context we must first refer to the source mechanism of the 1755 Lisbon earthquake.

In this regard we must refer to the source mechanism of large historical earthquakes in the SW Iberia margin. If at least two earthquakes occur at the same segment of the plate interface, we can infer slip rate across it.

The most destructive historical events are the Lisbon earthquake of 1 November 1775 and the Setúbal event of 11 November 1858.

The Lisbon earthquake is one of the largest historical earthquakes ever described. The Richter magnitude, M_L, is estimated to have been between 8.5 and 9 from isoseismal evidence and tsunami magnitude. The source zone was confirmed by recently modelled tsunami data (Baptista et al. 1998a,b). The source parameters require a thick schizosphere with 80 km, a length of ~200 km and a maximum displacement of 10–20 km. Using empirical seismotectonic data (Ribeiro 1994a; Wells and Coppersmith 1994; Zittelini et al. 1999), these parameters require a complex thrust source on the Marquês de Pombal structure but also along a large thrust structure to the SW, where the Horseshoe Abyssal Plain is underthrust below Africa. The location of the source is determined by tsunami modelling. These data suggest that the Lisbon earthquake involved three plates with the subduction of Eurasia below Iberia and below Africa. In this model the 1775 earthquake would be the first example of a seismic event in the interface between three plates. Usually, major interplate earthquakes involve only an interface between two plates.

The Setúbal event has a moment magnitude of 7.1 (Johnston 1989; Johnston and Kanter 1990). This event is included in the most important earthquakes at passive margins in the last three centuries. Its source is probably a blind thrust to WNW located off-shore Sines-Setúbal, to the north of the source of the 1755 event. So, it did not occur on the same segment of the incipient subduction zone of the SW Iberia margin and cannot be used to estimate the slip rate across it. It, nevertheless, demonstrates that the subduction is propagating northwards of the triple junction Africa–Iberia-Eurasia at a fast rate.

The slip rate for the West Iberian margin must decrease quite rapidly to the north, because the seismic activity decreases sharply to the north. The Iberia Plate, in the course of individualisation, rotates clockwise between Africa and Eurasia.

Given the uncertainties in the geometry of the incipient subduction zone SW of Cabo S. Vicente, the vertical separation across this transform could have been achieved in less than 3.2 Ma if the present rate of activity is constant.

We arrive at the conclusion that the Iberia plate will rapidly become individualised. It will act as a weak zone from where the subduction could propagate both northwards, inside the Eurasian Plate and southwards, inside the African Plate. Hence, in perhaps 100±50 Ma, the Atlantic will close again and the Atlantic Wilson cycle will be finished.

4.8 Obduction

In some orogens at some point of the Wilson cycle, the oceanic lithosphere is carried over the continental lithosphere in a process called **obduction** (Coleman 1971)

This is a tectonic anomaly, because usually the dense oceanic lithosphere is subducted below the buoyant continental lithosphere. Many models have been presented to explain this anomalous behaviour; only two are mechanically plausible (Moores and Twiss 1995) and have modern analogues (Fig. 4.16).

In *obduction synthetic to subduction* a passive continental margin or an arc collides with a subduction zone, stops it and rises isostatically after this event. This is the most frequent case and has many modern analogues in the SW Pacific and eastern Mediterranean (Mattauer et al. 1980). Some authors (Moores and Twiss 1995) restrict the term obduction to the next model, and consider the previous model ophiolite emplacement by collision.

In *obduction antithetic to subduction* an incipient intra-oceanic thrust forms in the subduction plate with movement opposite to the subduction zone. This thrust overrides the subduction zone and emplaces an ophiolite on top of the upper plate. Later on, the subduction zone decapitates the antithetic thrust system and subduction resumes. There are rare modern analogues the Pliocene Taitao ophiolite in Southern Chile (Forsythe et al. 1986) and the early tertiary Resumation Bay ophiolite in Alaska and, possibly, the Northland ophiolite of New Zealand (Malpas et al. 1992).

In antithetic obduction we have generation of a flake and a bifurcation at some stage of the process: the upper oceanic lithosphere is obducted at the same time that the lower oceanic lithosphere is subducted. For some authors, this process is impossible, because the continental margin is softer than the adjacent ocean, according to rigid plate theory. In the soft plate theory, it can be either softer or stronger, according to geodynamic setting.

In both models of obduction, we can have *hot obduction* if the obducted lithosphere is young and *cold obduction* if the lithosphere is old. Hot obduction is characterised by a low-pressure metamorphic thermal sole. Cold obduction is characterised by a footwall thrust with high-pressure metamorphism (Nicolas 1989).

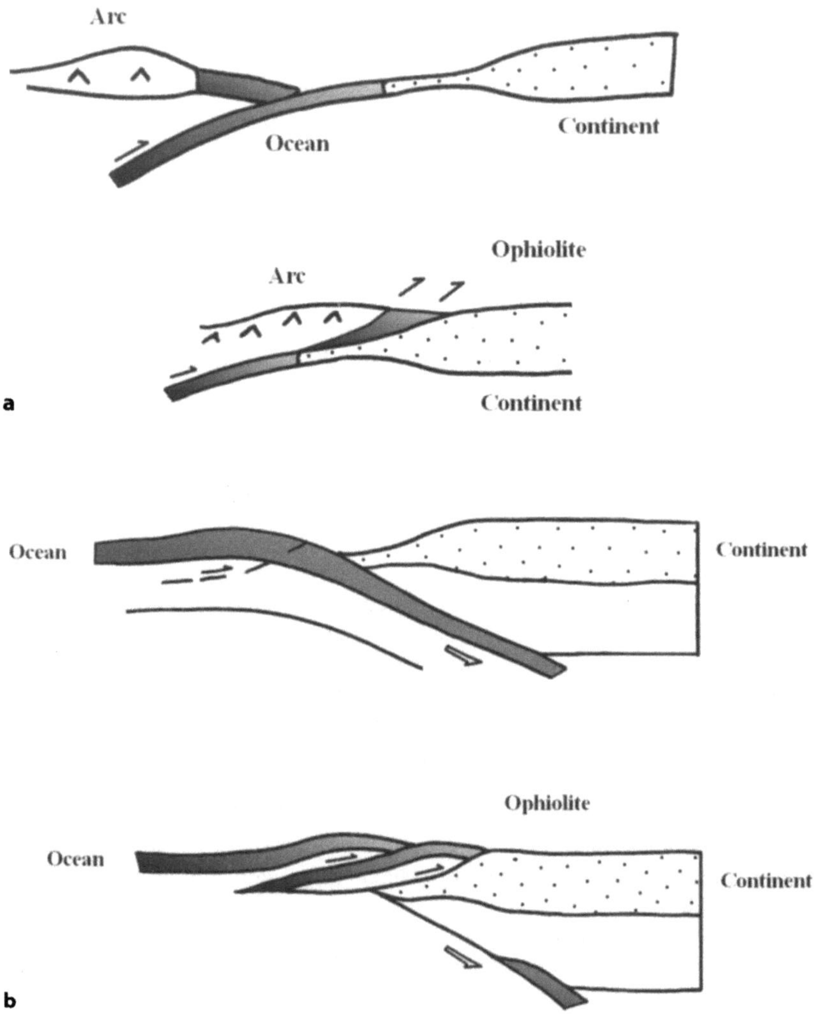

Fig. 4.16. **a** Obduction synthetic to subduction. A passive margin is changed to a subduction zone; when they collide, an ophiolite is obducted. **b** Obduction antithetic to subduction. An intraoceanic thrust is generated, antithetic to a subduction zone; continued convergence carries flakes of oceanic lithosphere over continent

It has been proved (Moores 1982) that major periods of ophiolite emplacement coincide with major discontinuities in sea-floor spreading. We think that soft plate tectonics can explain this correlation.

If there is a discontinuity in spreading such that the spreading rate varies in a period shorter than 10–20 Ma, the Maxwell relaxation time for the oceanic lithosphere, the elastic component of mechanical response dominates; stress is transmitted through both continental lithosphere and oceanic lithosphere with no dissipation and minor dissipation respectively. After some time, the stress concentration will be transmitted through the divergent plates and will affect adjacent plates. The weakest zones will be ridges of young oceans because they are hot and free of strain-hardening. A subduction zone will form in the young oceanic lithosphere and will produce synthetic hot obduction in one of the passive continental margins of the young ocean. This could explain obduction of the Oman ophiolite, perhaps the best example in the world.

In fact, many geodynamic models have been proposed for obduction of the Oman Ophiolite, based on tectonics, petrology, geochemistry and geochronology (for a revision, see Michard et al. 1989). However, the most plausible, because it is better constrained by all the field data (Nicolas 1989), can be synthesised as follows (Fig. 4.17).

Between Permian and Early Cretaceous, Arabia was a promontory of the passive margin of Gondwana, adjacent to the Neo-Tethys spreading ocean (Sengör 1979). Another passive margin is located to NE, in present geographic coordinates, attached to the Cimmerian continent; further to the NE the Paleotethys Ocean is closing by subduction to the SW, between Laurasia and this Cimmerian continent.

In upper Cretaceous there was a rearrangement of plate movements near Arabia because the South Atlantic began to open; the sense of motion of Arabia, attached to Africa because the Red Sea is younger, must have reversed with respect to the spreading axis of Neo-Tethys (Glennie et al. 1990). This reversal induced the closure of Neo-Tethys. The most likely place to establish the new plate boundary is at the mid oceanic ridge of Neo-Tethys itself. This is proved by some field data: the nappe of the Oman ophiolite thickens to the NE, along a very flat thrust; the shear zone formed during the detachment shows that it is a case of hot obduction, with ocean crust no older than 1–2 Ma. The radiometric ages of plagiogranites in the crustal section and the amphibolites of the thermal aureole overlap completely, showing the ridge was still active (Nicolas 1989).

Once the detachment is complete oceanic thrusting brings the young oceanic crust on top of the cold Arabian platform. Imbrication below the main nappe creates the thrust stack with all the members of the former passive margin of Arabia in tectonic inverted order promoting high-pressure metamorphism. The last phase of obduction probably includes elements of gravity gliding of the unstable thrust sequence, composed of heavy rocks thrust on top of light rocks (Shackleton 1987). The obduction is complete in late Campanian. A new subduction zone dipping NE nucleates at the base of the Makran accretion prism and magmatic arc, still active today.

We arrive at the conclusion that the hot obduction of the Oman ophiolite was, in geological terms, a rapid response of 20 Ma to the rapid change of plate velocity vectors that operate as boundary conditions to this process. This is a test of the value of 10–20 Ma for the Maxwell relation time for the oceanic lithosphere that we proposed, on the basis of the soft plate model. Thermomechanical modelling suggests that shear stresses in the order of 100 MPa were attained during thrusting of the ophiolite (Hacker 1991). This is probably an overestimation because the rheology parameters used in the study do not take hydrothermal seafloor metamorphism in the oceanic lithosphere into account.

If spreading is steady-state, with minor rate variations in periods longer than the Maxwell relaxation time for the oceanic lithosphere, there is time to dissipate slow

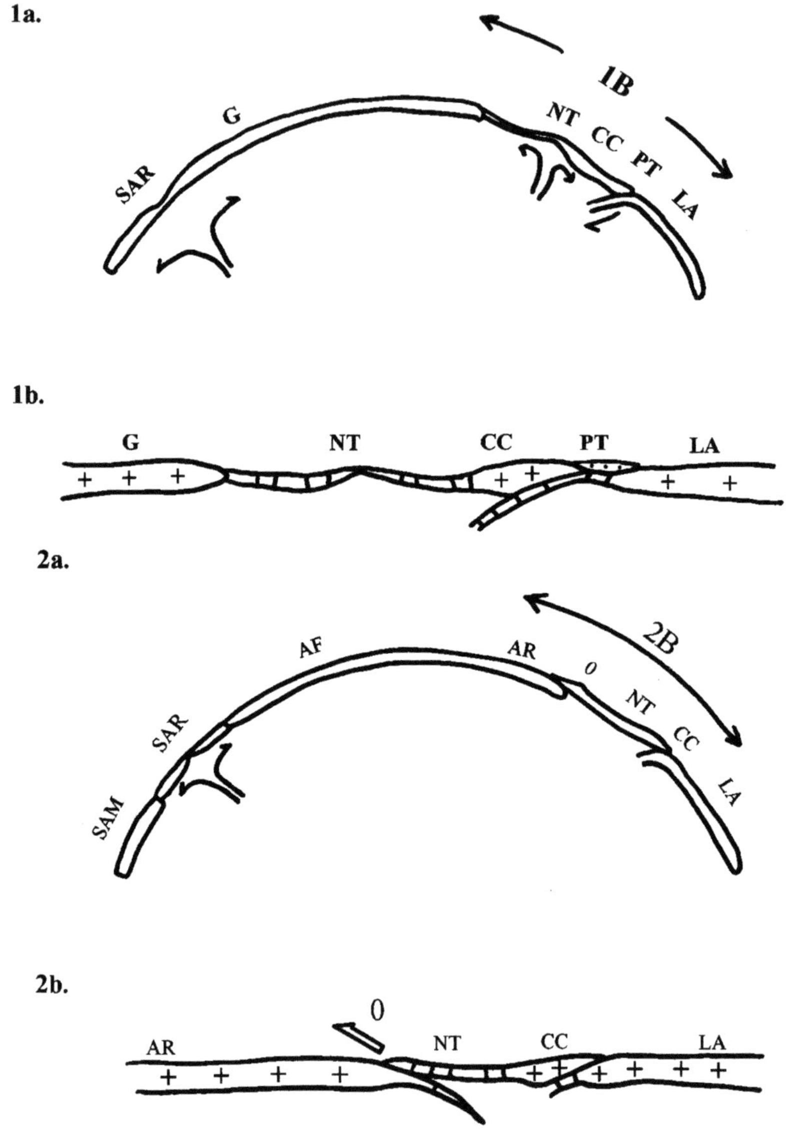

Fig. 4.17. Soft plate model for hot obduction as exemplified by the Oman ophiolite. **1a** Plate Tectonics situation at 120–98 Ma, Aptian–Albian time. The South Atlantic rift is in the infracontinental stage. **1b** Detail of the situation on the domain including Neo-Tethys, Cimmerian Continent and Paleotethys. **2a** Plate Tectonics situation at 85–75 Ma, Campanian time. The South Atlantic Ocean starts its spreading, splitting Gondwana in South America and Africa which is pushed away from the South Atlantic mid-ocean ridge. **2b** Detail of the situation as in **1b**. Paleotethys is closed, the Cimerian Continent is accreted to Laurasia and Neo-Tethys is obducted onto the Arabian Platform (based on Sengör 1979; Glennie et al. 1990). *AF* Africa; *AR* Arabia; *CC* Cimmerian Continent; *G* Gondwana; *LA* Laurasia; *SAM* South America; *SAR* South-Atlantic Rift; *O* Oman ophiolite; *NT* Neo-Tethys; *PT* Paleo-Tethys; *crosses* continental lithosphere; *vertical bars* oceanic lithosphere

stress variations through dominant viscous behaviour in oceanic segments of the plates; there is no cause for the formation of new plate boundaries.

Therefore soft plate tectonics can explain synthetic obduction better than standard theory, otherwise the correlation of obduction with discontinuities in seafloor spreading remains mysterious.

Hot obduction antithetic to subduction is explained if a ridge arrives at a subduction zone, as in the Taitao Ophiolite. A segment of the Chile Rise, possibly limited by transforms, is detached from the rest of the rise, subducted in the Chile Trench. This is why the Taitao ophiolite is located just south of the present triple junction of Nazca, Antartica and South America. In fact, the Taitao ophiolite was obducted in the past 3–4 Ma. The subducted ridge remained active and produced volcanism with calc-alkaline tendencies in the last 3 Ma at the same region (Lagabrielle et al. 1994). The heterogeneity of the ridge can explain why a segment, possibly the more buoyant because of being hotter due to maximum magmatic activity, can escape the general fate of the ridge, which is subducted elsewhere.

Cold obduction antithetic to subduction must be exceptional. Intraoceanic thrusting can trigger it (Moores and Twiss 1995). Intraplate deformation of the oceanic lithosphere is a consequence of soft plate theory. Lithosphere failure can occur after buckling in old oceanic lithosphere. If two conjugate thrusts nucleate, one can evolve to a subduction zone and the other will be antithetic to it; consuming the intervening block can produce a transient antithetic obduction and subduction. We will describe a possible modern analogue of this process in Section 5.3.

4.9 Pacific Stage

Once subduction starts an ocean is converted from Atlantic-type to Pacific-type. In Atlantic-type oceans spreading rates are low, ridges tend to follow the centre of the ocean basins and ridges have slow type morphology and structure. In Pacific-type oceans spreading rates are high, ridges may be displaced from the centre and ridges have fast type morphology and structure (Kearey and Vine 1996).

A very distinct feature of the Pacific Ocean is its asymmetry (Uyeda 1982). On the eastern American side the subduction is of "Chilean" type, shallow dipping and characterised by high compression, a pronounced outer rise, abundant calc-alkaline volcanism in the arc, and great interplate earthquakes. On the western, Asian side, the subduction is of "Mariana" type, steeply dipping extensional with back-arc spreading, little or no outer rise, few andesitic rocks in the arc and moderate intraplate earthquakes. This asymmetry has been attributed to differential rotation between the lithosphere and mantle, due to lateral mantle viscosity variations (see Sect. 3.11; Ricard et al. 1991). So westward subduction must produce eastward migration of back-arc extension, low structural relief, subduction hinge and a narrow active thrust belt with a pronounced fore deep; eastward subduction must produce westward migration of shallow fore deep, wide active thrust belt and high structural relief. This model could explain the differences in tectonic style between thrust belts related to west-dipping subduction (West Pacific accretionary wedges, Barbados, South Sandwich Islands, Carpathians and Apennines) and E to NNE dipping subduction (Andes, Alps,

Dinamides, Tamides, Zagros, Himalayas), extrapolating the present dissymmetry of the present Pacific to previous Wilson cycles of all Meso-Cenozoic oceans (Doglioni 1993). In both cases, changes in the style of subduction will produce inversion of the back-arc basin or intraplate compression in the hinterland, presently active in the Andean orogens, for example.

It remains to be proved if the model fits pre-Mesozoic orogens with proper reconstructions using paleomagnetic data and, keeping in mind, that the pole of net rotation of the lithosphere in respect to the mantle could have been shifted a lot through time from the present position to older positions.

Many tectonic aspects of the Pacific stage of the Wilson cycle are well known such as Arc-trench tectonics (Charvet and Ogawa 1994), related Mélange tectonics, and its influence on the stratigraphic record (Pickering and Taira 1994). Some details, such as the splitting of the Kashima seamount as it enters the subduction zone (Cadet et al. 1987) or the obstruction of subduction by a large seamount, inducing traction in the lower plate (Small and Abbot 1998) as referred to in Section 3.2, are extremely well documented. Other processes, such as *tectonic erosion* of the upper plate, are feasible but difficult to prove by direct arguments.

A major factor controlling the style of subduction-related orogens is the dip of the subduction plate. A steep dip causes little deformation of the overriding plate and a gentle dip causes shortening of the overriding plate (Coward 1994). Another factor is the strength of the subducting plate and the rate of subduction roll-back compared to the rate of plate convergence (Burchfiel and Royden 1982; Coward 1994). These factors will produce changes in both the upper and lower plates, in order to reconcile the steadiness of subduction with the extreme variability of episodic tectonic regimes in the upper plate, both in the Chilean and Mariana-type subduction zones.

A concept that was born in the circum-Pacific orogens is *terrane accretion* (Coney et al. 1980; Windley 1995), although highly criticised by some authors (Sengör 1990; Sengör and Dewey 1990), mostly based in the experience of Mesogea. We consider the terrane concept useful as long as each terrane is bounded by a former plate boundary; if we go to the excess of considering each regional fault block a terrane it loses significance, as some of the "microplates" that became as large in horizontal dimensions as deep in vertical dimensions. Both "microterranes" and "microplates" simply express the internal deformation of the lithospheric plates. We must keep in mind that there is a hierarchy in kinematics; at the top is strain with large intraplate movements and strain at plate boundaries and, second in order to these, moderate to minor intraplate movements and strains.

It was realised (Ben-Avraham et al. 1981; Nur and Ben-Avraham 1982a,b) that many exotic terranes around the Pacific would have their modern analogues inside this ocean, such as oceanic plateaux related to hotspot activity, aseismic ridges, old transform faults, remnant arcs and microcontinents. When those arrive at subduction zones they will resist subduction because they are buoyant, relative to the adjacent and more "normal" oceanic lithosphere. So the exotic terrane will be accreted to the continent above the subduction zone and this subduction zone must jump seaward (Van Andel 1985). We can envisage two possibilities for the tectonic contact between the exotic terrane and the upper plate (Fig. 4.18). In some cases, the upper plate can be thrust on top of the exotic terrane, such as North America over Salinia in central California, in *synthetic terrane accretion*. In other cases, the exotic terrane can be

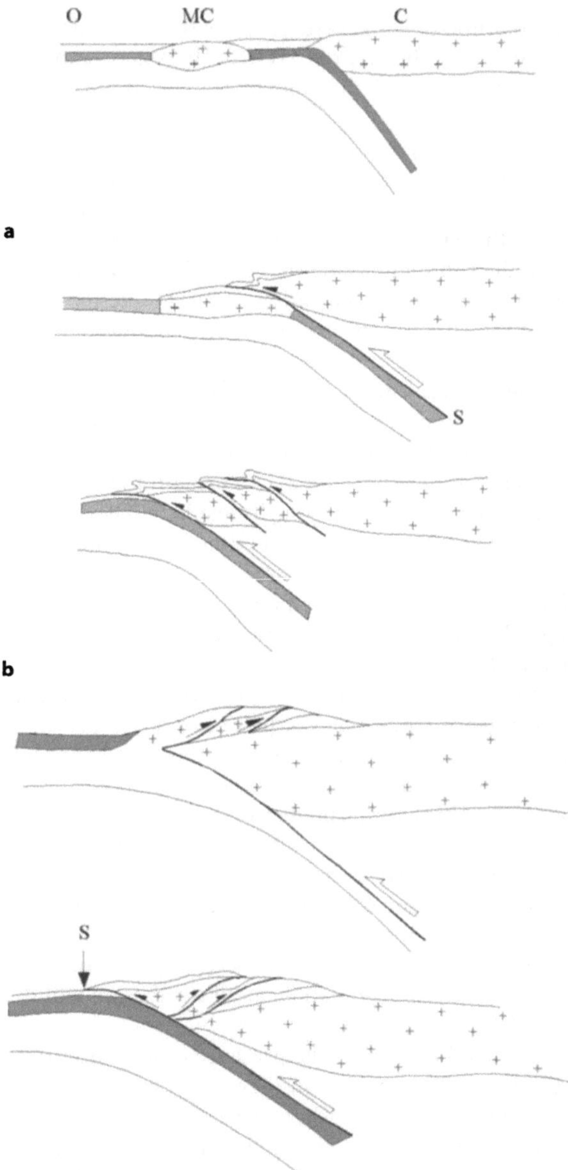

Fig. 4.18. Terrane accretion. A subduction plate contains normal a oceanic lithosphere and an anomalous "terrane" with a distinct composition and/or rheology. **a** Synthetic terrane accretion. The subduction zone jumps to a new position and the accreted terrane is underthrust below the plate above the subduction zone. **b** Antithetic terrane accretion. The accreted terrane is cleaved in a flake: the upper part of this flake is thrust on top of the plate above the subduction zone; the cover plate is subducted. *C* Continent; *MC* microcontinent; *O* ocean; *S* suture

thrust on top of the plate above the subduction zone in *antithetic terrane accretion*, with generation of a tectonic flake; this situation can probably be inherited from a previous *backthrust* and *backstop*, visible in some arcs (Charvet and Ogawa 1994).

Antithetic terrane accretion should be more frequent than synthetic terrane accretion, because the buoyancy of the exotic block must resist its dragging down the subduction zone; but the opposite is happening in the modern analogue, the Koshima seamount, so this is also a possible situation. In some cases, such as the Erastosthenes seamount microcontinent, caught between the Cyprus arc-trench system and the Levantine passive margins, the ultimate fate of the microcontinent is uncertain (Ziegler et al. 1998).

There are large variations in the tectonic style of subduction-related orogens, such as the contrasts between the Cordilleran orogens in North America and the Andean orogens in South America (Windley 1995). A common feature, however, is a gross symmetric structure with a synthetic thrust belt in the oceanic side, a middle rotation zone and an antithetic thrust belt in the continental side (Roeder 1973). This symmetry is acquired before any central collision takes place and must be subduction-related. In other words, Benioff subduction (B-subduction) or oceanic subduction characterises the oceanic side; and Ampferer subduction (A-subduction) or continental subduction characterises the continental side (Bally 1981). For some authors, continental subduction is limited to some hundreds of kilometres of underthrusting (Molnar and Gray 1979) and a detachment of brittle upper crust above ductile lower crust or the entire brittle crust (Hatcher and Hooper 1992) can explain the observed profiles. The entire orogen can be floating in the underlying lithosphere (Oldow et al. 1990). For other authors, there is no limit to continent subduction that can underthrust thousands of kilometres of continental lithosphere (Matte et al. 1997).

A modern analogue of this style will be the transverse ranges in California or "Big Bend". A balanced profile through this region shows N-directed subduction of the Pacific Plate; a mid-crustal detachment; and N-directed thrusting of ranges above the North American Plate (Namson and Davis 1988). In this case, we are dealing with an *active flake* with some 53 km convergence since 2–3 Ma; there is a difference with antithetic terrane accretion in the sense that there is no exotic terrane being accreted but it is the margin of North America itself that is flaked off the craton.

In all types of convergent orogens, resulting from subduction and eventually, collision, the amount of convergence is controlled by the velocity of convergent plates; in frontal convergence the plate boundary is nearly at right angles to the convergence vector. A more general case is oblique convergence; this introduces a strike-slip component of movement besides the dip-slip one, inducing transpression and a variety of partitioning tectonic regimes (see Sect. 2.6). If strain is partitioned in thrust and strike-slip components, both components must be compatible at all depths; we will discuss this problem again in Section 4.12.

A general feature of subduction-related orogens is the presence of arcs or curved orogenic belts, convex towards the lower plate; this is an inevitable consequence of subduction along an inclined plane (Frank 1968; Le Pichon et al. 1973) on a spherical earth. Again, this feature is previous to any collision stage. The flexible but inextensional model of Frank gives radii of curvature that are too small. However, if plates are not inextensible, but permanently deformable, the observed curvatures agree with the curved geometries of subduction zones inferred from seismicity, sometimes with great detail, such as in the Izu-Bonin Arc (Isacks and Barazangi 1977; Bevis and Isacks 1984; Fowler 1990).

We will delay further discussion on strain distribution in subduction-related curved orogenic belts until we consider collision-related arcs in the next section.

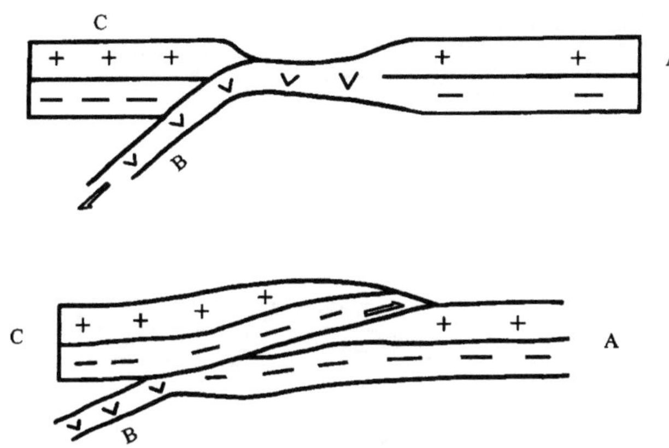

Fig. 4.19. Synthetic collision between plate *AB* (continental *A*, and oceanic *B*) and plate *C* (continental). The mountain belt is situated above *A* and the root zone is situated well below *C*

4.10 Collision Stage

If the rate of subduction on a Pacific-type ocean is greater than the rate of spreading, the result will inevitably be ocean closure and finally, a collision of the plates bordering the ocean. If the closing ocean has only one subduction zone, the passive margin opposed to the convergent boundary will be dragged towards it and we will have collision of one continent attached to passive margin and the opposing continent or arc. If the closing ocean has two subduction zones, we will get bilateral collision of two continents or two arcs. The result of the collision process will be the generation of a supercontinent and a major reconfiguration of plate boundaries. Therefore, collision will result at the end of one Wilson cycle and the beginning of another.

We can anticipate that the modalities of collision are highly variable and the tectonic style will be controlled by them. The major factor will be the degree of continental overriding (Sengör 1990) and the presence or absence of flip in the polarity of the convergent boundary (Roeder 1973). Both factors express that once they have collided, the buoyant opposing plates, either continents or arcs, can only suffer a small amount of subduction, just because they are buoyant, relative to the subducted oceanic lithosphere.

When collision begins the continent or arc above the subduction zone overrides the opposing continent. The collision can proceed in two ways.

In one modality the overriding has the same sense of the subduction zone, a process of synthetic collision (Fig. 4.19). A prism of crustal or lithospheric accretion will form by forward propagating thrusting in the lower plate below the suture, as in the Himalayas, or by backward propagating thrusting in the upper plate above the suture (Coward 1994).

In the other modality the overriding is opposed to the sense of subduction zone and a tectonic flake (Fig. 4.20) is generated by tectonic wedging or by tectonic delamination (Fig. 4.21).

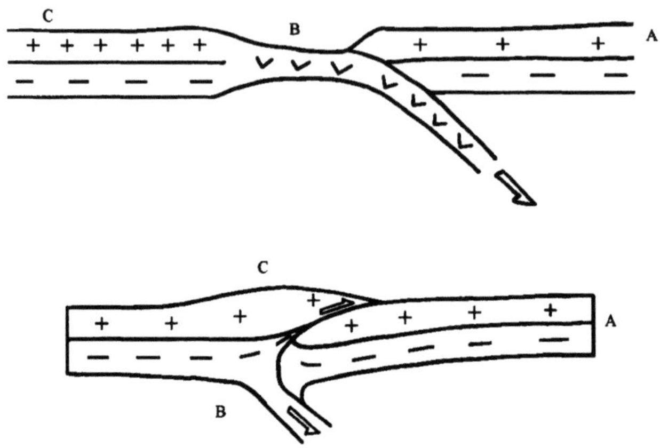

Fig. 4.20. Flake formed by antithetic collision between plate *A* (continental) and plate *BC* (oceanic *B* and continental *C*). The mountain belt is situated just above the roof zone

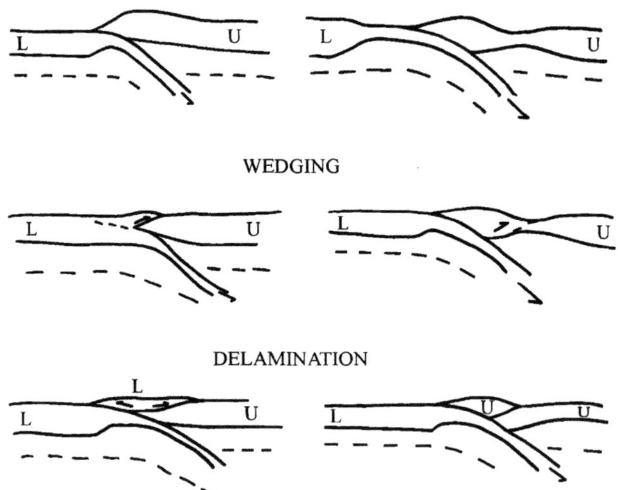

Fig. 4.21. Processes of flake generation by wedging and/or delamination of the lower *L* or upper plate *U*. *L*, Lower plate, *U*, upper plate

The concept of tectonic flake was introduced independently by Oxburgh (1972) and Laubscher (1971) to explain the following facts in the Eastern Alps.

A continental plate A, Eurasia, is overridden by an oceanic plate B, Tethys, which is itself overridden by another continental plate C, Adriatic Plate. The upper continental plate C, Adriatic, exposes only an upper crustal section, just 4–12 km thick. Why were they not brought to the surface by thrusting, deeper levels of Plate C?

One possibility is that the lower part of plate C was dragged below plate A by a subduction zone dipping on that sense. This is the flake model.

The other possibility is that the thrust plate boundary never reaches the deeper levels; but in this case the thickening caused by duplication of crust C over the crust of plate A should produce a mountain range well above plate A but a mountain root well below plate C, which is not the cause. So the first interpretation by the flake model explains the facts, which is not the case with the second interpretation.

Both hypotheses should be tested by geophysical methods (seismic, gravimetric, magnetic profiling).

A tectonic flake is an example of a bifurcation in physical space. A superficial segment of plate C goes up, thrust over plates B and A. The other deeper segment goes down under plates B and A. As long as the bifurcation plays a role, points that are adjacent inside plate C can follow a divergent path and end up either above or below plate A. This process is known in metal plasticity.

The concept of tectonic flake has been confirmed by different approaches in recent years, such as seismic profiling, numerical and analogue modelling (Burg and Ford 1997). Seismic reflection profiling of Proterozoic and Phanerozoic orogens (Beaumont and Quinlan 1994) has shown that reflectivity of deep levels can be due to shear zones, fluid-filled faults and lithological boundaries (Klemperer and BIRPS Group 1987). In many cases, the surface trace of deep reflector coincides with major thrusts, so much of reflectivity is due to tectonic processes. In particular, the flake concept was confirmed in the orogen where it was proposed for the first time, the Alps (Schmid et al. 1996). So the bilateral symmetry of many orogenic structures can be explained by flake generated at the late collision stage of a former one-sided subduction related orogen. This does not exclude that other orogens can result from collision above a closing ocean, limited on both sides by subduction zones (Roeder 1973). Both cases must be discriminated on the basis of geological and geophysical criteria (Soesoo et al. 1997), especially the peculiar magmatism induced by bilateral subduction.

Numerical modelling (Beaumont and Quinlan 1994) of compressional orogens has been conducted in two-dimensional plane-strain by finite-element methods. The rheology is Coulomb plastic with an angle of friction of $\phi=15°$ for the upper crust, to simulate Byerlee's law; below the brittle-plastic transition it is viscous with a combination of "wet" quartz and "wet" feldspar flow laws for different strain rate and temperature parameters to simulate lower crustal conditions. Flexural rigidity of 10^{22} Nm is assumed for the elastic schizosphere to include isostatic effects in evolving orogens. The mantle behaves in an essentially rigid manner; so this one end-member of lithosphere response in compression (Fig. 4.22).

The originality of the model consists in the kinematic basal boundary condition. The lithospheric mantle detaches from the crust and underthrusts the adjacent mantle lithosphere at a singularity point, *S*. The Lagrangian grid allows the specification of velocity, strain rate and finite strain for 150 km of convergence by continuum mechanical considerations that are valid everywhere except near the singularity.

Deformation migrates from the singularity upwards in the crust and spreads laterally according to the assumed rheological properties. This induces the generation of a triangular pop-up with apex in the singularity; a tectonic wedge develops above the subducting plate, the pro-wedge, equivalent to synthetic subduction (Roeder

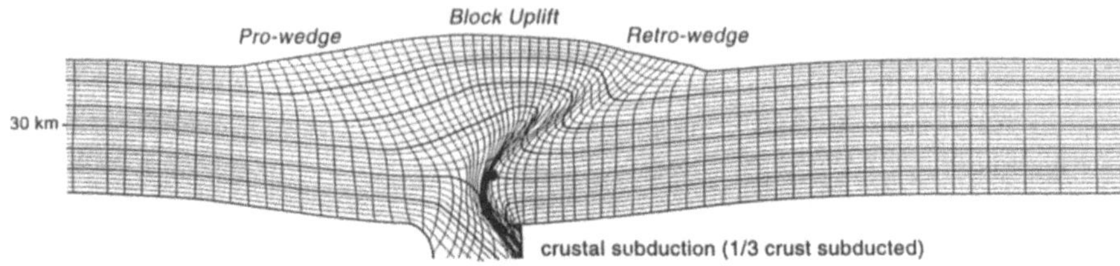

Fig. 4.22. Langrangian grid showing deformation of the crust generated by a two-dimensional plane strain finite element model of a small compressional orogen (after Beaumont and Quinlan 1994, Fig. 6, with kind permission from Blackwell Science Ltd.). This model represents a cold, single layer crust where one-third of the crust is subducted. The crust is modelled with the rheology of wet feldspar. Decoupling of the lower one-third of the left-hand crust occurs at the singularity (*black spot*). The model is shown after 2 Ma and 40 km of convergence

1973) and another tectonic wedge develops above the fixed plate, called retro-wedge, equivalent to antithetic subduction.

The style of deformation simulates qualitatively observed structures in many orogens such as: coupling between step-up thrusts and subhorizontal detachments, localisation of deformation by subduction of plastosphere at or on more levels and the role of surface denudation in controlling the tectonic style. This last particular aspect has been extensively modelled in the Southern Alps of New Zealand (Beaumont et al. 1992). Many variants of the basic model can be obtained by changing controlling factors such as strength of basal coupling between crust and mantle, the number of rheological layers, the thickness of subducted lower crust and the extent of surface denudation.

The numerical flake model (Beaumont and Quinlan 1994) has obvious limitations: it is two-dimensional and therefore cannot take into account lateral components of movements that generally exist in three-dimensional orogens; the amount of convergence is below 150 km, so it cannot be extrapolated to more advanced collisional models. However, the model explains qualitatively many aspects of collision orogens: the approximate bilateral symmetry that can include two asymmetric end members depending on the amount of subduction with predominant synthetic or antithetic wedges; in both cases the wedges develop simultaneously, and this explains simultaneous antithetic obduction/subduction or overthrusting/underthrusting in the intracontinental stage of collision. The initial model was extended to include oblique convergence of the crust driven by basal forcing (Ellis et al. 1995)

The qualitative conclusions of numerical modelling have been confirmed by analogue modelling; the fault pattern inferred by numerical experiments and analogue experiments, such as in sand-boxes (Malavielle 1984), are similar (Burg and Ford 1997); so both methods support each other in simulating reality. In fact, we can say that numerical modelling of lithosphere deformation using the thin sheet approximation, to which we made reference in Section 2.6, leads to the smoothing out of deformation into a bulk homogeneous strain of the lithosphere. This is far from the reality studied by field structural geologists, with heterogeneities at all scales. This state of affairs is recognised by field specialists when they comment on results obtained by

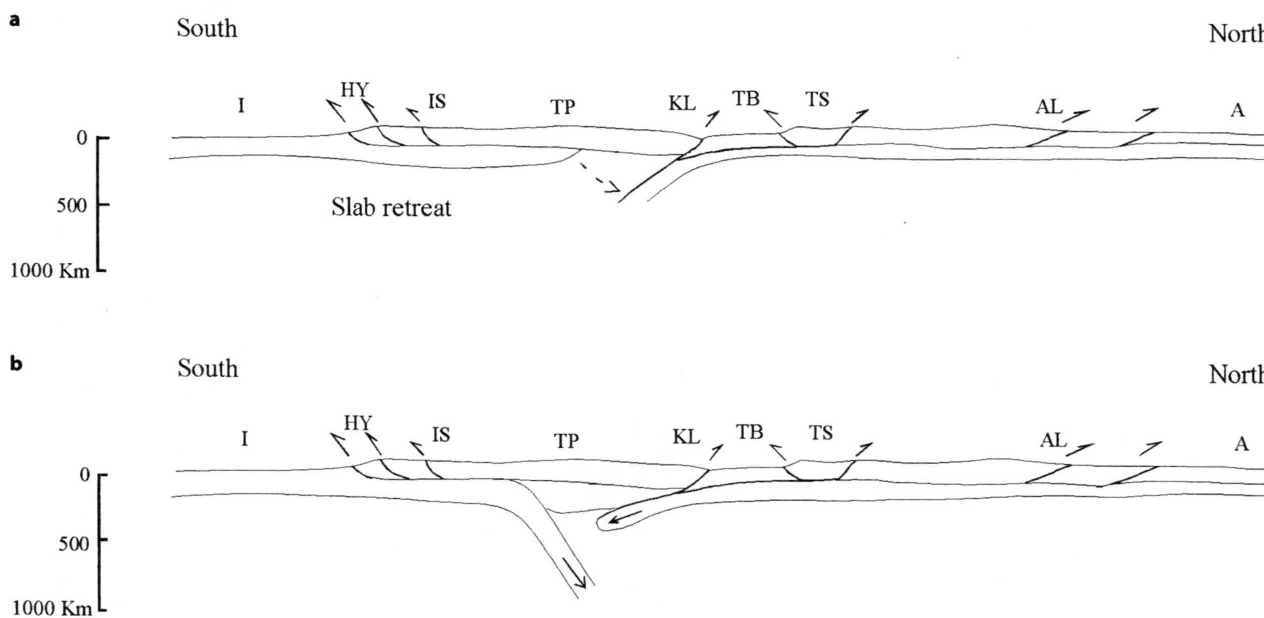

Fig. 4.23. North–South profiles of the India-Asia collision at the lithosphere scale. **a** Convective removal of thickened lithospheric mantle of Asia beneath the northern Tibetan Plateau (Willet and Beaumont 1994). **b** Continental subduction of 1000 km of India after oceanic subduction of Tethys and limited continental subduction of Asia (Matte et al. 1997; Mattauer et al. 1999). *A* Asia, *Al* Altai; *Hy* Himalaya; *I* India; *IS* Indus Suture; *Kl* Kun Lun; *TB* Tarin Basin; *TP* Tibetan Plateau; *TS* Tien Shan

theorists: "it is theoretically possible for a set of three equations to describe all the displacements throughout the European Alps, but such equations would probably contain as many terms as the Alps have cubic millimetres of actual rock" (Ramsay 1969, p. 51); "it frustrates geologists to know that their observations and measurements become local, mechanical anecdotes and noise when it comes to understanding collisional orogeny in terms of lithospheric deformation" (Burg and Ford 1997, p. 5). The state of the art can be synthesised as follows: there is a large gap between the complex reality of tectonics and the simple theoretical models of lithosphere deformation of the 1980s. However, the numerical models of the 1990s and the analogue models are narrowing this gap and there is indisputable progress in theoretical tectonics. And this progress results from inferring, from field relationships, the boundary conditions more adequate to impose in numerical and analogue experiments. This has happened in tectonics, a synthetic approach, as it happened in structural geology, an analytic approach to the same objects. "I think that the most important task of the geologist-mechanic is to recognise general and specific boundary conditions. They can only be read from field relations...the purpose of field mapping by the geologist-mechanic is to discover the boundary conditions of process" (Johnson 1970, p. 24).

Although the flake concept seems well supported in some orogens, it is still being debated for some collision orogens, what is the predominant sense of subduction, even in active orogens, such as central Asia (Fig. 4.23) and, *a fortiori*, for old orogens where kinematic reconstructions are less accurate. So is the Indian lithospheric mantle being subducted beneath Tibet, as most authors assume and then there was a flip in subduction and Asia is now being subducted beneath Tibet, at the level of the entire lithosphere (Matte et al. 1997; Mattauer et al. 1999), or only at the level of the lithospheric mantle? (Willet and Beaumont 1994). We can say "to flip-or to flake-that is the question". However, the presence of active faults with opposed senses of movement limiting active flakes, such as in transverse ranges of southern California (see Sect. 4.9), shows that these faults can develop synchronously.

Another example of active plate is the southern branch of the Banda Arc (Snyder et al. 1996). This arc is due to internal deformation of a triple junction of the convergent plates Eurasia, Pacific and Australia. Between 8.0 and 0.2 Ma, the arrival of the Australian continental margin at the subduction zone of the Java trench induced the formation of a dual, doubly vergent orogen by continent-arc collision; southward thrust was active in the Timor Trough in direct continuation of the Java Trench, and it put the Outer Band Arc on top of the Australian continental margin. Simultaneously, northward thrusting initiated below Wetar, thereby putting the inner volcanic Banda Arc on top of the Back-Arc Basin Banda Sea. This convergence between thrust boundaries with opposed vergences caused intense deformation and uplift between the pro-wedge of the Timor Trough and the retro-wedge of Wetar thrust. From 0.2 Ma up to now, seismotectonics and GPS monitoring (Genrich et al. 1996) show minor left lateral strike-slip movement between the outer Banda Arc in Timor and major N–S convergence across the Wetar thrust. Deep seismicity is located below the outer Banda Arc and is directed northward. Seismicity below the Wetar thrust is presently shallow, so the active flake will probably change in the near future to a flipped subduction zone of Banda Sea below the Banda Arc, accreted to Australia.

These observations, taken altogether, suggest that flakes evolve dynamically into other structures, in space and in time; in space singularities can evolve to and from triangle zones (Davison 1994); in time a flake can evolve either into pop-up, if the sense main thrusting is maintained or flip, reversing this sense.

So we must now ask an important question: what are the factors that control the generation of a one-side accretion prism at the crustal scale by synthetic collision or of a tectonic flake?

A mechanism of syn-collisional rock exhumation and associated normal faulting has been proposed recently (Chemenda et al. 1995) to explain subduction and rapid exhumation of continental crust during continuing plate convergence, as it occurs in the orogens resulting from continental collision such as the Himalayas (Matte et al. 1997; Mattauer et al. 1999), the Alps (Michard et al. 1993) and the Uralides (Matte et al. 1993; Chemenda et al. 1997).

The field relationships are particularly clear in the linear N–S Urals and were extrapolated to depth by deep seismic profiling (Berzin et al. 1996); the main Uralian Fault, a tectonic melange of peridotites and serpentinites, separates the Devono-Carboniferous volcanic arc of the Magnitogorsk syncline to the E, with very mild deformation and metamorphism, from the Maksutov Complex, with middle Devonian high-pressure metamorphism reaching perhaps coesite assemblages (Matte et al. 1993). The Maksutov Complex is itself thrust to the west on top of allochthonous and parautochthonous units belonging to the East European platform and the oceanic domain to the east of it, obducted into the Kraka Allochthon (Brown et al. 1995). From field relationships we need a large west verging thrust to emplace the Maksutov Complex on top of the East European Platform and thrust units derived from it and a large normal fault to the east to put the high grade rocks of the Maksutov Complex side by side with the unmetamorphosed rocks of the Magnitogorsk volcanic arc. These opposed senses of movement should occur along the Main Uralian Fault Zone, the suture of the Urals and are required by the eduction mechanism of emplacement.

Physical modelling suggests that the buoyant continental crust and erosional unloading could be the mechanisms for the exhumation of previously subducted oceanic lithosphere (Chemenda et al. 1995). However, in most of continental collision belts, we do not have any tectonometamorphic evidence for rebound of the continental lithosphere. On the other hand, the erosion is too slow to induce the rapid exhumation of high-pressure rocks required by non-re-equilibration of their mineral assemblages. In order to avoid these objections we propose an alternative model, as follows. It has been noted that: "conceptual and mechanical difficulties of subduction of substantial portions of continental lithosphere are in part overcome by the simple observation that in all cases we deal with a presumably attenuated continental crust of the lithosphere of a former passive margin" (Bally 1981, p. 18). This explains the usual limited amount of A-subduction of continental lithosphere, although this fact has been challenged by another recent interpretation that we will mention below.

The Galicia Bank provides a modern analogue of attenuated continental crust along a passive margin related to the Atlantic opening of the Mesozoic that will become involved in the convergence of Eurasia and North America, a process that has already started. We have seen in it that a thinned continental crust, 10–3 km thick, overlies a wedge of serpentinised mantle 4 km thick thinning to zero kilometres towards the normal continental crust of the Iberian massif. Hence, the normal peridotite mantle was converted during continental thinning, in serpentinite probably, because listric normal faulting in the crust above provides an easy access to descending fluids coming from the ocean and submarine sediments. These conclusions can be generalised to other crustal structures in passive margins of Atlantic-type oceans.

When this thinned continental lithosphere reaches a subduction zone in the opposite margin of a closing ocean, the constraints for the operation of eduction process are fulfilled. This happens under even more favourable conditions than in an oceanic domain, because the thinned continental crust above serpentinite is more buoyant than oceanic crust. All the segments composed by thinned continental crust and serpentinised mantle refuse to subduct. They will be detached from underlying peridotite mantle and thrust over the previous continental passive margin. They will also be detached from above the oceanic rocks pinched in the subduction zone converted into a suture by continued convergence. This model of subcontinental mantle eduction (Ribeiro and Fyfe, in press; see Sect. 2.3) obviates the objection to the subduction and exhumation of the normal continental crust, and can be applied to continental collision orogens such as the Urals, Alps and Himalayas. We can speculate further that the nature of the former passive margin can control the operation of continental eduction, a particular case of mantle eduction. Two extreme cases can be considered (Kuznir and Park 1986).

A slow rate of extension will favour a wide passive margin zone with thinned continental crust dipping away from it (Fig. 4.24a.1). When this passive margin enters a subduction zone the conditions for mantle eduction are potentially favourable and the alternative model we proposed will be entirely applicable (Fig. 4.24a.2).

A high rate of extension will cause a narrow zone of extension and abrupt contact of normal continental crust with oceanic crust (Fig. 4.24b.1). When a continent or microcontinent adjacent to a closing ocean of this type enters a subduction zone, the conditions for mantle eduction are not fulfilled because in this second case the subcontinental mantle will be composed of peridotite and not of serpentinite. The mechanical problem of subducting such a continental lithosphere will induce different dynamic processes such as flake tectonics, with emplacement of continental crust and eventually underplated material on the hanging plate above the subduction zone (Fig. 4.24b.2), as in the NW Iberia Variscides (Ribeiro et al. 1990).

The amount of continental subduction was considered to be minor in the earlier years of plate tectonics. However, the recent discovery of deep subducted slabs in the deep mantle (see Sect. 3.11) has opened up the possibility that they could be composed, not only of the oceanic, but also of the continental lithosphere attached to it. In the India-Asia collision, more than 1000 km of continental lithosphere of Great India would have been subducted below the Himalaya subduction–collision orogen, according to some authors (Mattauer et al. 1999). This interpretation is based on paleomagnetic reconstructions of the location and size of Great India. If this model is true, the problem of rapid exhumation of high-pressure metamorphic rocks becomes even more acute: why do some continental rocks quickly return to the surface after being subducted to depths of 100 km, while others in the same slab, go down at least 1000 km? We think that synthetic subduction collision can explain the deep subducted slabs while the antithetic subduction collision process will favour the resurfacing of slabs that never went below some 50 km, if we accept that tectonic overpressure can affect the pressure-temperature-time paths of metamorphic rocks (Mancktelow 1993) on which the subduction-exhumation evolution of high pressure rocks is based.

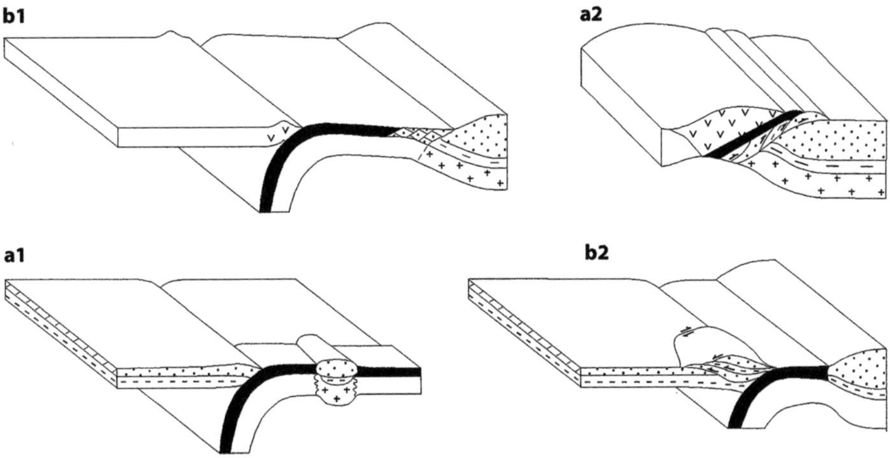

Fig. 4.24. **a.1** Passive margin with thinned continental crust, on the right side, belongs to a plate which is being subducted; In the upper plate, on the left side, a volcanic arc develops above the subduction zone. **a.2** The passive margin collides with the upper plate; oceanic crust and serpentinitic mantle are obducted on the passive margin with the development of footwall duplexes of thinned continental upper and lower crust; the volcanic arc is put in normal fault contact with the obducted ophiolite. **b.1** Subduction zone with oceanic plate on the right side and continental plate on the left side. The oceanic plate includes a microcontinent with continental lithosphere and peridotite upper mantle in the otherwise oceanic lithosphere with serpentine upper mantle. **b.2** The microcontinent arrives at the subduction zone and resists subduction, generating a flake of continental material by antithetic collision. Continental upper crust (*dotted*); continental lower crust (*detached*); perioditic upper mantle (*crosses*); oceanic crust (*black*); serpentinitic upper mantle (*blank*); volcanic arc (*v symbol*)

One-sided thrusting or flaking is the major scale structure inside a collisional belt. In both cases, second order structures can form in response to more local situations. Inside *thrust wedges* (Davis et al. 1983) the structures vary as a function of the geometry of the wedge and basal shear strength. Extensional faults directed foreland or hinterland can form near the surface at the same time as internal compressive deformation by folding and backthrusting.

As a result of collision the orogen is uplifted and a flexural basin forms at the point of contact with the foreland (Coward 1994). Migration of uplifted areas causes hinterland basins. The uplifted area will collapse outwards, if the lithosphere below the orogen is weak, or be sustained if the lithosphere below is strong (Molnar 1986).

In the final phases of collision and after collision the welded continents continue to deform by the process of lateral expulsion or extrusion (McKenzie 1972a). The more rigid plate will push the less rigid plate out of the high convergence area towards the subduction zone limiting the remnants of the closing ocean inside which the collided belt start to form (Fig. 4.25). Hence, both the shape of more rigid, continental promontories and the location of more long-lived subduction zones will control the location and shape of expulsed areas. The process can be symmetric as in the

FLAKE

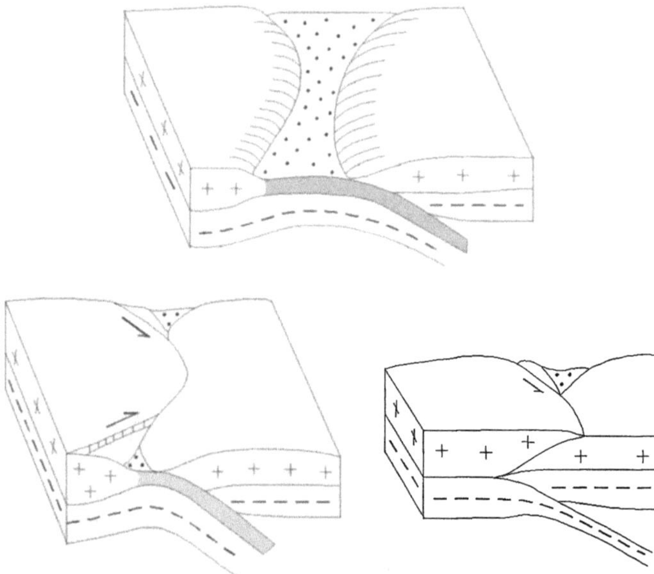

Fig. 4.25. Convergence of irregular margins induces lateral escape from promontories to reentrants where the oceanic lithosphere is still being consumed. (Modified from Oxburgh 1974)

Eurasian north of the Arabian Plate, expulsed to the west in the Aegean subduction zone and to the east in the Makran subduction zone. It can also be asymmetric as in Eurasia north of the Himalayas because there are subduction zones to absorb the expulsed blocks only on one side of the less rigid continent.

Independently of the rheological properties of the collided plates (viscous or plastic, as seen in Sect. 2.7), one is softer than the other, and some authors call the process *indentation, expulsion* or *escape tectonics* (for a review see Coward 1994).

The indentation is a type of bifurcation in physical space; in the flake the bifurcation occurs in a section of the collided orogen, in indentation the bifurcation occurs in a plane view of the orogen. The blocks in front of the indenter may go to one side of it, or the other , depending on the initial and boundary conditions. The geometry of these blocks depends on the shapes of the indenter (Tapponier and Molnar 1977; Molnar and Tapponier 1975). In the case of a triangular indenter the "undecided" domain separating the blocks that go one or the other side is reduced to a point! Nature solves the problem by a fractal distribution of blocks separated by conjugate strike-slip faults, well visible on maps, and reproduced by experiment (Tapponnier et al. 1982; Davy and Cobbold 1988; Cobbold and Davy 1988). Further away from the collision orogen, and in the foreland side of the indented zone with strike-slip tectonics, we can find extensional structures caused by secondary tension, orthogonal to the front of the collided range (Tapponier and Molnar 1976; Molnar and Tapponier 1975), such as the Baikal rift zone, related to Himalayan collision.

Collision orogens are either linear or arcuate in the plane view; there is extensive terminology for curved orogenic belts (Sengör 1990, pp. 126–127). Linear collision orogens are produced by frontal collision, where bulk strain is planar with an upward escape of material or by oblique collision with a transpressive tectonic regime (Sanderson and Marchini 1984; Dewey et al. 1998). In curved orogenic belts the predominant crustal thickening, two-dimensional tectonic regime must be combined with tectonic processes that cause the curvature of the belt, producing the observed three-dimensional strains. In fact, the origin of curved orogenic belts (COB) can be studied by strain distribution inside them (Ries and Shackleton 1976; Ribeiro 1974; Hindle and Burkhard 1999) and by palaeomagnetism (Kissel and Laj 1989). Concerning strain distribution, we must look at arcs as folds around vertical axes.

In *primary arcs*, the curvature exists since the beginning in the shape of convergent plates; the strain is the same around the arc, due to uniform transport direction.

In *secondary arcs*, a linear orogenic belt becomes curved by:

1. bending, caused by maximum forward motion in the centre of the arc, where strain increases; this produces an orocline.
2. shear zone of opposed sense of movement at the flanks of the arc; strain increases from hinge to flanks.
3. buckling of the entire arc with the flanks moving towards one another producing the Piedmont Glacier type with divergent transport directions; strain follows the distribution characteristic of tangential longitudinal strain or orthogonal flexure.

These elementary processes can be combined in various ways for the genesis of specific arcs. The Jura Arc (Hindle and Burkhard 1999) formed as a Primary Arc with a minor component of "Piedmont Glacier"-type divergence in transport directions. Palaeomagnetism proved that the Ibero-Armorican Arc was Primary since the beginning, but curvature accentuated dramatically during orogeny, by a combination of pure shear with orthogonal flexure (Ribeiro 1974; Pérez Estaún et al. 1988; Ribeiro et al. 1995) and shear buckling (Matte and Ribeiro 1975). This example shows that COB can be the result of processes during subduction accentuated by collision; it is probable that the salient of subduction zones will be promontories during the collision stage.

Some arcs and intra-arc basins, such as in the western Mediterranean, are due to mantle diapirism with upward bending that produces a thinned continental crust, but not a back-arc basin (reviewed in Llyboutry 1998).

Most COB are *centripetal* – with vergence towards the convexity of the arcuate structure; the superficial layers can move outwards in a free fashion. This is valid for subduction and collision arcs.

However, some COB are *centrifugal* – with vergence towards the concave side of the arc. This can be due to the generation of a flake, with centripetal vergence synthetic to subduction and centrifugal vergence antithetic to subduction (Fig. 4.26), such as in the Ibero-Armorican Arc (Ribeiro 1974). We can extend the flake mechanism (Quinlan and Beaumont 1984) to an arc with increasing curvature, such as the Ibero-Armorican Arc. In this case, the singularity will probably migrate around the arc in a spiral way. This could explain the final disposition of major units, similar to that of

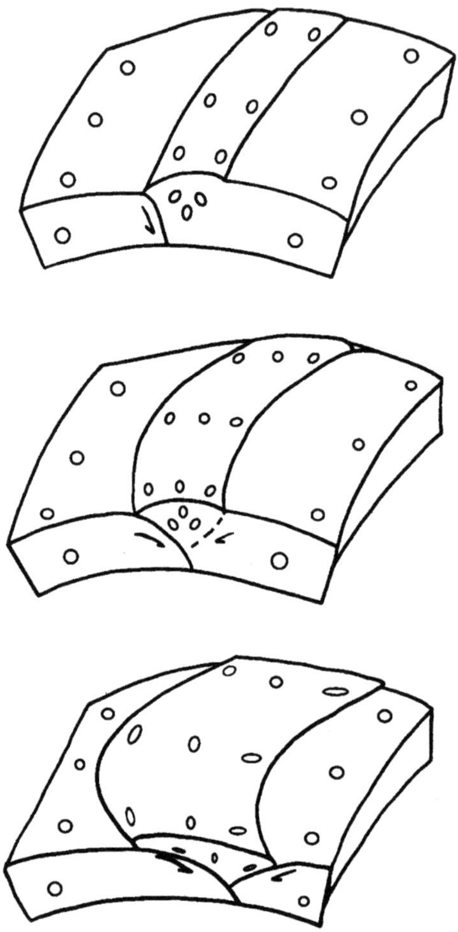

Fig. 4.26. Curved orogenic belt with centripetal vergence produced by a flake mechanism; the vergence becomes centrifugal in the pro-wedge of the outer arc and centripetal in the retro-wedge of the inner arc. This solves the space problem posed by the centripetal vergence in a synthetic convergence

the leaves of a photographic iris in the Cantabrian Zone, in the centre of the arc (Pérez-Estaún et al. 1988) and will solve the space problem of this centrifugal arc.

An alternative view, as in the Himalayas, is the flip between E-directed subduction in the earlier stages of orogeny and W directed subduction in the last stages.

The difference between linear and curved orogenic belts seems to be controlled primarily by the kinematics of plate boundaries, which in turn are controlled by boundary conditions of convergent plates. For arcs in the continental lithosphere it has been suggested (McNutt and Kogan 1987; McNutt et al. 1988) that the plane view shape of COB is related to curvature of the flexed plate in cross section. Steep dips of highly curved plates will imply a low equivalent elastic thickness of weaker plates and shallow dips of linear plates will imply a higher equivalent elastic thickness of stronger plates.

Collision orogens differ in plane view but also in cross section width. In Alpine-type orogens, such as the Alps and Caledonides, deformation is located in a narrow band near the suture. In Variscan-type orogens, such as central Asia to the north of the Tibetan Plateau, the Variscides and some Pan-African orogens, there is a very wide deformation zone well inside the plates bordering the suture. Both orogens are the product of plate tectonic processes in a Wilson cycle of opening and closing of oceans of variable width, but they differ in the amount of intraplate deformation outside the suture, eventually multiple: it is minor in Alpine-type and major in Variscan-type orogens. This led some authors (Rey et al. 1997) to suggest that this difference is mainly due to low heat-flow, around 150 °C/km, in Variscan-type orogens, across the two types of lithosphere. Both types show different orogenic evolutions: higher-pressure metamorphism and less granitoids in the Alpine type. They also differ in the pre-orogenic evolution: fast spreading ridges that produce wider oceans and less important extensional in Alpine-type, versus slow spreading, diffuse and long-lasting rift episodes and subordinate ophiolites in Variscan-type orogens. In this case, we must conclude that the difference in heat-flow regime must have lasted during the whole Wilson cycle, with a more diffuse and lower strain-rate extension in the opening stage of Variscan-type orogens. In the Variscides, the domain with high heat-flow is very wide and attached to a plate that is moving rapidly according to palaeomagnetic data. This requires long and broad incubation at the lithospheric level (Tackley 2000), even if the ultimate origin lies in a deep mantle superplume (McNutt 1998), rather than a hotspot (Ribeiro 1999). We arrive at the conclusion that the lithospheric subcontinental mantle is probably highly heterogeneous, leading to a variety of tectonic styles. In fact, deep reflection seismic profiles tend to show that mantle reflectors are abundant and they may correspond to continental crust material of various compositions and origins that was incorporated into the subcontinental mantle during orogeny, and influence the tectonic style of the next Wilson cycle (Ziegler et al. 1998). Another possibility consists of the fact that heat flow will control the rheology behaviour of lower crust, strong in Alpine-type orogens and weakened in Variscan-type orogens, leading to coupling and decoupling of crust and mantle respectively (Royden 1996) and to two contrast tectonic profiles of the orogen.

The latest stages of orogeny lead, in many cases (Burg and Ford 1997, p. 9) to extension or collapse when convergence rates between plates show a marked decrease (England and Houseman 1989); body forces generated in high elevation regions are sufficient to develop extension (Tapponier and Molnar 1976; Bird 1991). However, other processes can operate in orogenic collapse; the increase in the potential energy of the mountain belt can be achieved by removal of the mountain root. In fact, the mantle lithosphere is denser than the underlying asthenosphere (Houseman et al. 1981; England and Houseman 1989; Molnar et al. 1993). Therefore, we can induce that convective removal of the lithospheric root and delamination of the mantle lithosphere takes place. The thickened lithosphere becomes gravitationally unstable, the root is detached and sinks and is overridden by hot asthenosphere. The orogen collapses by extension and returns to a flat topography above a neoformed flat Moho (Nelson 1992). The process must have occurred in old mountain chains but must take some MA because lithospheric roots are still visible in recent mountain chains and, surprisingly, in some pre-Mesozoic orogens such as the Urals (Berzin et al. 1996). Lithosphere delamination is presently active below the Alboran Sea and Rift-Betic mountains as inferred from seismology (Seber et al. 1996) from the presence of deep earthquakes reaching 600 km and potassic magmatism reaching the surface. For some authors (Llyboutry 1998), delamination as a foundering mechanism is impossible, but continued subduction of ductile lower crust detached from a floating upper crust is viable, and controls the end of orogeny.

Orogeny can be thought of as a thermodynamic process that drives mountain belts progressively far from equilibrium and again towards equilibrium by collapse and erosion. Hence, it is a clear example of a dissipative dynamic process.

4.11 Intraplate Orogens

The best evidence that plates deform internally is the presence of intraplate orogens. In the oceanic lithosphere plate boundary zones such as the central Indian Ocean and the central Atlantic Ocean will become well defined plate boundaries as time elapses. The situation is different in continents because intracontinental zones of deformation may rarely evolve to new plate boundaries, but they will accumulate intraplate strains and will evolve to intraplate orogens.

Intraplate deformation is minimal at cratons (Park and Jarozewski 1994). Nevertheless, even cratons deform, mostly by vertical movements. Cratonic structures include basins, uplifted areas such as shields, fault zones and intracratonic systems of folds. Basins and uplifts of large horizontal dimensions result from vertical movements. Deep fault zones are long-lived and multiphase, acting as weakened zones inside cratons. Cratons are crossed by widespread rifts and aulacogens of extensional origin. Compressional structures are rare, because the lithosphere is stronger in compression than in extension. Intracratonic folds, of drape type generally can occur in some inverted areas, or in restraining bends in strike-slip fault systems.

To quantify the state of strain inside cratons we must keep in mind that some bulk strain must exist outside discrete displacements on faults because the stresses in cratons are high (Zoback et al. 1989), requiring some strain in order to store such high stresses. Slow strain rates would favour pressure solution and generation of stylolites, which are in fact sometimes observed in the cover of cratons (Park and Jarozewski 1994).

The mechanics of many cratonic structures is still the subject of considerable uncertainty. Vertical movements are still one of the least understood topics in geodynamics (Park and Jarozewski 1994). They can arise by processes unrelated to plate tectonic movements such as isostatic response to various processes, underplating above mantle plumes, density variations due to deep phase changes; other processes are plate-related such as lithosphere flexuring. Only a few cratonic structures may be due to tangential movements such as stretching of the lithosphere and subsequent minor inversion of stretched domains, or along weaker fault zones.

Another type of intraplate deformation is expressed by chains of basement uplifts within cratons marginal to orogenic belts (Rodgers 1987). Examples are the NW European platform north of the Alpine Chain, with Meso-Cenozoic deformation, or the North American Craton with Upper Palaeozoic deformation to the west of the Appalachian-Quachita orogen or with Meso-Cenozoic deformation to the east of the

Cordilleran orogen, the active Sierras Pampeanas in NW Argentina and the High and Middle Atlas of Morocco. Therefore, the deformation has the same age as the Wilson cycle responsible for the generation of the orogen but it is located outside the orogen, well inside the platform adjacent to this orogen. Marginal basement uplifts were first explained by vertical movements, as in the case of cratonic structures. Reverse faults were considered to be due to drape effects over nearly vertical faults or as the result of gravity sliding at the margins of uplifted blocks. This explanation can be discarded on the basis of two arguments. First, it does not explain the synchronicity with tectonic pulses in adjacent orogenic belts. Second, seismic profiling and drilling has shown that some of the thrusts limiting the uplifts are low angle and accumulated significant displacements, such as in the Wind River Mountains (Brewer et al. 1982).

The other explanation consists in lateral compression of the entire continental crust (Rodgers 1987). In this case, various models have been advanced for the relationship between the source of intraplate and interplate compressive stress.

In one genetic model (Ziegler 1983; Bally 1984) these orogens are located above ramps of dêcollements at various levels such as cover/basement unconformities or inside a basement at rheological contrast such as between the upper brittle crust and lower ductile crust or deeper down, at the Moho. This model has been applied to the extensive orogens inside the Iberia Plate, such as the central Cordillera, the Celtiberian Chain, the Coast Ranges of Catalonia (Banks and Warberton 1990; Ribeiro et al. 1990), or to the Alleghenian foreland of North America, with reactivation of former late Precambrian rifts, generating fault-propagation folds and pop-ups in the south-central United States, thin-skinned imbricate structures in the Oklahoma aulacogen and strike-slip and normal faults in Illinois, Michigan and Kentucky.

In another interpretation (Coward 1994) it is noted that these detachments must cross zones of extension in the foreland of the same age as collision events inside the orogen, which is mechanically improbable. So the deformation is related to transform direction inside the European Plate that connects transform faults along the Mid-Atlantic Ridge to convergence boundaries inside the Alpine orogen; or in other words the driving forces are not only collision-resistant forces but also ridge push.

Both interpretations are not incompatible and probably apply to different domains inside NW Europe and other forelands that show both detachments at the crustal level and whole lithosphere pure shear-type compression (Ziegler et al. 1995, 1998).

The geometry of inversion structures in foreland uplifted blocks is quite different from orogenic structures (Fig. 4.27). These are mainly one-sided thrusts, climbing up the passive margin in one side of the orogen. The asymmetry of boundary conditions given by a wedge of sediments above the basement favours the generation of thin-skinned thrusts with the same vergence towards the foreland. In contrast, the uplifted blocks form nested pop-ups at different scales with some hierarchy in organisation; the symmetry of the boundary condition inside the platform favours coaxial shortening with the presence of the two conjugate sets of reverse faults on thick-skinned blocks, with no preferred polarity. This was the basis for the classical distinction between the "Alpinotype" and "Germanotype" tectonic styles (Sengör 1979).

The localisation of intraplate deformation must be explained by the presence of weakened heterogeneities that concentrate deformation; these are inherited from previous cycles, such as rifts, aulacogens, deep fault zones, old sutures. In some cases,

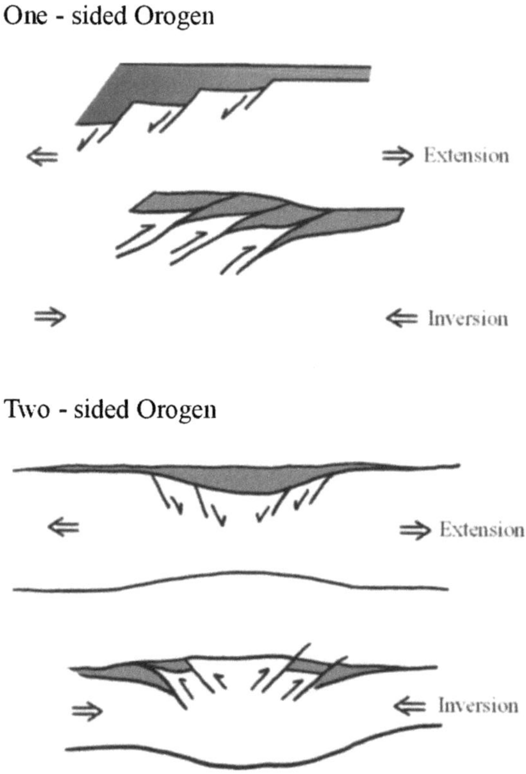

Fig. 4.27. One-sided and two-sided orogens. In one-sided orogens the asymmetry of boundary conditions generates one-sided vergent thrusts by inversion of normal faults. In two-sided orogens the bilateral symmetry of boundary conditions generates bivergent thrusts. Basement *white*, cover *shaded*

such as in the foreland uplifted blocks of western United States in Wyoming, intraplate deformation was possibly controlled by zones of magmatic activity that weakened the lithosphere inside the plate just before the inversion episodes.

The strains that caused the inversion of platform basins are usually low. For example, in the Sulaiman Ranges in western Pakistan the shortening is only 23 km in the thick-skinned interpretation of inversion instead of 250 km for the thin-skinned interpretation (Coward 1996), over some 300 km across the deformed orogen. In Iberia the strain inside the pop-up that extends along the central Cordillera towards the Lusitanian Basin is in the order of 5% shortening.

However, in other, more evolved, intraplate orogens the strains can be higher, such as in central Asia. For some authors (Beekmann 1994) major zones of deformation occur between various sub-plates such as Gobi, Pamir-Tibet and East-Asia between the Eurasia and India Plates. It is inevitable that, if plates deform internally, there will be a transition from plate-related orogens to intraplate orogens with a fuzzy boundary between both.

4.12
Dynamics of the Lithosphere: Inferences from Tectonics

Active tectonics allows us to apply the uniformitarian principle even if substantive uniformitarianism must be relaxed the present provides a model for understanding the past. However, the study of active tectonics, even supplemented by all geodynamic information, is restricted to near surface direct observation; the project of continental scientific drilling (Zoback and Emmermann 1994) will allow the direct monitoring of deformation related to the seismic cycle, near the San Andreas Fault, for the first time in earth sciences. However, in eroded orogens, we can have direct access to finite and progressive history of tectonometamorphic processes; we can observe the processes operating at depth at the present time. So, we may also say that the "past is a key to the present", reversing uniformitarianism (see Sect. 4.3).

So let us look at the inferences that we can make from our revision of plate tectonics in the past concerning the problems that we faced in the study of active tectonics and geodynamics.

The fundamental problems, as stated in Chapter 3, are the rheology of the lithosphere and the driving mechanism for soft plate tectonics that we will consider next under the heading of dynamics of the lithosphere.

The observations on ophiolites and on continental and arc-derived rocks confirm that the view of soft plate tectonics is a better approximation than the standard theory, for each phase of the Wilson cycle and for each tectonic setting. However, the mechanics of lithosphere deformation can be directly observed in deeper levels of eroded orogens and they may contribute two disputed questions concerning the driving mechanism: (1) are plates driven from below or from the sides? (2) are plates passively driven by the asthenosphere or does the driving force come from plate interaction with the passive asthenosphere? The second question can be directly addressed by observation of the fossil asthenosphere/lithosphere boundary in ophiolites (Nicolas 1989). The sense of shear in tectonic harzburgite near the Moho shows that the asthenosphere drags the lithosphere. The answer to the first question is more indirect, as follows.

In old orogens we observe a general law: all types of faults and shear zones – extensional, contractional and strike-slip – become flat at intralithosphere dêcollements and detachments. Strain compatibility requires a gradient of deformation in the same sense (Ramsay 1969). Even shallow upward faults must die out at horizontal decoupling zones (Coward 1983; Mattauer 1986). This is due to the fact that the ductility of the lithosphere increases with depth. If there are less ductile levels, such as a more rigid mantle below a ductile lower crust, then the tectonic movement is transferred to the level immediately above and below it, generating intralithospheric dêcollements and detachments. This explains the success of models of underthrusting below orogens (Beaumont and Quinlan 1994) as the main process, combined perhaps in some cases with side-driven deformation (Ellis et al. 1995). We must anticipate what happens at a larger scale near the lithosphere-asthenosphere boundary: the movement in the convective layer must become parallel to the dip of the boundary, pure dip-slip extension at the mid-oceanic ridges and predominant dip-slip contraction at oceanic subduction zones.

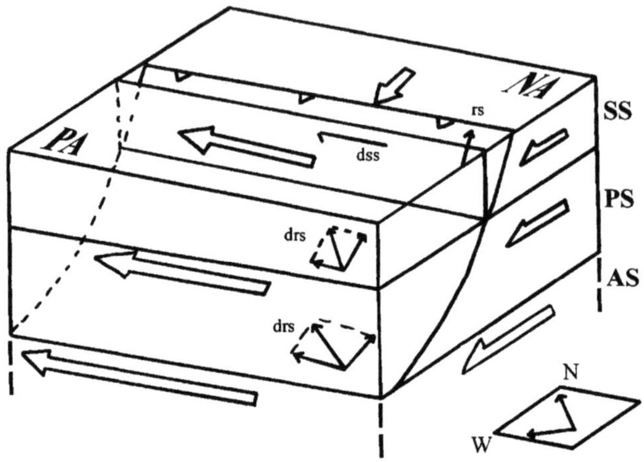

Fig. 4.28. Kinematics of the San Andreas Plate boundary system. *SS* Schizosphere; *PS* plastosphere; *AS* asthenosphere; *NAM* North-American Plate; *PA* Pacific Plate; *SAF* San Andreas transform; *wide vectors* absolute plate motion; *narrow vectors* slip direction on fault planes; *dvs* dextral reverse slip; *dss* dextral strike-slip; *rs* reverse slip

This strongly suggests that plates are driven from below (Molnar and Gipson 1996), confirming the same inference from active tectonics. The main sense of relative movement inside orogens is clearly defined by facing relationships even in the case of flake tectonics and backthrusts. However, a problem remains in the case of strike-slip and transforms; they become flatter at depth, as seen in many interplate and intraplate faults, but with which polarity? Again, strain compatibility requires that strike-slip shear zones and faults must become listric beneath the block or plate which is moving faster, relative to deeper levels in the case of intraplate dêcollements or, in the case of plates, moving faster to the asthenosphere . For plate movements, absolute velocity vectors are available in the hotspot reference frame, less mobile than the plate themselves, because it is anchored below the asthenosphere. We can test this idea in the case of the San Andreas transform, which concentrates dextral movement between the Pacific and North American Plates (Fig. 4.28). Hypocentral depths of earthquakes and source mechanism of the Loma Prieta seismic event of 17 October 1989 (Oppenheimer 1990; USGS 1990) show that the transform dips steeply to SW and the fault slip is dextral with an equal component of thrust. Seismic data (Holbrook et al. 1996) show a step in the Moho 10–15 km SW of the San Andreas Fault and confirms this geometry. The absolute plate kinematic model (Chase 1978; Argus and Gordon 1991) shows that North America is moving slowly to SW and the Pacific Plate is moving faster to NW. Applying the strain compatibility test, we infer that the San Andreas must dip below the Pacific Plate with transpressive movement by combination of the two absolute velocity vectors for Pacific and North America Plates. Most of the seismicity occurs in the upper 10–15 km of the schizosphere; it is partitioned in a dextral strike-slip component along subvertical planes oriented close to the vector of relative motion, N 34.5° W between the Pacific and North American Plates

(Zoback et al. 1999). The dip-slip component is variable and is extensional when releasing dextral bends of the subsiding Golden Gate platform, just SW of San Francisco. It is thrusting to NE on restraining sinistral bends, causing uplift of the Santa Cruz mountains. Sporadically major earthquakes nucleate deeper, such as the Loma Prieta event, with a hypocentre depth near 18 km; and oblique slip mechanism of dextral strike-slip and thrust to NE, without decoupling and revealing the probable kinematics of the plastosphere.

The previous deformation model at the lithospheric level is further supported by independent evidence combining kinematics and seismic anisotropy in the upper mantle (Teyssier and Tikoff 1998).

The previous model is purely kinematic in the sense that it uses geometrical constraints in plates rotating relative to each other, and to the asthenosphere, on a spherical earth. It does not take into account material properties that are invariant, relative to a rotation on the surface of the spherical earth but are controlled by pressure and temperature conditions as a function of depth. An example is the sense of subduction between the Australian and Pacific Plates. Near the mid-Pacific Ridge the Australian Plate, older than the Pacific Plate, is colder and is subducted beneath it in the Macquarie Trench. Further north the sense of subduction is inverted through the Alpine Transform Fault of New Zealand and the Pacific Plate is subducted below the Australian Plate across the Tonga Kermadic Trench; there the polarity of subduction agrees with the purely kinematic model.

If we accept that the plates are driven from below, what is the role of lateral expulsion, where plates must be driven by the sides? It must be minor, in the case of Tibet (England and Molnar 1997). The asymmetry in the potential energy field caused by the presence of the plateau and the subduction zones on the east side of Asia must be enough to induce minor lateral expulsion to the east of the China block. In the case if the Ibero-Armorican arc the constrictive strains must be caused by higher potential energy in the hinge of the arc combined with sideways motion, subparallel to increasing curvature of the orogenic belt as we proposed in Section 4.10. Therefore, the mechanics of lithosphere deformation tend to show that thickening mechanisms are more important than escape mechanisms, contrary to earlier views on this problem (Dewey et al. 1998). In fact, active lateral expulsion supported by geodetic measurements is only demonstrated by dextral motion of Anatolia relative to Eurasia and absorbed by subduction in the Aegean Trench (see Sects. 2.7 and 3.10).

4.13 Plate Tectonics Through Time and Space

The concept of the Wilson cycle introduces a periodic component into earth history. This cyclicity exists at a plate level for each ocean but can be integrated at a higher, global level, in the supercontinent cycle (Nance et al. 1988; Murphy and Nance 1992; Condie 1997; Windley 1995).

At the plate scale the driving forces control the evolution of each ocean during the Wilson cycle (see Sect. 3.11). Opening and closure of oceans occurred during 400–500 Ma and continents tend to accumulate over zones of major mantle downwelling accreting to form a supercontinent because plate movements are coupled. These coincide with first order geoid laws (Anderson 1989). Continental insulation preserved this unique supercontinent for some 100 Ma. However, heat is not easily lost by conduction, so the mantle beneath the supercontinent heats up again and downwelling must be interrupted and a new downwelling zone must develop outside the supercontinent. So it will heat up, weaken and be broken apart by new rifts.

The supercontinent cycle is also expressed in external geodynamics; periods of aggregation will cause regressions of sea level and glaciations of the supercontinents situated near a geographic pole. Periods of maximum dispersal will cause transgressions, maximum hydrocarbon generation and warmer climates. Hence, the supercontinent cycle can even control biological evolution itself.

Major periods of supercontinent formation occurred at 1.1 Ga with Rodinia; at 600 Ma with Gondwana and at 250 Ma with Pangea. The next period must occur in approximately 250 Ma or a bit more, if the duration of supercontinent cycles is slowly increasing from one cycle to the next. Note that the beginning of closure of the Atlantic in the SW Iberia margin fits nicely into this story! We may ask for how much time the supercontinent cycle as been operating in the past – and will be operated in the future. This leads us to the problem of plate tectonics through the earliest periods of earth history.

Plate tectonics is the present regime operating on earth. We should ask some questions that can help us to understand "why" this is so, before or after the "how" this is so. In this respect, we must look at plate tectonics through the earth's history of 4600 Ma and the possible operation of some kind of plate tectonics in other planets of the solar system.

Plate tectonics, as we understand it, extends at least to the beginning of Proterozoic – 2.5 Ga. After some discussion, specialists reached a consensus on the status of plate tectonics in the Archean 2.5–4.0 Ga (for a review, see Windley 1993, 1995). Plate tectonics operate since the beginning of Archean but with subtle differences during Proterozoic–Phanerozoic. Rocks and structures are the same but the rate of process evolved through time. Continental growth curves (Taylor and McLennan 1996) show that around 2.5 Ga continents become stable, with large passive margins and able to drift and collide. This is also marked in the chemistry of sediments, reflecting the decline of heat production through time and the long-term evolution of hydrosphere and atmosphere. This history of earth shows a combination of a cyclic component, the Wilson cycle, and the sagittal component related to continuous growth of continents and continuous decay in heat production. This view of continents as "scum of the earth" (Allégre 1999) has been challenged by a rival view of continent recycling but the evidence for major crustal recycling in the Archean is weak (Sylvester 2000).

Archean tectonics (Myers and Kroner 1994) differs nevertheless from present plate tectonics in some respects. Thermal evolution is demonstrated by distinct Archean rock associations like komatiites in greenstone belts and tonalite, trondhjemites and granodiorites (TTG) suites characterise the higher heat flow in Archean than at later times. We know from theory and experiment that the convection regime changes as a function of the Rayleigh number; below 1700 no convection occurs, between this value and 20,000 we have cylindrical shells in monomodal regime, between 20,000 and 100,000 we have bimodal flow and above 100,000, a spoke pattern. Present mantle convection involves values of 10^6–10^7. It is possible that in Archean the lithosphere was thinner and the convection shells had an aspect ratio smaller than today with lo-

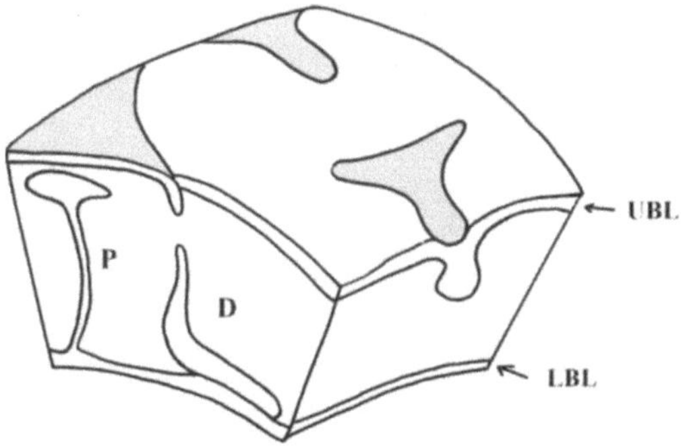

Fig. 4.29. Blob tectonics on early Earth and on present Venus. *D* Delaminated UBL. *LBL* Lower boundary layer; *P* plume; *UBL* upper boundary layer, lithosphere

wer horizontal/vertical length dimensions, because the Rayleigh number was higher.

For other authors (Davies 1992), the present plate tectonics regime initiated during the Middle Proterozoic (0.9–1.4 Ga), on the basis of the following reasoning. The oceanic lithosphere is buoyant until 20 Ma; after that it becomes negatively buoyant. Before the present regime the mantle was hotter, the oceanic crust would have been thicker and neutral buoyancy would have been achieved later. However, a hotter mantle convected faster, so the average age of oceanic plates at subduction was less than the present 100 Ma. In the past, the average oceanic lithosphere would have only just become negatively buoyant as it reached a subduction zone. Therefore, plate tectonics could have operated, but more slowly, or a different tectonic mode has operated such as subcrustal delamination (Davies 1999).

The existing data on the surface of planets of the solar system shows that some of them have active volcanism or presently inactive volcanism, such as Io, Mars, Venus or in the distant past, such as the Moon (Morrison 1993). No planet has plate tectonics like the present Earth; very recent data show that magnetic lineations are present in the old crust of Mars; this suggests that seafloor spreading was active 4 Ga ago (Connerney et al. 1999). Venus shows some form of tectonics, more similar to that of the Archean on Earth, with smaller convection cells, that have been called blob tectonics (Fig. 4.29) more similar to the plume mode on Earth than to the plate mode (Davies 1999). If we compare the rollover time for convection for different planets (Elder 1976), we gain a planetary perspective. The giant planets such as Jupiter are vigorous because they have hardly cooled. Considering terrestrial planets, Venus is the more active, followed by the Earth, Mars and the Moon is almost inactive. So plate tectonics on Earth since the 2.5 Ga is intermediate between very active convection and very slow convection. This is a "dry" theory, in the sense that the effect of fluids is not a controlling factor. However, some authors (Fyfe 1988; Mian and Tozer 1990) give the impression that plate tectonics is made easier by liquid water that facilitates mechanical failure of the lithosphere. In Venus no plate tectonics are possible because free water is absent even if the surface temperature is higher. This strongly favours the view that the Earth is a water-cooled convective planet in the plate mode regime and the volcanic dry planets exhibit only the plume mode regime.

The Earth as a Dynamic Open System

5.1
Introduction: Impact Tectonics

Tectonic phenomena that affect the lithosphere can occur at a wide variety of rates (Price 1975) from 100 to 10^{-18} s^{-1} for the processes described in previous chapters. The Soft Plate approach was proposed mainly for processes on the slow side of the spectrum of strain rates but also has implications on the fast side; in fact, viscoelastic rebound has deep implications in earthquake faulting. Can we extend the range of behaviour of the lithosphere by studying processes that operate at an even faster rate than earthquake faulting, if such processes exist? Only by exploring the whole range of possible behaviours can we propose a coherent theory for the central problem in geodynamics, the rheology of the lithosphere.

In fact, there is a geological process that operates at a faster strain rate, $\sim 10^6$–10^9 s^{-1}, than earthquake faulting. It is impact cratering; and it can be used to probe planetary lithospheres, besides the case of the earth (Thomas et al. 1997).

The physical conditions for impact cratering are known from shock metamorphism (see Grieve 1990; Grieve and Pilkington 1996 for a synthesis). Pressures rise from 10 to 100 Gpa and temperatures from 1000 to 10,000 °C; under these conditions crystals successively develop planar features, diapletic glasses by disorganisation of the lattice, melts and vaporisation. This field is well beyond the fast geological processes such as earthquake faulting and volcanism, with temperatures below 1000 °C but with pressures between 1 and 0.1 Gpa. Theoretical energy considerations confirm the observations: a stony chondritic body at 25 km s^{-1} gives a peak pressure of ~900 Gpa at the impact point in granite.

The direct tectonic effects of impact are known from observation of known impact craters and from analogies with nuclear explosions (Melosh 1989; Price and Cosgrove 1990). Different stages in the cratering mechanism such as contact, compression, excavation, modification, in this order, produce: the crater itself, tectonic inversion attaining recumbent folding below the crater rim, radial fractures and a circumferential graben outside the crater rim. Therefore, compressive stresses produce the recumbent folding and tensional stresses, during the modification stage, they produce the tensional structures by collapse. All these effects are scaled to the energy of the impactor, depending on its mass and velocity. In the case of Chicxulub the impactor has a 12 km diameter, the multi-ring crater has a 195 km diameter and the Moho at 35 km is displaced 3 km by normal faults (Morgan et al. 1997).

Other direct effects of impact on the lithosphere are well known and documented. They include, besides morphology and tectonics, a petrological signature in the form of shock metamorphism, already referred to, geophysical signature (Grieve and Pilkington 1996) and geochemical signature.

The importance of impact cratering on Earth has been growing since the discovery of the Meteor Crater (Shoemaker 1960; Gilbert 1983). However, space exploration has imposed the idea that impact cratering is a major process in shaping the evolution of the solar system (for a review, consult Melosh 1989), if not life itself. Our goal is not to discuss impact cratering per se, but rather the implications for tectonic evolution of the Earth by the impact processes. We must say right away that there is no consensus on that question: on one end of the spectrum some authors think that plate tectonics is just a minor aspect of impact tectonics (Shaw 1994) and, on the other end, there are authors that deny any tectonic influence of impact, besides their direct effects (Melosh 1989; Courtillot 1995). So the discussion of this theme is mandatorily speculative when compared with previous themes.

Our question is rather: can impacts, namely mega-impacts, fracture the lithosphere, at least partially, and induce magmatism or related phenomena such as mantle plumes? Can mega-impacts even induce continental breakup, shaping the Wilson cycle in a critical way?

To address this problem we can use the data coming from experiments on dynamic rock fragmentation and from numerical simulations, apart from the indirect evidence from geological records.

Dynamic fracture (Atkinson 1987, p.16) is caused by unstable crack propagation under stresses above the critical crack-tip stress intensity, at velocities that may approach the shear wave velocity of the host rock. It is the opposite of subcritical crack propagation, with slow velocities of 1 mm/s due to stress corrosion.

Dynamic rock fragmentation (Grady and Kipp 1987) was applied to well shooting showing different mechanical behaviour as a function of explosive or propellant loading rates: at a fast rate we get explosive crushing in the vicinity of the well, at an intermediate rate we get multiple radial fracturing, at a slow rate we get one fracture propagating by hydraulic fracturing (Fig. 10.11 in Grady and Kipp 1987, p. 449), similar to the one produced by static pressurisation (Fig. 5.1).

Computational simulations (Boslough et al. 1996) were stimulated by two facts: monitoring of the propagation of seismic rays on Earth shows that surface and body waves converge at the antipode of the seismic source, by focusing due to axial symmetry. This explains impact-induced antipodal volcanism on some planets. On the other hand, energy considerations show that a large impact could induce shock heating in the upper mantle, contrary to views that upward motion in the mantle should be adiabatic, precluding additional melting. The simulations use shock physics theory in the impact source region and seismological theory over the entire interior.

The results of these simulations showed that maximum energy is transferred in the mantle along the impact-antipodal axis with intense focusing within the asthenosphere at the antipode. Strong axial focusing also occurs directly beneath the impact site. We conclude that excavation and fracturing of the lithosphere and triggering of magmatism in the asthenosphere are possible consequences of impact.

Observation, physical theory of shock, experiments and computational simulations lead to the same conclusion: mega-impacts can fracture and excavate the lithos-

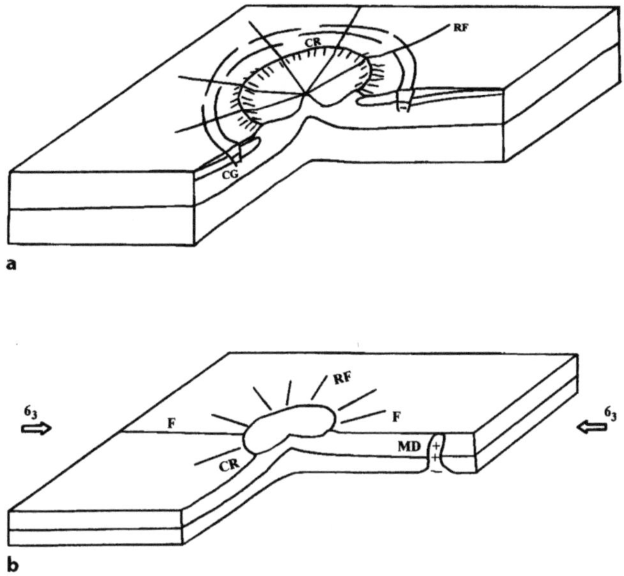

Fig. 5.1. **a** Synthesis of structures produced by impact. From the core of the impact, with a central uplift, to the outer side, we see a crater rim (*CR*) with a reverse fold produced during the excavation stage; outside the crater rim we see a circumferential graben (*CG*) produced during the collapse stage: and radial fractures (*RF*). The domain inside the crater rim is affected by explosive crushing at fast loading rates. The domain outside the crater rim is affected by multiple fracturing at intermediate loading rates (after Grady and Kipp 1987; Price and Cosgrove 1990). **b** At a late stage of evolution, one of the radial fractures (*F*) propagates at a slow loading rate by hydraulic fracturing well beyond the domain with radial fractures; along this slow loading rate fracture a magma diapir (*MD*), triggered by impact, pierces the surface. The main fracture is the most favourably oriented to crack propagation, relative to maximum compressive horizontal stress, σ_3

phere, with rock comminution at the centre of the structure, radial fracture around this centre, and normal faulting below and outside the crater rim.

At the impact site itself pressure, P, and temperature, T, can lead to melting and vaporisation of all geological materials. However, the shock wave attenuates rapidly with distance and away from the impact site, the effects at the strain rate of $\sim 10^6$–10^9 s^{-1} show that the lithosphere is extremely brittle. The lithosphere is soft for geological strain rates, but hard for impact strain rates at the other end of the spectrum. We can oppose Soft Plate to Hard Impact Tectonics.

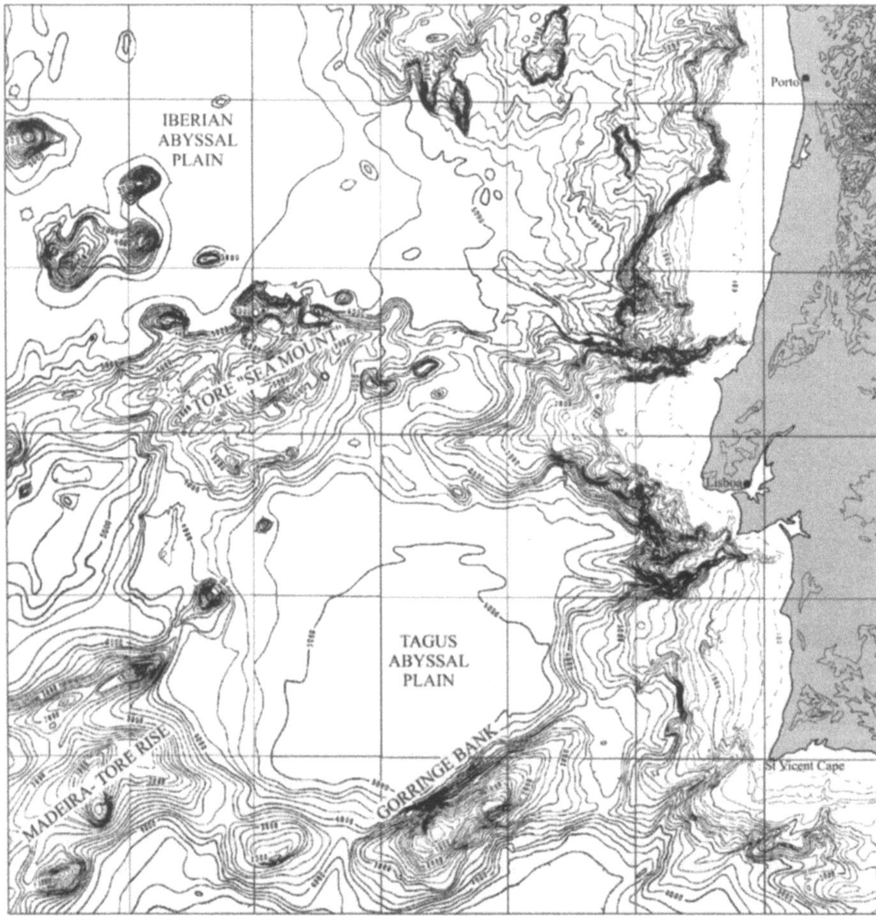

Fig.5.2. Bathymetric map of the West Iberia margin (After Lallemand et al. 1985). Approximate scale is 1/4,186,000; Mercator projection. Contours at 200-m interval

5.2
Tore as an Impact Crater in the Ocean

5.2.1
The Tore "Seamount"

The Tore "Seamount" has been mapped and described (Laughton et al. 1975). This peculiar geomorphic feature is situated at the northern end of the Madeira-Tore Rise, 1400 km long in NNE direction (Fig. 5.2).

The Tore "Seamount" has a toroidal shape, hence its name. It has a rim, 2 km deep on average, and a centre 5 km deep; the "Seamount" is surrounded by the abyssal plains, Tagus and Iberia, 5 km deep on average. The rim has an elliptic shape in the plane, with the major axis oriented N22±5°E, approximately 120 km long and a short axis approximately 85 km long. The "seamount" was surveyed in more detail (Lallemand et al. 1985), but the main geomorphic features persist on the new bathymetric map. The discoverers of Tore "Seamount" considered that its unusual shape could have three possible origins.

First, Tore could be a giant caldera. Giant calderas of this size are only known in continental settings or above subduction zones, because they are generated by an explosion or by the collapse of a viscous magma of rhyolitic to andesitic composition. This tectonic setting can be discarded in the case of Tore, because the morphologic feature is part of the evolution of the opening and spreading North-Atlantic, in the oceanic lithosphere of basaltic composition and much less viscous than andesite-rhyolite magmas.

Second, the Tore could be a meteorite crater, but then we have to consider the coincidence of its occurrence on the line of the Tore-Madeira Rise.

Third, Tore could be due to a chance mixture of tectonic trends. Although some of the segments of the "seamount" are parallel to tectonic trends in this part of the North-Atlantic, the authors state: "However, the southwest part of the encircling ridge has a trend NW–SW which is not observed elsewhere" (Laughton et al. 1975, p. 803). The authors prefer this explanation but recognise the lack of more data to propose a more definite answer to the problem of the origin of Tore.

In retrospect, we must credit these authors for admitting, in 1975 ,the possibility of an impact crater as the origin of Tore; this happened at a moment where very few people considered impact cratering a matter of science.

The search for the "smoking gun" responsible for the Cretaceous/Tertiary extinction (Alvarez et al. 1980) triggered the search for an impact site, and Tore was explicitly admitted as a possibility (Alvarez et al. 1982). Other authors supported this hypothesis (Ribeiro 1993/1994; Cabral 1995), because absence of evidence is not evidence of absence.

5.2.2 Age of the Tore "Seamount" and Relationship with the Tore-Gulf of Cadiz Lineament

Even if we cannot constrain the origin of the Tore "Seamount", we can constrain the age of the geomorphic feature. For this purpose we shall use the available geophysical information (Silva 1995), which includes namely an aeromagnetic map of the West-Iberia margin (Fig. 5.3).

On this map we recognise a dipole field on Tore "Seamount"; the magnetic anomaly is displaced slightly to the north of the deep centre of Tore; this is the effect of the inclined magnetic field at the time of formation of the "seamount". This dipole field is imposed on N-S magnetic anomalies; these are ascribed to the Mo event, in the magnetostratigraphic scale, corresponding to an age of 118–119 Ma. Further south in the Madeira Tore Rise South of Tore the axis of the rise and the Mo magnetic anomaly become subparallel.

Another peculiar feature consists of an alignment of dipoles describing an arc that changes its orientation from almost E–W to almost N–S as we move in a clockwise di-

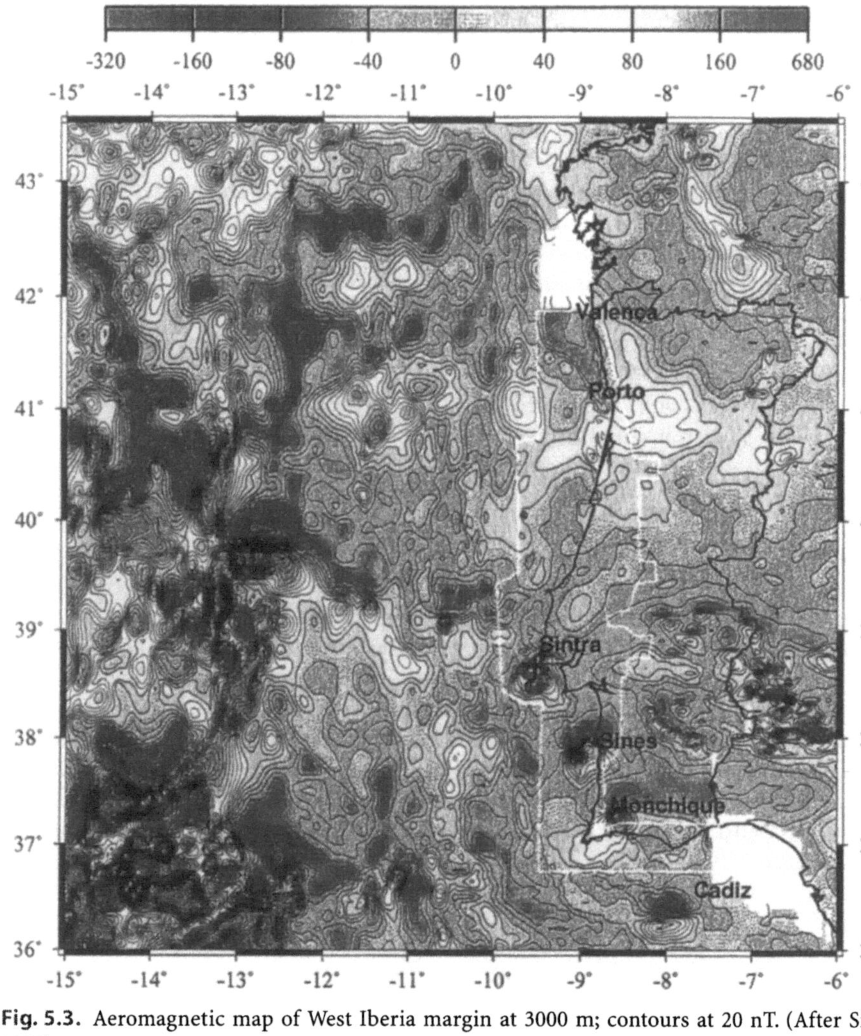

Fig. 5.3. Aeromagnetic map of West Iberia margin at 3000 m; contours at 20 nT. (After Silva 1995)

rection. The western end coincides with the Tore dipole. We then have three offshore dipoles of unknown origin. Onshore we have three dipoles coinciding with the Sintra, Sines and Monchique Massifs. These are subvolcanic complexes with alkaline affinities related to the opening of the Atlantic, with ages ranging between 82 and 65 Ma (see Sect. 5.4). Further south, offshore the Algarve, we have a dipole in the Cadiz Gulf that probably corresponds to an intrusion in the same time interval, because thermometamorphic biotite has been found on slate-greywacke samples of carboniferous flysch, dredged at this site (Ribeiro et al. 1979). As the Variscan metamorphic gradient decreases to SW in the South-Portuguese Zone this thermal event must be post-Palaeozoic.

These data prove a minimum age for Tore "Seamount" of 118–119 Ma, the age of the surrounding oceanic lithosphere and suggest a maximum age of 82 Ma, the oldest intrusion on the Tore–Sintra–Sines–Monchique–Gulf of Cadiz lineament which we will call Tore-Gulf of Cadiz lineament for short.

5.2.3 Ejecta at the Cenomanian-Turonian Boundary in Praia da Vitória, Nazaré, Central Coast of Portugal

Additional evidence for the origin of Tore came in an unexpected way, with the discovery of an impact ejecta horizon near the 91 Ma Cenomanian–Turonian boundary, at a site 0.5 km south of Praia da Vitória and 10 km NNE of Nazaré (Monteiro et al. 1998; Monteiro, in prep.).

This site belongs to the Lusitanian Basin, the present western continental margin of Iberia, related to the opening of the North Atlantic; the ejecta is found at the top of a stratigraphic section corresponding to a carbonate platform of Cenomanian–Turonian age (Soares 1966).

The ejecta is noticeable by the presence of brecciated limestone with exotic blocks; it forms a corridor 75 m long and 10 m wide aligned N80°E; this direction points to the Tore "Seamount" (Fig. 5.4).

The clasts of the ejecta vary from millimetres to 80 cm. Each clastic boulder corresponds to a polymictic breccia and dark coloured fragments with iron sulphides, quartz fragments and calcitic nodules; solid hydrocarbons occur as surface stains and impregnations; other clasts are composed of limestone and chert, and show a preferred orientation of long axes. Some clasts show surface figures – facets, striae, polishing, percussion works and penetration features with random orientation for each clast. This is known from ejecta ballistic curtains (Ocampo et al. 1996). The ejecta have all the characteristics of the impact-related diamictites (Rampino et al. 1997).

Textures diagnostic of impact cratering are present: diaplectic glasses; a few quartz grains display planar deformation features as in "shocked quartz", sometimes in two sets; impact melt textures in glassy fragments, microscopic spherules and irregular shard-like particles. All these features are diagnostic of shock metamorphism.

A clay layer, with variable thickness from some mm up to 40 cm, occurs on top of the limestone; there is an onlap of a poorly sorted sandstone of not well-constrained upper Cretaceous age on the Cenomanian–Turonian carbonate platform. This layer is geochemically anomalous to platinum group elements, including a peak of Iridium with 200 ppt analysed by neutron activation. This is ten times the average continental crust and exceeds some of the peak Ir anomalies of the world for Cenomanian–Turonian sections (Orth et al. 1993), a period of minor extinction, probably stepwise; other anomalous values include Au, Os and Ru.

Ten km south of Praia da Vitória, at Nazaré, there is a section (Reis et al. 1997; Calapez 1998) with a sharp discontinuity between very brecciated Cenomanian limestone and Turonian sandstones, 5 m thick with wave-ripple laminated material, parallel lamination, cross-bedding, fossil debris and collapse structures. These indi-

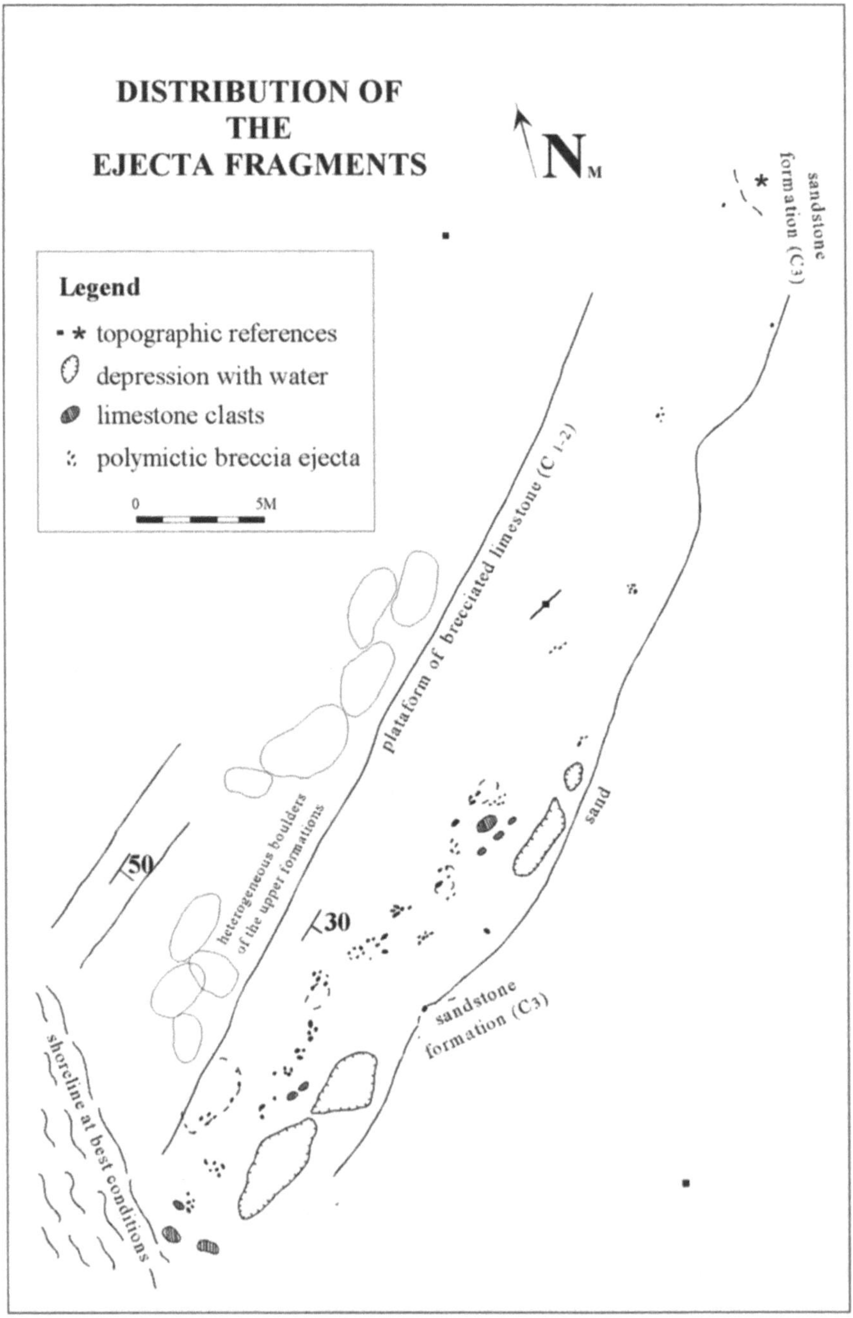

Fig. 5.4. Detailed map of ejecta deposits at Praia da Vitória (after J.F. Monterio, in prep.). Location of Praia da Vitória is shown in Fig. 5.10, north of Nazaré

cate shallow-water high energetic environment of tempestite-tsunami origin. This formation is absent in the Praia da Vitória because of the onlap of upper Cretaceous sandstones on top of the Cenomanian–Turonian platform carbonates. Subvertical mafic dykes, oriented NW–SE, cut the Turonian and are probably of Late Cretaceous age.

What is the source of the ejecta at Praia da Vitória and the tsunami event at Nazaré, both at 91 Ma? The tsunamigenic nature of the Turonian sandstones at Nazaré requires an oceanic event; this is confirmed by a source west of the Praia da Vitória ejecta, simply because ejecta disappears to the east at the same stratigraphic level. The question of origin of the ejecta in the oceanic versus continental lithosphere is more debatable; the scarcity of shocked quartz, the presence of sulphides and chert fragments favour an oceanic origin; the limestone clasts suggest a shallow-water environment, but we cannot exclude that they were generated by a secondary impact in shallow waters, the main impact taking place in the deeper ocean (J.F. Monteiro, in prep.).

5.2.4 Possible Link Between Tore "Seamount" and Praia da Vitória Ejecta

If we consider the age of ejecta at Praia da Vitória, 91 Ma and the interval of ages for the genesis of Tore, 118–82 Ma, we notice that the age of the ejecta falls in the interval age for Tore. Therefore, there is compatibility in the age of these events; however, there is no proof for the genetic linkage between them but a suggestion of a possible cause and effect linkage. This calls for more geophysical, geological and geochemical data on Tore and all the domain between Tore and the West-Coast of Iberia. International co-operation is planned for this purpose.

We can speculate on the age of Tore, in order to refine its limits. The Tore–Sintra–Sines–Monchique–Gulf of Cadiz lineament seems to perturb the magnetic anomalies west of Tore until anomaly M9 (118 Ma). Further west, the magnetic anomalies do not seem perturbed, but their dating is imprecise because we fall into the quiet magnetic zone. If we reconstitute the palaeogeography of Atlantic Ocean at the time of the last perturbed anomaly along the Tore-Gulf of Cadiz lineament we conclude that it must have nucleated sometime before Anomaly 34, 84 Ma, because the younger oceanic lithosphere is not affected by it. We must allow some time interval for eventual propagation between Tore and the youngest perturbed anomaly. This reasoning suggests the highly probable synchronous generation of both Tore and the impact that produced the 91-Ma-old ejecta at Praia da Vitória and tsunami at Nazaré.

5.2.5 Possible Direct Impact Tectonics Effects at Tore

Now we must review the possible origin of Tore as an impact crater in the light of our present knowledge about this geological process.

In a later section, we will try to explain that the location of Tore at the northern end of the Madeira-Tore rise is not a product of chance (see Sect. 5.4). Once we remove this difficulty for the impact hypothesis, we must reconsider its feasibility.

With this perspective, we must take a new look at Tore morphology on the basis of what we now know about the tectonic effects of impact. The 5-km-deep centre of Tore can be explained by intense comminution at the crater centre; the central peak will be recovered by sediments younger than the impact and this can be tested by seismic profiling. The rim of the crater is cut by discontinuities with saddle geometry, which suggest a radial arrangement of fractures, as we see around the centre in known impact craters. One of the radial fractures continues along the Tore–Gulf of Cadiz lineament. This may be due to propagation of the fracture more favourably oriented in relation to a regional stress field; the curvature of the lineament can be due to the changing stress field.

5.3 Tore as a Strain Marker in Mature Oceanic Lithosphere

5.3.1 The Elliptic Shape of Tore "Seamount"

If Tore "Seamount" is an impact crater, then we must study the implications of its elliptic shape. In fact, impact craters are usually circular in shape. Observation and experiment show (Melosh 1989, pp. 81–82) that craters are circular down to angles of incidence less than 10°. The process of oblique impact is not well understood and the asymmetry of the process is only evident in the ejecta distribution.

What was the original shape of the impact crater at Tore? A process of impact with an angle of incidence below 10° is possible if we consider the present elliptic shape, independently of any other geometric relationship. However, if we consider the pattern of magnetic anomalies around Tore, we can discard the hypothesis of an initial elliptic shape. In fact, the orientation of the major axis of the ellipse at Tore, N22±5°E is oblique to the N–S magnetic anomaly around M_0, visible to the north of Tore but parallel to younger magnetic anomalies to the west, such as Anomaly 34, 84 Ma, and younger (Fig. 5.5).

To explain this geometric relationship we have, a priori, two possible explanations.

In the first case, the elliptic shape is original, due to an oblique impact with an angle of incidence below 10° and contained in a NNE plane parallel to the younger than M_0 magnetic anomalies to the west of Tore. This would be a highly improbable coincidence, in the order of 6.2×10^{-3}, estimated on the basis of the probability product, $p_1\times p_2$, of an impact below 10°, which is $p_1=10°/90°$, and of an impact parallel, within 10°, to the azimuth of younger magnetic anomalies, $p_2=10°/180°$. Both probabilities assume uniform distributions in the sphere.

In the second case, the original shape is circular, due to an impact above 10° of incidence, and it becomes elliptic after deformation by a strain with the major axis parallel to the younger magnetic anomalies to the west of Tore. In other words, the ellipse expressed by Tore is a finite strain ellipse accumulated since the impact event; the corresponding probability is 99.38%. In this explanation the state of strain at Tore is

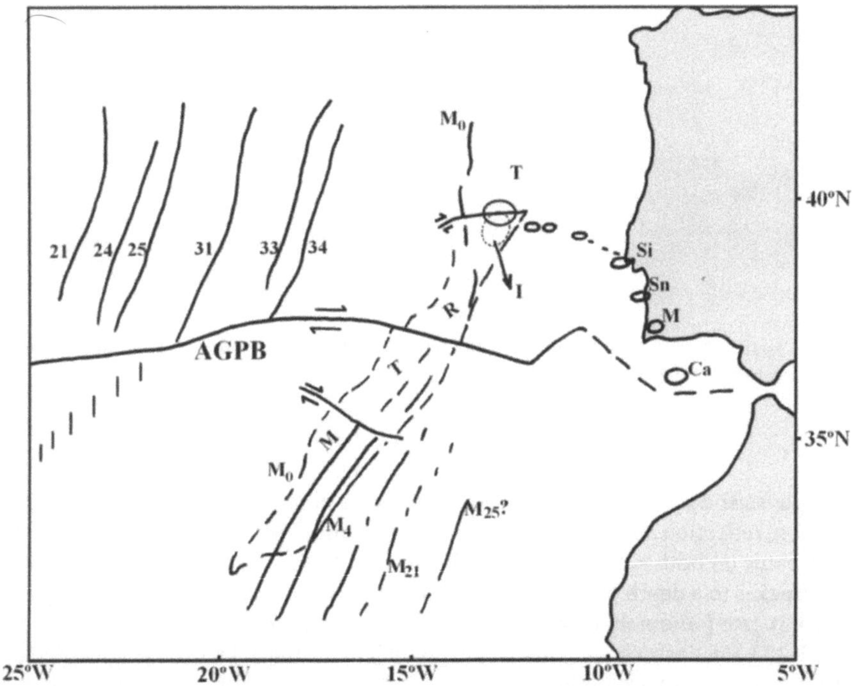

Fig. 5.5. Geodynamic interpretation of magnetic anomalies. Magnetic anomalies M25 to M0 and 34 to 21. Magnetic dipoles. *Ca* Cadiz, *M* Monchique, *Sn* Sines, *Si* Sintra, *T* Tore, *I* possible impact crater at Tore, with elliptic shape; the magnetic T is displaced northwards because no reduction to pole was processed. *AGPB* Azores Gibraltar Plate Boundary, *dotted* where diffuse. *MTR* Madeira Tore Rise, with *hatched edge*. (After Roeser et al. 1992)

simply a consequence of the asymmetric model of deformation for the oceanic lithosphere in an Atlantic-type ocean, which we developed in Section 3.3.2, with co-axial deformation since approximately 91 Ma and the present time. In fact, we know that trajectories inferred from magnetic anomalies for Iberia, since the Cenomanian–Turonian boundary, 91 Ma, are smooth and compatible with the present Euler Pole for rotation between North America and Eurasia, due to North Atlantic spreading. The impact must have occurred near or after the end of rotation of Iberia, between 118 and 84 Ma (Roest and Srivastava 1991), otherwise the deformation at Tore would not have been co-axial in the last 91 Ma.

5.3.2
The Origin of Madeira-Tore Rise

The position of Tore at the northern end of the Madeira-Tore Rise (MTR) stimulates our interest in the genesis of this morphotectonic feature which extends for 1200 km in a NNE direction.

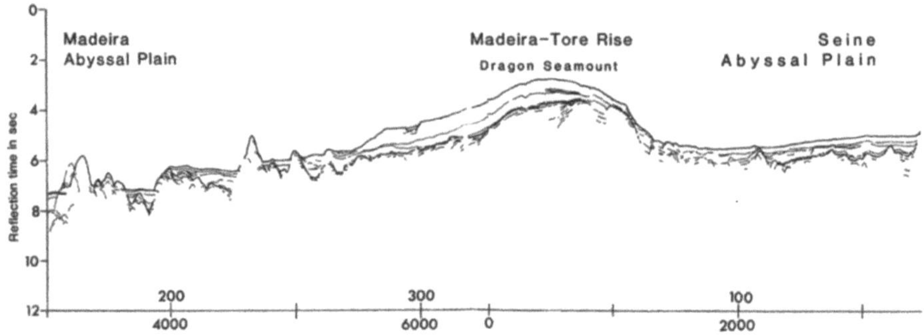

Fig. 5.6. Seismic reflection profile of the Madeira-Tore Rise in the Dragon Seamount at 36°N. (After Roeser et al. 1992; courtesy of Hans Roeser, BGR)

Geophysical data (Pierce and Barton 1991) were obtained for the MTR, including refraction, reflection and gravity data. It was concluded that the oceanic crust is typically oceanic on both sides of the MTR, however, it has an anomalously high velocity and is thicker to a depth of 17–18 km beneath the rise, but isostatically compensated.

A conjugate J anomaly ridge exists on the North American side, between M_4 and M_0, 108–113 Ma, Early Cretaceous. A hotspot origin for the MTR can be discarded because hotspot tracks are oblique to magnetic anomalies and the MTR is parallel to them. Another possibility is a period of major magmatic accretion at the Mid-Atlantic ridge at this time, above the normal value of the magmatic/tectonic ratio for this slow-spreading ridge. This static interpretation for the MTR has been challenged by other data.

A seismic reflection profile (Roeser et al. 1992) in the vicinity of Dragon Seamount shows clearly that the sedimentary column is deformed above the MTR as well as the most recent sediments and the sea bottom itself. They define an anticline with a half-wavelength at ≈150 km, asymmetric with a gentle western limb and a faulted eastern limb. This suggests, as in the case of SW Iberia margin, an anticline above a ramp of a thrust fault with movement to SE, subperpendicular to the MTR (Fig. 5.6).

If we look at seismicity in the MTR and adjacent domains, we notice that the eastern flank of the MTR coincides with an alignment of instrumental epicentres. A major strike-slip event with M_s=7.9 occurred on 26 May 1975, west of the MTR, with a major dextral focal mechanism (Lynnes and Ruff 1985). The present stress field shows a maximum horizontal compressive stress oriented NW–SE, approximately orthogonal to the trend of the MTR. Therefore, these seismotectonic data are compatible with a major SE-directed thrust along the eastern flank of the MTR that slightly displaces the oblique Tore-Cadiz lineament, with a minor component of dextral slip (Fig. 5.7).

We conclude that the MTR is then an active buckle of oceanic lithosphere (Fig. 5.8). Along the major part of the MTR the rise is parallel to the M_0 anomaly; but at the northern end, around Tore, the rise has a NNE trend parallel to the major axis of the ellipse defined by Tore rim but the magnetic anomaly M_0 becomes N–S oriented. These data suggest that the MTR active buckle and thrust nucleated in an inherited heterogeneity at 118 Ma, responsible for the M_0 anomaly where they are parallel

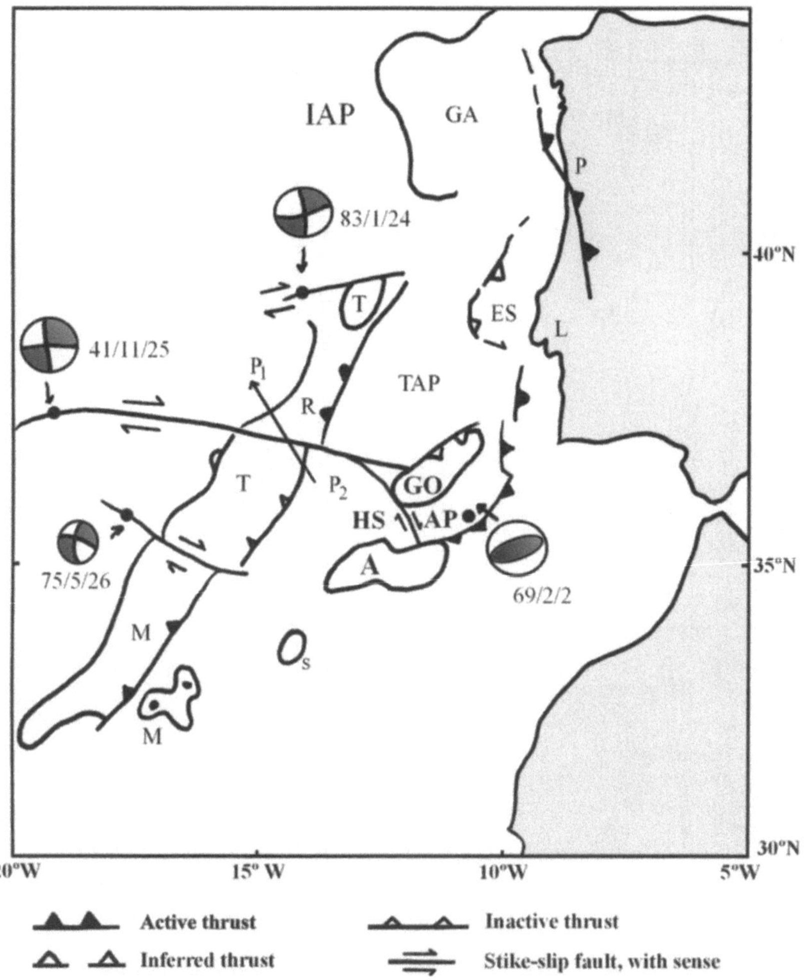

Fig. 5.7. Geodynamic setting of the Madeira-Tore-Rise. Tore "Seamount" at the northern end of the Madeira-Tore Rise (*MTR*). Active tectonics and focal mechanism of large earthquakes are shown (M>7). P_1—P_2 The location of seismic profile across Josephine Seamount. Abysmal plains: *HSAP* horseshoe, *IAP* Iberia, *TAP* Tagus. Banks: *Ga* Galicia, *Go* Goringe. Ocean Island: *M* Madeira. Seamounts: *A* Ampère, *S* Seine, *T* Tore. Spur: *ES* Extremadura

and nucleated at the Tore impact crater at the northern end of the MTR. The impact crater behaves as a weak point in the oceanic lithosphere, probably because the impact process weakened the oceanic crust or even the entire lithosphere.

This dynamic interpretation for the MTR is further strengthened by the fact that the conjugate J anomaly in the North American Plate does not have the obvious morphologic expression as the MTR, because it is not tectonically reactivated under the present stress regime (Tucholke and Ludwig 1982).

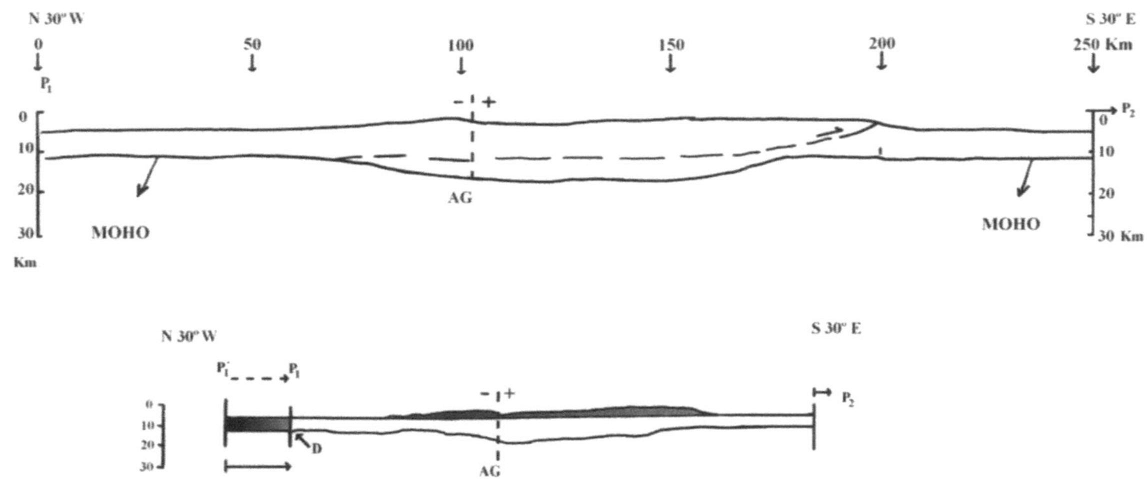

Fig. 5.8. Balanced cross section across the Madeira-Tore Rise in the Josephine Seamount. The location of the profile is shown in Fig. 5.7. The depth to the Moho is after Pierce and Barton (1991). The geometry of the thrust is based on the surface trace shown in Figs. 5.5 and 5.7 and its extrapolation to depth. The excess area (*dotted*) above the datum at 5 km is due to moving pinpoint P'_1 to P_1 above dêcollement at depth *D*, at 12 km. *AG* Azores Gibraltar Plate Boundary Fault

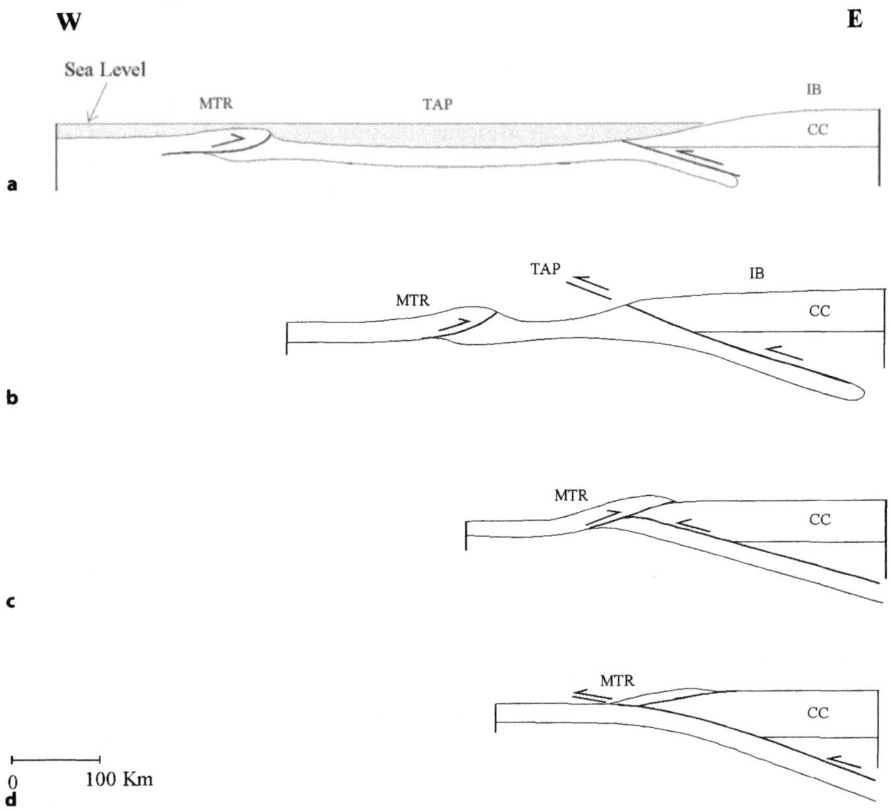

Fig. 5.9. Future geodynamics evolution West Iberia margin illustrated by E–W profile at 35°5′N. **a** Present situation: intra-oceanic thrusting at MTR; incipient subduction of West Iberia margin. **b** Subduction acceleration at West Iberia margin; the oceanic floor of the Tagus Abysmal Plain is consumed at the subduction zone. **c** The opposed thrust at the Madeira-Tore Rise and West Iberia subduction zone come into contact; MTR is cold obducted on top of Iberia. **d** The subduction zone cut across the MTR thrust and oceanic lithosphere to the west of MTR is consumed. *IB* Iberia, *MTR* Madeira-Tore Rise, *TAP* Tagus Abysmal Plain

If we look at the present geodynamic setting of the West-Iberia margin and the MTR, we must conclude that the entire domain with mature oceanic lithosphere, older than 100 Ma, is in a transition between passive and active margin stages of the Wilson cycle. All the domains between the incipient West Iberia subduction zone and the footwall to the east dipping intraoceanic thrust in the eastern flank of the MTR will be subducted below Iberia. In the near future the W-directed thrust will face the E-directed thrust of the West Iberia subduction zone and we get the conditions for cold obduction of a segment of the MTR above continental Iberia by a flake mechanism. This situation would be transient and the general subduction regime will take over again, closing the North Atlantic in 100±50 Ma (Fig. 5.9). The uncertainty depends on the initiation, or not, of a conjugate subduction zone in the North American side of the Atlantic.

The activation, by thrusting, of the eastern flank of the Madeira-Tore Rise, by buckling, proves that these structures are not only due to present retreating subduction boundaries, in response to Late Miocene collision, as proposed by Royden (1993). The Atlantic lithosphere is in a mature state and ready to subduct along previous heterogeneities related to passive margin evolution, independently of their relation to neighbour continental collision systems.

5.3.3
Finite Strain at Tore: Estimation and Partitioning

If the initial shape of the impact crater is circular, as we tried to demonstrate in the previous section, its final elliptic shape can be used as a finite strain indicator in mature oceanic lithosphere. This would be the only finite strain marker available in this type of lithosphere until now, hence its critical importance in the Soft Plate Tectonics theory.

The ellipse at Tore has a major axis oriented N20±5° and is 120±6 km long and the minor axis is 85±5 km long. The axial ratio, R_s, of the strain ellipse is $1.41\ ^{+0.17}_{-0.14}$. This strain ellipse results from coaxial strain because the major axis stays approximately parallel to magnetic anomalies younger than ≈91 Ma in the North Atlantic and are subparallel to the axis of the active buckle of the MTR. As the oceanic lithosphere cools, this coaxial deformation results from a shortening of 29% parallel to the minor axis and no significant longitudinal strain parallel to the major axis; so the ellipse derives from a circle of ≈120 km diameter. The coaxial shortening must be partitioned into two mechanisms: homogeneous shortening and buckling of the lithosphere.

To estimate the homogeneous component of shortening we must use the axisymmetric model of oceanic lithosphere deformation. We assume that the non-coaxial deformation component due to spinning of the Euler axis between Eurasia and North America is negligible. Knowing the Euler Pole for the Eurasia-North America plate pair from NUVEL-1 (de Mets et al. 1990), the coordinates of the centre of Tore strain ellipse (39.4°N, 12.8°W) and the coordinates of the Mid-Atlantic Ridge, we can estimate the angular distance from to Tore to Mid-Atlantic Ridge along a minor circle around the Euler Pole, which is θ=20±2°. This gives an axial ratio

$$R = \frac{1+e_1}{1+e_2} = \frac{1}{\cos\theta} = \frac{1}{0.94} = 1.06$$

with shortening of 6% along the minor axis of the strain ellipse. This underestimates the homogeneous component of shortening, because the Euler Pole EU-NA probably migrated northwards since 91 Ma. Given the age of Tore, ≈91 Ma, and the estimation of shortening we can extrapolate this result backwards and for the oldest oceanic lithosphere, ≈120 Ma, we obtain an accumulated shortening of 8%, if we use a linear approximation, which is valid for these low strains.

To estimate the buckling component of shortening we must use a cross section of the MTR. Assuming plane strain in the coaxial regime of deformation, we can balance the section by the excess area method (for a review see Ramsay and Huber, Vol. 2, 1987); if the depth to décollement is known, we can estimate the amount of shortening (Hossack 1979).

Table 5.1. Data on observed and estimated finite strain at Tore

	Maximum (%)	Average (%)	Minimum (%)
Observed shortening	37	29	21
Estimated homogeneous shortening	7	6	5
Estimated shortening due to buckling	18	15	12
Total estimated shortening	25	21	17

The section is shown in Fig. 5.6. The seismic profile crosses the Azores-Gibraltar Plate boundary; we neglect the minor dextral strike-slip displacement of the Madeira-Tore Rise at this location, below 20 km, when compared with the 150 km width of the MTR. The rounded nature of the section across the buckle of the MTR favours a model with deformed hanging wall and footwall above and below a gently curved thrust, as seen in some field examples (Ramsay 1992), rather than the classic fault bend model (Suppe 1983), with deformed hanging wall above a thrust with ramp flat-planar geometry; this conclusion is supported by the seismic image of the Moho in the section.

The excess area can be estimated from the geomorphological expression of the active buckle above the datum represented by the depth of abyssal plains at 5 km. The difficulty in the application of the method of balancing the section consists in the estimation of the depth of dêcollement. However, by using the crustal structure model of the MTR across the Josephine Seamount (Peirce and Barton 1991), we can estimate a maximum and minimum value for the shortening by inferring a minimum and maximum depth for the dêcollement, independently of any inherited crustal thickening prior to buckling. The maximum depth must be 12 km, because this is the depth of the Moho on both sides of the MTR, showing that the dêcollement should be at this level or above it, otherwise the Moho would be shallower in the hanging wall on the west side of the MTR. The corresponding shortening is 18%. If the dêcollement is higher than the Moho, the amount of shortening will be increased accordingly, but this is unlikely, given the rheological layering in the oceanic lithosphere. If some of the relief of the MTR is inherited, the amount of shortening is decreased accordingly: 1 km of inherited relief reduces the shortening to 12%.

Table 5.1 summarises the data on finite strain observed and estimated at Tore; the estimated values fall in the range of observed values, taking into account the error range for each value. We conclude that values for the shortening due to deformation in mature oceanic lithosphere are in the range of 6±1% for the homogeneous component and 15±3% for the buckling component; the amount of volume loss is unknown because both estimations assume constant volume/area for ellipse/ellipsoid.

These results have obvious implications for deformation of the oceanic lithosphere on a global scale. The buckling component of shortening must be restricted to the length of the MTR, but the homogeneous component of ≈8% must be extrapolated to the whole slow-spreading Atlantic Ocean, because there should be compatibility between the width of magnetic anomalies at the MTR and elsewhere in the Atlantic Ocean.

Table 5.2. Results of finite strain estimation for Tore and for the plate boundary zones of the central Indian Ocean

Plate pair	Distance to Euler Pole	Angle of rotation	Strain regime
Capricorn--India	2°	40±5°	Buckling
Eurasia--North America	40–50°	20°	Homogeneous shortening

From the strain estimation we can infer the strain rate for ocean lithosphere deformation in the domain of Tore, in order to compare it with other strain rate values in other oceanic domains. The critical parameter to be estimated is the duration of the homogeneous shortening and buckling regimes. The geodynamic interpretation of the tectonic regimes in the West-Iberia margin (Sect. 4.7) shows us that the change in stress field from the collision regime Eurasia-Africa to incipient subduction–obduction in the West-Iberia margin and Madeira-Tore Rise must have occurred near the Miocene-Pliocene boundary, at ≈5 Ma or near the Pliocene–Pleistocene boundary, at ≈2 Ma. The estimated strain rates are, on average, 6.9×10^{-16} to 7.16×10^{-16} s^{-1}, for homogeneous shortening, and 2.3×10^{-14} to 5.7×10^{-14} s^{-1} for the buckling regime. This confirms our previous views that intraplate deformation proceeds at strain rates two orders of magnitude lower than interplate strains.

We can now compare the results of finite strain estimation for Tore and for the plate boundary zones of the central Indian Ocean. We can use the well-defined Euler Pole between Capricorn and India and the angle of rotation around this pole for the domain where buckling was attained. For Tore we can estimate the same parameters, keeping in mind that the Euler Pole for Eurasia-North America migrated between 90 Ma and the present position. The results suggest that for an angle of rotation, measured in the strained state of 30°±10°, we change the strain regime from homogeneous shortening to buckling. This corresponds to a linear shortening between 6 and 23%, around a mean value of 13% (Table 5.2).

5.4 Impact Tectonomagmatism at Tore

If Tore is a mega-impact crater we must consider its possible geodynamic effects.

If one process influences another of the same or different nature, we must carefully characterise the link between the cause and its effects. We must face two possibilities, clearly defined for instance in the case of seismicity (Simpson 1986).

If the causative process originates another process, which was completely absent when the first operated, we speak of induction; this would be the case of seismicity induced by operations in big oil fields (McGarr 1991) or of shear heating due to seismic rupture that generates pseudotachylites in a domain where no melt was present before the rupture. If the causative process merely accelerates another process that was already going on before the causative process started, we speak of triggering; this is the case when seismicity is triggered by impounding a high reservoir, if the lubrication, by high pressure fluid, of a previous active fault is sufficient to anticipate a seismic event that would occur naturally later on. It is also the case if a natural earthquake triggers another natural earthquake in a different active fault, where the process of accumulation was already advanced (Scholz 1998); or if a ductile shear zone promotes melting, due to increased permeability, in a wall rock that is already near melting conditions.

In the case of impact cratering we must distinguish between the direct tectonic effects induced by the impact (referred to in Sect. 5.1 in general terms, and in Sect. 5.2 as applied to Tore) and the indirect effects, tectonic and magmatic, triggered by the impact that we will now consider.

The Tore-Gulf of Cadiz lineament (Fig. 5.10) has been studied recently on its onshore segment (Terrinha 1998; Terrinha et al. 1999). It corresponds to a discontinuous fracture zone, with variable expression at the present surface. When it is well defined, such as south of the Sintra Massif, and 20 km west of Lisbon, it is a subvertical strike-slip fault, oriented NW–SE with a subhorizontal offset of ≈0.5 km in the Mesozoic cover of the Lusitanian Basin (Ribeiro et al. 1979).

Where the lineament cuts Mesozoic rocks, we can observe that it is independent of extensional faults and transfer zones related to the opening of the Lusitanian Basin, from Trias to Turonian. The lineament is a neoformed fault of post-Turonian age in most of the onshore segment. However, south of Monchique, the lineament is inherited from a N–S fault that was active during the opening of the Algarve Basin, in Trias–Jurassic times, called the Portimão Fault (Terrinha et al. 1999).

Along the Tore-Gulf of Cadiz (TGC) lineament, we find the massifs of Sintra, Sines and Monchique and mafic dykes south of Sintra and south of Monchique.

These magmatic rocks belong to an alkaline province with variable composition (Rock 1982). The plutonic rocks have ages between 82 and 72 Ma and the volcanic and hypabyssal rocks have ages between 77 and 72 Ma.

A model for emplacement of the plutonic rocks has been proposed (Terrinha 1998; Fig. 5.11); each massif is a magma diapir of deep origin. Magma ascends at releasing bends of the TGC strike-slip fault zone below the plastosphere; it migrates horizontally at a lower crustal detachment, necessary to explain the listric geometry of extensional faults during the Mesozoic; above, in the schizosphere, the massifs are emplaced at the intersection of the lineament and E–W extensional faults probably inherited in the Variscan basement. So the plutonic rocks of Sintra, Sines, Monchique and possibly Gulf of Cadiz are tectonically controlled by the TGC lineament. The plutons occur only along the lineament and, in the case of Sines, they are even dragged dextrally along it.

There is some evidence of fracture propagation along the lineament: an older magmatic diapir was emplaced at 82 Ma in Sintra; younger diapirs occurred at 72 Ma in Sintra, Sines and Monchique. This is compatible with propagation from NW to SE along the lineament. Taking into account the geochronological and magnetostratigraphic evidence, we must note that the magnetic dipoles along the TGC lineament indicate that the magmatic intrusions took place in periods of normal magnetic polarity, such as the Cretaceous Normal Polarity superchron, between 119 and 83 Ma, or a younger normal polarity interval between 78 and 72 Ma. Most of the Sintra diapir would have been emplaced during the older interval, whereas the younger diapir in

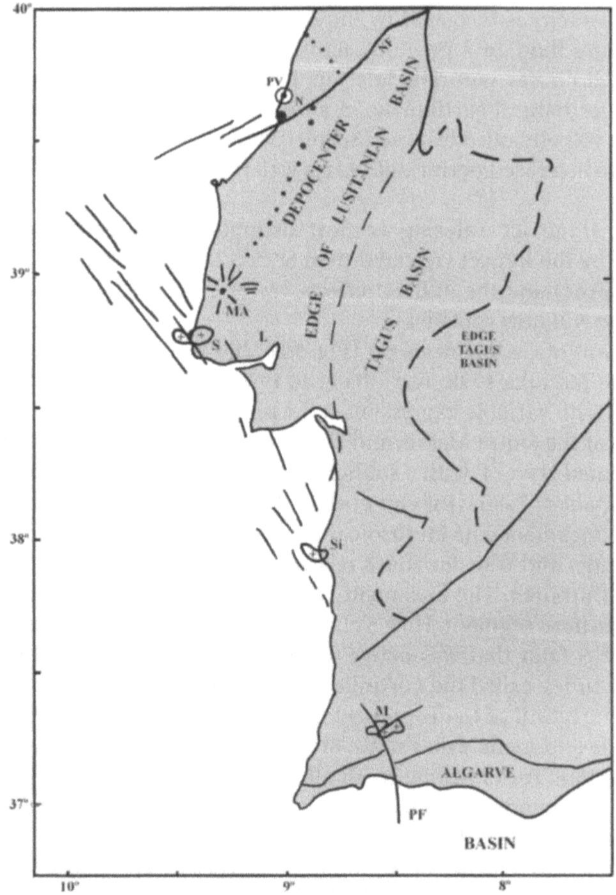

Fig. 5.10. Tectonics of the Tore-Cadiz lineament on the continental domain. Basins: Mesozoic–Lusitanian Basin, with depocentre (*dashed*) and Algarve Basin. Cenozoic–Tagus Basin. Magmatic rocks – subvolcanic massifs: *M* Monchique, *Si* Sines, *S* Sintra, *MA crosses*. Mafra – volcanic centre with radial mafic dykes. *L* Lisbon, *N* Nazaré, *PV* ejecta site at Praia da Vitória. *Thick lines* Faults. *NF* Nazaré Fault, *PF* Portimão Fault. Other faults – distributed shear along the Tore-Cadiz Lineament. (Modified from Terrinha 1998)

Sintra, and the other diapirs in Sines and Monchique, should belong to the younger interval in this model.

There are also minor intrusions outside the lineament, but they obey a different control; the oldest one is Mafra: a stock of gabbro from which a radial mafic dyke swarm irradiates; this suggests a state of lithostatic stress with negligible remote tectonic stress. The isotopic ages, ~100 Ma by K/Ar methods (Ferreira in Ribeiro et al. 1979), must be due to excess argon; in fact, there is no palaeogeographic indication in the carbonate platform of any disturbance prior to Turonian. The Mafra stock is loca-

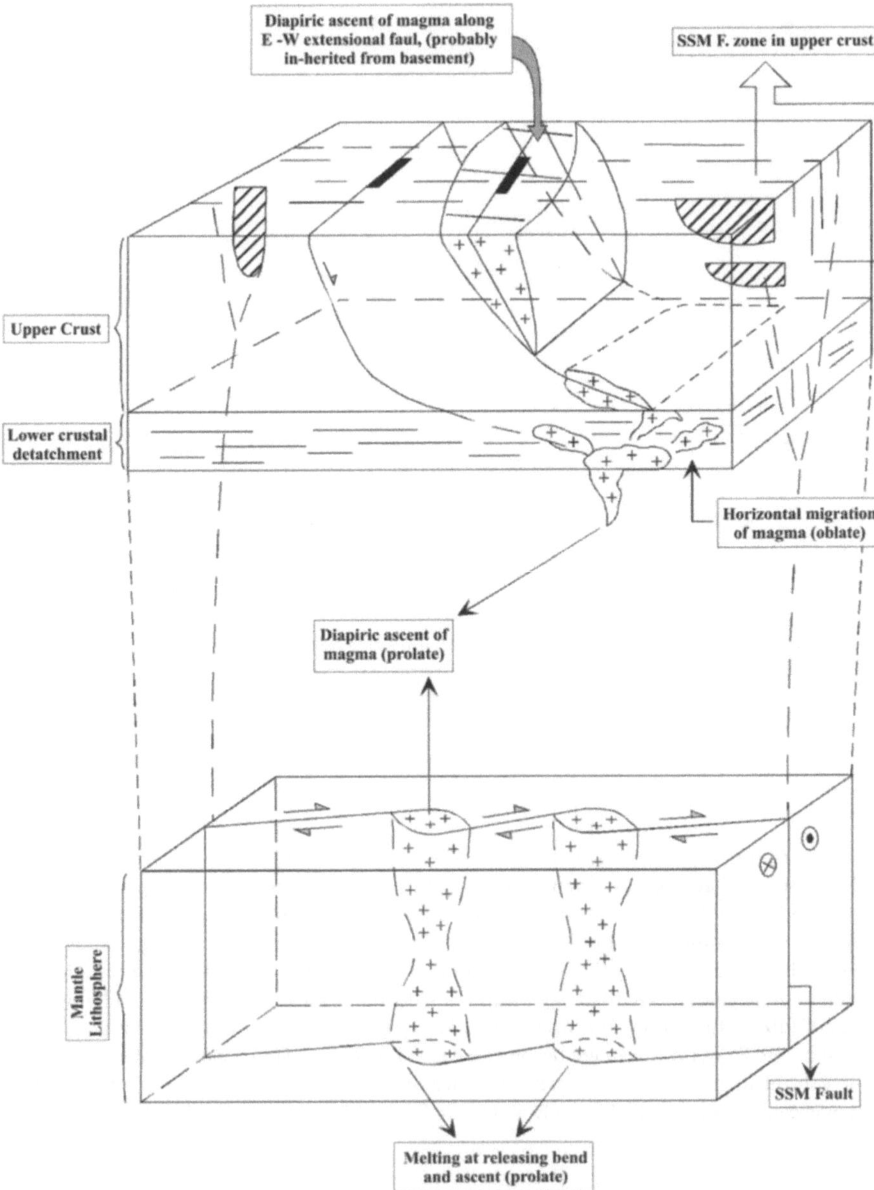

Fig. 5.11. Speculative model for the emplacement of the Sintra, Sines and Monchique Massifs. It is suggested that magma generation and ascent are controlled by releasing bends located along the deep-seated strike slip fault, joining the Massifs and shown in Fig. 5.10. A horizontal ductile detachment (possibly in the lower crust) inhibits upward propagation of the fault to shallower crustal levels, where E–W faults, active during the stretching of the Lusitanian Basin, controlled emplacement and shape of the Massifs. (After Terrinha 1998)

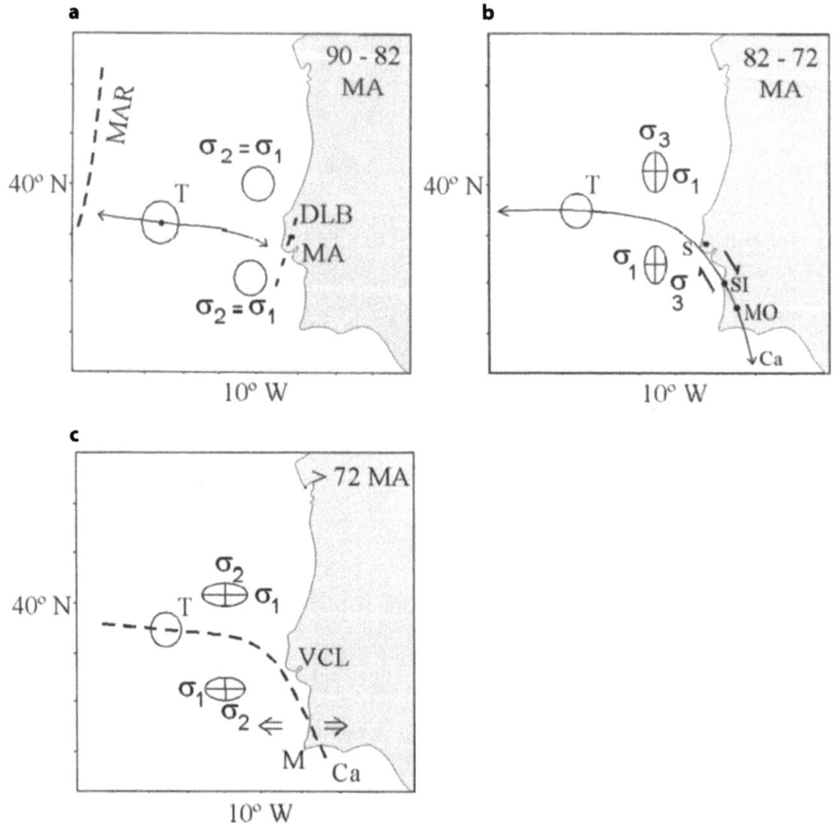

Fig. 5.12. Geodynamic evolution of the Tore-Cadiz lineament. **a** Between 90 and 82 Ma. Impact at the Tore (*T*) that triggers magmatism at Mafra (*MA*) situated on the depocentre of the Lusitanian Basin (*DLB*). The lineament is mainly tensile and propagates westwards until it is stopped at the location of the Mid-Atlantic Ridge (*MAR*) at that time; and propagates eastwards. σ_2 and σ_1 have approximately the same magnitudes; σ_3 is vertical. **b** Between 82 and 72 Ma. The lineament propagates eastwards, triggering magmatic diapirs at *S* Strike, *Si* Sines, *M* Monchique, and *Ca* Cadiz, in that order. The stress field corresponds to N–S compression with σ_3 N–S, σ_2 vertical σ_1 and E–W, the lineament becomes a dextral strike-slip fault zone. **c** 72 Ma and younger. Mafic volcanism generated the volcanic complex of Lisbon and dykes around Monchique (*M*). This magmatism is controlled by an E–W tensile field, with σ_1 E–W; σ_2 N–S and σ_3 vertical

ted in the deepest axis of the Lusitanian Basin, well outside the TGC lineament. The nearly lithostatic state of stress is compatible with the small E–W extension in the faulted margins of the Lusitanian Basin, but not with the stress field that induced propagation of the TGC lineament.

A period of mafic volcanism, the so-called Lisbon Volcanic Complex, follows the magmatic events referred to above and its age must be 72.5±3 Ma (Ferreira in Ribeiro et al.1979)

We conclude that there is a close spatial and temporal connection between the impact event at Tore and the tectonomagmatic events along the TGC lineament. In Section 5.1, we showed that a mega-impact could induce fracturing of the lithosphere and trigger magmatism in the asthenosphere. On the basis of this relationship, we propose the following scenario for the Upper Cretaceous events in the West-Iberia margin (Fig. 5.12).

Near the Cenomanian-Turonian boundary at 91 Ma, a mega-impact occurred at Tore, generating a ≈120 km diameter crater; the body that caused it must have been in the range of 5–10 km in diameter, using the scaling relationship between size of the crater and size of the body. The ejecta of this impact is preserved at Praia da Vitória. This impact caused a large tsunami, preserved in the stratigraphic record at Nazaré. The impact triggered the ascent of magma at Mafra, probably from an underplate magma chamber in the axis of the Lusitanian Basin that produced earlier mafic magmatic episodes near the Jurassic-Cretaceous boundary 144 Ma. The impact also induced a radial dynamic fracturing in its vicinity; the most favourably oriented fracture propagated slowly both to the west and to the east. To the west, this slow propagation proceeded during some million years until it stopped when the fracture reached the hotter Mid-Atlantic ridge at that time. To the east, the fracture propagated across the stretched margin of West Iberia and was deviated to a NW–SE direction when it reached the normal continental crust of Iberia. The impact triggered magmatism along the fracture, with a delay of some 10–20 Ma for diapiric ascent of the lithosphere above the source in the asthenosphere; Sintra reached the surface first, Sines and Monchique followed. There was a probably a large mantle thermal anomaly at upper Cretaceous because the magmatic province extends all over Iberia and conterminous domains (Rock 1982). Along the lineament, previous fracture zones were opened, giving access to mafic magma, south of Sintra and Monchique, both in the main dextral shear zone, NNW–SSE, and on Riedel shears, N–S oriented.

The curved trace of the TGF lineament suggests that it is due to heterogeneities between the oceanic, stretched continental and normal continental lithosphere. It probably indicates a changing palaeostress trajectory; the lithostatic stress field at Mafra, around 90 Ma, changes later to N–S compression giving a dextral component of strike-slip along the lineament during diapir emplacement. This can be due to remote tectonic stresses at the convergent boundaries of the Iberia Plate during Upper Cretaceous, either by N–S subduction along the northern Iberian margin or by early collision in the Betics southern margin. This remote stress became relaxed during extrusion of the Lisbon volcanic complex, around 72 Ma, with maximum expression at the intersection of TGC with the axis of the Lusitania Basin.

The TGC lineament is a singular feature which remains unexplained in the context of the plate tectonics but can be understood in the context of impact tectonics. It is satisfying to observe that the Discovery Scarp is a 500-km-long thrust fault nucleating in an impact crater in Mercury (Jones 1999).

Therefore, we favour the hypothesis, on the basis of field evidence, that impact cratering can <u>induce</u> fracture of the lithosphere and can <u>trigger</u> magmatic activity, possibly in the asthenosphere but perhaps also in the lithosphere itself.

5.5 Conclusion

In conclusion, we found evidence for generalised deformation in the oceanic lithosphere at a strain rate some two orders of magnitude lower than in plate boundary zones, but some two orders of magnitude above stable continental interiors (Table 5.3). The entire oceanic lithosphere shows evidence of incremental strain. This includes homogeneous shortening perpendicular to plate motion and extension at right angles; posterior buckling and simultaneous faulting start leading to subduction initiation and then slab pull produces elongation parallel to plate motion.

The direct evidence for finite strain is restricted more to particular domains, where it has been proved. These include young oceanic lithosphere such as in the Azores but also mature oceanic lithosphere such as in Tore "Seamount". However, indirect evidence is widespread in oceanic lithosphere of all ages from rift to abyssal plains; the extrapolation from the observed strain rates leads to the inevitable conclusion that all oceanic lithosphere is able to accommodate considerable strains; these are lower if the lithosphere fails in rapid tension and higher if it fails in long-lived compression. A realistic plate tectonic theory should include intraplate deformation as a fundamental mechanism in the dissipative dynamic process. This is a fundamental step in the study of the Earth as an open dynamic system.

The Earth is a dynamic system. All the geospheres interact with each other; they become self-organised in a system operating on a global scale. Some authors (Keilis-Borok 1990; Scholz 1990; Bak 1996) even think that the lithosphere is a self-organised critical (SOC) system, by the play of internal and external geodynamic subsystems: each subsystem can drive or be driven by another one. In self-organised systems, feedback mechanisms are essential; intraplate deformation is such a fundamental feedback process to maintain the system near a critical state.

Reservoir-triggered seismicity (Simpson 1986) shows that a column of water above 100 m can inhibit seismicity in compressive stress environments. This suggests that ocean water depths of up to 10 km should drive active compressive structures from instability and make a significant contribution to aseismic deformation in the oceanic lithosphere.

The Earth is also an open system. Impacting is a very important process on a global scale. However, the extra-terrestrial forcing can be coupled with the inner forcing of Earth itself; non-linear coupling with resonance can explain the coincidence in time between mega-impact showers and superplumes that punctuate the Earth's history (Shaw 1994).

The Earth is the only planet with plate tectonics instead of blob tectonics. It is also the only planet with a bimodal lithosphere. In fact, throughout Earth's history, plate

Table 5.3. Strain rates

Orogens and PBZ	$\dot{\varepsilon} = 10^{-12}s^{-1}$ to $10^{-14}s^{-1}$
Oceanic interiors	$\dot{\varepsilon} = 10^{-15}s^{-1}$ to $10^{-17}s^{-1}$
Cratons	$\dot{\varepsilon} = 10^{-18}s^{-1}$

tectonics contributed to the segregation of the continental crust from an undifferentiated primitive outer layer, therefore, the duality between ocean and continental lithosphere is peculiar to the Earth (Taylor and McLennan 1996). The presence of liquid water controls the form of convection required by plate tectonics; so the Earth is a water-cooled, convective planet. Presumably, plate tectonics will stop on Earth, in some 2000 Ma time, when surface heat flow and dehydration will make the oceanic lithosphere as rigid as the continental lithosphere in cratonic interiors. We conclude that fluids play a major role in the driving mechanism for plate tectonics from its beginning to its end.

References

Agar SM (1990) Fracture evolution in the upper ocean crust: evidence from DSDP hole 504B. In: Knipe RJ, Rutter EH (eds) Deformation mechanisms, rheology and tectonics. Geological Society, London, Special Publication, vol 54, pp 41–50

Agar SM (1994) Rheological evolution of the ocean crust: a microstructural view. J Geophys Res 99:3175–3200

Aki K (1981) A probabilistic synthesis of precursory phenomena simpson. I. In: David MI, Richards PG (eds) Earthquake prediction an international review. II. American Geophysical Union IV, Series. American Geophysical Union, Washington, DC, pp 566–574

Aki K (1984) Asperities barriers, characteristic earthquakes and strong motions predictions. J Geophys Res 89:5867–5872

Alexander RJ, Harper GD, Bownam JR (1993) Oceanic faulting and fault controlled subseafloor hydrothermal alteration in the sheeted dike complex of the Josefine Ophiolite. J Geophys Res 98:9731–9759

Allègre C (1999) L'Écume de la Terre, 3ème édn. Fayard, Paris, 312 pp

Allègre C, Jaupart C (1985) Continental tectonics and continental kinetics. Earth and Planetary Sci Lett 74:177–186

Allen P A (1997) Earth Surface Processes. Blackwell, London, 404 pp

Allen P A, Allen JR (1990) Basin analysis principles and applications. Blackwell, London, 451 pp

Allis, RG (1981) Continental underthrusting beneath the Southern Alps of New Zealand. Geology 9:303–307

Alvarez LW, Alvarez W, Asaro F, Michel HV (1980) Extraterrestrial cause for the Cretaceous-Tertiary extinction. Science 208:1095–1108

Alvarez W, Alvarez LW, Asaro F, Michel HV (1982) Current status of the impact theory for the terminal Cretaceous extinction. Geol Soc Am Spec 190:305–315

Anderson D (1975) Accelerated plate tectonics. Science 187:1077–1079

Anderson DL (1984) The Earth as a planet: paradigms and paradoxes. Science 223:347–355

Anderson DL (1989) Theory of the earth. Blackwell, London, 366 pp

Anderson DL, Dziewonski M (1984) Seismic tomography. Sci Am 251:58–66

Anderson EM (1905) Dynamics of faulting. Trans Edinb Geol Soc 8:387–402

Anderson EM (1951) The dynamics of faulting and dyke formation with application no Britain, 2nd edn. Oliver and Boyd, Edinburgh, 206 pp

Andeweg B, De Vicente G, Cloetingh S, Giner J, Martin AM (1999) Local stress fields and intraplate deformation of Iberia: variations in spatial and temporal interplay of regional stress sources. Tectonophysics 305:155–164

Angelier J (1994) Fault slip analysis and palaeostress reconstruction. In: Hancock PI Continental deformation (ed) Pergamon Press, Oxford, pp 53–100

Angelier J, Lyberis N, Le Pichon X, Barrier E, Huchon P (1982) The tectonic development of the Sea of Crete; a synthesis. Tectonophysics 86:159–196

Argus DF, Gordon RG (1991) No-net rotation model of current plate velocities incorporating plate motion model NUVEL. Geophys Res Lett 18:2039–2042

Argus DF, Gordon RG (1996) Test of the rigid-plate hypothesis and bounds on intraplate deformation using geodetic data from very long baseline interferometry. J Geophys Res 101:13555–13572.

Argus DF, Gordon RG (1999) Current Sierra Nevada-North America motion from very long baseline interferometry: implications for the kinematics of the western United States. Geology 19:1085–1088

Argus DF, Gordon RG, Demets C, Stein S (1989) Closure of the Africa-Eurasia-North America plate motion circuit and tectonics of the Gloria Fault. J Geophys Res 94:5585–5602

Atkinson BK (1984) Subcritical crack growth in geological materials. J Geophys Res 89:4077–4117

Atkinson BK (1987) Introduction to fracture mechanics and its geophysical applications II. In: Atkinson B (eds) Fracture mechanics of rock. Academic Press, London, pp 1–26

Auzende JM, Olivet JL, Charvet J, Le Lann A, Le Pichon X, Monteiro JH, Nicolas A, Ribeiro A (1978) Sampling and observation of oceanic mantle and crust on Gorringe Bank. Nature 273:45–49

Avilles CA, Scholz CH, Boatwright J (1987) Fractal analysis applied to characteristics segments of the San Andreas Fault. J Geophys Res 92:331–344

Badal J, Corchete V, Payo G, Canas JA, Pujades L (1993) Study of Iberian subcrustal lithosphere and asthenosphere: methods and models. In: Mezcua J, Carreno E (eds) Iberian lithosphere heterogeneity and anisotropy. Instituto Geografico Nacional, Madrid, Monografia no 10, pp 57–73

Bak P (1990) Is the world at the border of chaos? Ann N Y Acad Sci 581:110–118

Bak P (1996) How nature works: the science of self-organized criticality. Springer, Berlin Heidelberg New York, 212 p

Bak P, Chen K (1991) Self-organized criticality. Sci Am 254:26–33

Bak P, Chen K (1995) Fractal dynamics of earthquakes. In: Barton C, La Pointe PR (eds) Fractals in the Earth Sciences. Plenum Press, New York, pp 227–236

Bak P, Tang C (1989) Earthquakes as a self-organized critical phenomenon. J Geophys Res 94:15685–15687

Bak P, Tang C, Wiesenfeld K (1988) Self-organized criticality. Phys Rev A 38:364–374

Baker ET, Massoth GJ, Feely RA (1987) Cataclysmic hydrothermal venting on the Juan the Fuca Ridge. Nature 316:149–151

Baker GL, Gollub JP (1990) Chaotic dynamics: an introduction. Cambridge University Press, Cambridge, 182 pp

Baker P, Stout P, Kastner M, Elderfield (1991) Large-scale lateral advection of seawater through oceanic crust in the central equatorial Pacific. Earth Planet Sci Lett 105:522–533

Bally AW (1981) Thoughts on the tectonics of folded belts. In: McClay K, Price NJ (eds) Thrust and nappe tectonics. Special Publication Geological Society 9, London. Blackwell, Oxford, pp 13–32

Bally AW (1984) Tectogenèse et sismique réflexion. Bull Soc Géol Fr 2:279–285

Banda E, Torne M (1995) Iberian Atlantic Margins Group investigates deep structure of ocean margins and the Iberian margins. American Geophysical Union, vol 76, EOS Transactions, pp 25–29

Banks CJ, Warburton J (1990) Mid-crustal detachment in the Betic system of southeastern Spain. Tectonophysics 191:275–289

Baptista MA, Heitor S, Miranda JM, Miranda P, Victor LM (1998a) The 1755 Lisbon tsunami; evaluation of the tsunami parameters. J Geodyn 15:143–157

Baptista MA, Miranda PMA, Miranda JM, Victor LM (1998b) Constrains on the source of the

1755 Lisbon tsunami inferred from numerical modelling of historical data on the source of the 1755 Lisbon tsunami. J Geodyn 25:159–174

Barriga FJAS, Fyfe WS, Landefeld LA, Munha J, Ribeiro A (1992) Mantle eduction: tectonic fluidisation at depth. Earth-Sci Rev 32:123–129

Barth GA, Mutter JC, Madsen JA (1991) Upper mantle seismic reflections beneath the EPR. Geology 19:994–996

Barton CC, Larsen E (1985) Fractal geometry of two-dimensional fracture networks at Yucca Mountain, Southwestern Nevada. In: Stephanson O (ed) Proc Int Symp on Fundamentals of Rock Joints, pp 77–84

Barton CC, LaPointe PR, Malinverno A (1992) Fractals and their use in the earth sciences and in the petroleum industry. AAPG Short Course Manual, Houston, TX, 339 pp

Barton CHC, Cameron BG, Montgomery JR (1986) Fractal scaling of fracture and fault maps at Yucca Mountain, southern Nevada. EOS 67:871

Bastos L, Osório J, Barbeito A, Hein G (1998) Results from geodetic measurements in the western part of the African-Eurasian plate boundary. Tectonophysics 294:261–269

Beaumont C, Quinlan G (1994) A geodynamic framework for interpreting crustal-scale seismic-reflectivity patterns in compressional orogens. Geophys J Int 116:754–783

Beaumont C, Fuklsack PH, Hamilton J (1992) Erosional control of active compressional orogens. In: McClay K (ed) Thrust tectonics. Chapman and Hall, London, pp 1–18

Beavan J, Moore M, Pearson C, Henderson M, Parsons B, Bourne S, England PH, Walcott D, Blick G, Darby D, Hodgkinson K (1999) Crustal deformation during 1994–1998 due to oblique continental collision in the central Southern Alps, New Zealand, and implications for seismic potential of the Alpine fault. J Geophys Res 104:25233–25255

Beekman F, Bull JM, Cloetingh S, Scrutton RA (1996) Crustal fault reactivation facilitating lithospheric folding/buckling in the central Indian Ocean. In: Buchanan PG, Newland DA (eds) Modern developments in structural interpretation, validation and modelling. Geol Soc Spec Publ 99:251–263

Beekman W (1994) Tectonic modeling of thick-skinned compression intraplate deformation. PhD Thesis, Free University Amsterdam, Amsterdam, 152 pp

Ben-Avraham Z, Nur A, Jones D, Cox A (1981) Continental accretion and orogeny: from oceanic plateaus to allochthonous terranes. Science 213:47–54

Bercovici D (1995) A source-sink model of the generation of plate tectonics from non-Newtonian mantle flow. J Geophys Res 100:2013–2030

Bergé P, Dubois M (1992) Chaos deterministe expérimental et attracteurs étranges. In: Dalmedico AD, Chabert JL, Chemla K (eds) Chaos et déterminisme. Éditions du Seuil, Paris, pp 115–169

Bergman EA, Solomon SC (1984) Source mechanisms of earthquakes near mean-ocean ridges from body waveform inversion: implications for the early evolution of oceanic lithosphere. J Geophys Res 89:11415–11441

Bergman EA (1986) Intraplate earthquakes and the state of stress in oceanic lithosphere. Tectonophysics 132:1–35

Bergman EA, Solomon SC (1990) Earthquake swarms on the Mid-Atlantic Ridge: products of magmatism or extensional tectonics? J Geophys Res 95:4943–4965

Berzin R, Oncken O, Knapp JH, Pérez-Estaún A, Hismatulin T, Yunusov N, Lipilin A (1996) Orogenic evolution of the Ural Mountains: results from an integrated seismic experiment. Science 274:220–221

Bevis M, Isacks BL (1984) Hypocentral trend surface analysis: probing the geometry of Beniolf zones. J Geophys Res 89:6153–6170

Biddle KT, Rudolph TW (1988) Early Tertiary structured inversion in the Stord Basin, Norwegian North Sea. J Geol Soc Lond 44:201–219

Biot MA (1965) Mechanics of incremental deformations. Wiley, New York, 504 pp
Bird P (1991) Lateral extrusion of lower crust from under high topography, in the isostatic limit. J Geophys Res 96:10275–10286
Birot P (1964) La Méditerranée et le Moyen.Orient. I. Généralités Péninsule Ibérique, Italie. Presse Universitaire de France, Paris, 550 pp
Blundell D, Freeman R, Mueller S (1992) A continent revealed: the European geotraverse. Cambridge University Press, Cambridge, 275 pp
Bohannon RS, Howell DG (1982) Kinematic evolution of the junction of the San Andreas, Garlock, and Big Pine faults, California. Geology 10:358–363
Boillot G, Beslier MO, Comas M (1992) Seismic image of under-crusted serpentinite beneath a rifted margin. Terra Nova 4:25–33
Bonatti E, Michael PJ (1989) Mantle peridotites from continental rifts to ocean basins to subduction zones. Earth Planet Sci Lett 91:297–311
Boslough MB, Chael EP, Trucano TC, Crawfoo DA, Campbell DL (1996) Axial focusing of impact energy in the Earth's interior: a possible link to flood basalts and hotspots. In: Ryder G, Fastowsksy D, Gartner S (eds) The Cretaceous-Tertiary event and other catastrophes in earth history. Geol Am Spec Publ 307:541–550
Bosworth W, Strecker MR, Blisniuk PM (1992) Integration of East African paleostress and present-day stress data: implications for continental stress field dynamics. J Geophys Res 97:11851–11865
Bott MP, Kusznir NJ (1984) Origins of tectonic stress in the lithosphere. Tectonophysics 105:1–14
Boucher C, Altamimi Z, Sillard P (1999) The 1997 International Terrestrial Reference Frame (ITRF97), IERS Tech. Note 27, IERS Cent Bur, Obs de Paris.
Bourne SJ, England PC, Parsons B (1998) The motion of crustal blocks driven by flow of the lower lithosphere and implications for slip rates of continental strike-slip faults. Nature 391:655–659
Bratt SR, Bergman EA, Solomon D (1985) Thermo-elastic stress: how important as cause of earthquake in young ocean crust? J Geophys Res 90:10249–10260
Brewer JA, Allmendinger RW, Brown LD, Oliver JE, Kaufman S (1982) Cocorp profiling across the Rocky Mountain Front in southern Wyoming, part 1: laramide structure. Geol Soc Am Bull 93:1242–1252
Brikowski T, Norton D (1989) Influence of magma chamber geometry on hydrothermal activity at mid-ocean ridges. Earth Planet Sci Lett 93:241–255
Brocher T M, Karson JA, Collins JA (1985) Seismic stratigraphy of the oceanic Moho based on ophiolite models. Geology 13:62–65
Brown D, Puchkow V, Alvarez-Marron J, Perez-Estaun A (1995) The structural architecture of the footwall to the Main Uralian Fault, southern Urals. Earth-Sci Rev 915:1–21
Brown GC, Mussett AE (1993) The inaccessible earth. An integrated view to is structure and composition. Chapman and Hall, London, 276 pp
Brown SR, Scholz SH (1985) Broad bandwidth study of the topography of natural rock surfaces. J Geophys Res 90:12575–12582
Brum A (1990) Neotectónica e Sismotectónica da Região Vidigueira-Moura. MSc Thesis, Dep Geologia, Fac Ciências Lisboa, 204 pp
Buck WR (1986) Small-scale convection induced by passive rifting: the cause for uplift of rift shoulders. Earth Planet Sci Lett 77:362–372
Buck WR, Poliakov ANB (1998) Abyssal hills formed by stretching oceanic lithosphere. Nature 350:272–275
Buforn E, Udias A, Colombás MA (1988) Seismicity source mechanisms and tectonics of the Azores - Gibraltar plate boundary. Tectonophysics 152:89–118

Bull JM (1992) Seismic reflection images of intraplate deformation, central Indian Ocean, and their tectonic significance. J Geol Soc Lond 149:955–966
Bull JM, Scrutton RA (1992) Seismic reflection image of intraplate deformation, Central Indian Ocean and their tectonic significance. J Geol Soc Lond:149:955-966
Bull JM, Martinod J, Davy P (1992) Buckling of the oceanic lithosphere from geophysical data and experiments. Tectonics 11:537–548
Burchfiel BC, Royden L (1982) Carpathian foreland fold and thrust belt and its relation to Pannonian and other basins. Bull Am Assoc Petrol Geol 66:1179–1195
Burg JP, Ford M (1997) Orogeny through time: an overview. In: Burg JP, Ford M (eds) Orogeny through time. Geol Soc Spec Publ 121:14–17
Burnham CW, Davis NF (1971) The role of H2O in silicate melts, I, P-V-T relations in the system $NaAlSi_3O_8$–H_2O to 10 kilobar and 10,000 °C. Am J Sci 270:54–79
Burnham CW, Davis NF (1974) The role of H2O in silicate melts, II, thermodynamic and phase in the system $NaAlSi_3O_8$–H_2O to 10 kilobar and 10,000 °C. Am J Sci 274:902–940
Burov EB, Diament M (1995) The effective elastic thickness (Te) of continental lithosphere: what does it really mean? J Geophys Res 100:3905–3927
Byerlee JD (1978) Friction of rocks. Pure Appl Geophys 116:615–626
Cabral J (1989) An example of intraplate neotectonic activity, Vilariça basin, northeast Portugal. Tectonics 8:285–303
Cabral J (1995) Neotectónica em Portugal Continental. Mem Inst Geol Mineiro 31:265 pp
Cabral J, Ribeiro A (1989) Incipient subduction along the West-Iberia continental margin. Abstracts, 28 h International Geological Congress, Washington, DC, vol 1, 223 pp
Cadek O, Ricard Y (1992) Toroidal/poloidal energy partitioning and global lithosphere rotation during Cenozoic time. Earth Planet Sci Lett 109:621–622
Cadet JP, Kobayashi K, Lallemand S, Jolivet L, Aubouin J, Boulègue J, Dubois J, Hotta H, Ishii T, Konishi K, Niitsuma N, Shimamura H (1987) Deep scientific dives in the Japan and Kuril Trenches. Earth Planet Sci Lett 83:313–328
Calapez P (1998) Estratigrafia e Paleobiologia do Cenomaniano-Turoniano, o significado do eixo da Nazaré-Leiria-Pombal. PhD Thesis, Coimbra, 379 pp
Calvert A (1995) Seismic evidence for a magma chamber beneath the slow spreading Mid-Atlantic ridge. Nature 377:410–414
Campbell AC, Palmer MR, Klinkhammer TS, Edmond JM, Lawrence JR, Casey JF, Thompson G, Humphris S, Rona P, Karson JA (1988) Chemistry of hot springs on the Mid-Atlantic Ridge. Nature 335:514–519
Cann JR (1974) A layered model for oceanic crust developed. Geophys J R Astron Soc 39:169–187
Cann JR, Strens MR (1982) Black smokers fuelled by freezing magma. Nature 298:147–149
Cann JR, Strens MR (1989) Modeling periodic megaplume emissions by black smoker systems. J Geophys Res 94:12227–12237
Cannat M (1993) Emplacement of mantle rocks in the seafloor at mid-ocean ridges. J Geophys Res 98:4163–4172
Cannat M, Mével C, Stakes D (1991) Stretching of the deep crust at the slow spreading southwest Indian ridge. Tectonophysics 190:73–94
Cannat M, Mevel C, Maia M, Deplus C, Durand C, Gente P, Agrimier P, Belarouchi A, Dubuisson G, Humier E, Reynolds J (1995) Thin crust, ultramafic exposures, and rugged faulting patterns at the Mid-Atlantic Ridge (220–240 N). Geology 23:49–52
Carlson RL, Herrick CN (1990) Densities and porosities in the oceanic crust and their variations with depth and age. J Geophys Res 95:9153–9170
Carr MH (1987) Water on Mars. Nature 326:30–35

Carter N, Tsenn MC (1987) Flow properties of continental lithosphere. Tectonophysics 186:27-65

Carter NL, Kronenberg AK, Ross JW, Wilischko DW (1990) Control of fluids on deformation of rocks. In: Knipe RJ, Rutter EH (eds) Deformation mechanisms, rheology and tectonics. Geological Society, London, Special Publication, vol 54, pp 1–13

Cathles L (1990) Scales and effects of fluid flow in the upper crust. Science 248:323–329

Cazenave A, Feigl K (1994) Formes et Mouvements de la Terre. Satellites et géodésie. BELIN CNRS Editions, Paris, 159 pp

Chamot-Rooke N, Jestin F, De Wood B, Phedre Working Group (1993) Intraplate shortening in the central Indian Ocean determined from a 2100-km long north-south deep seismic reflection profile. Geology 21:1043–1046

Chapple WM (1968) The equations of elasticity and viscosity, and their application to faulting and folding. In: Riecker RE (ed) NSF advanced science seminar in rock mechanics for college teachers of structural geology. Air Force Cambridge Res Lab, Bedford, MA, pp 225–284

Charvet J, Ogawa Y (1994) Arc-Trench Tectonics. In: Hancock P (ed) Continental deformation. Pergamon Press, Oxford, pp 180–199

Chase C (1978) Plate kinematics: the Americas, East Africa, and the rest of the world. Earth Planet Sci Lett 37:355–368

Chemenda A, Matte P, Sokolov V (1997) A model of Paleozoic obduction and exhumation of high-pressure/low-temperature rocks in the southern Urals. Tectonophysics 276:217–227

Chemenda AI, Mattauer M, Malavielle J, Bokun AN (1995) A mechanism for syn-collisional rock exhumation and associated normal faulting: results from physical modelling. Earth Planet Science Lett 152:225–232

Chen WP, Grimson L (1989) Earthquakes associated with diffuse zones of deformation in the oceanic lithosphere some examples. Tectonophysics 166:155–158

Chen YJ (1992) Oceanic crustal thickness versus spreading rate. Geophys Res Lett 19:753–756

Chen YJ, Morgan WJ (1990) A nonlinear rheology for mid-ocean ridge axis topography. J Geophys Res 95:17583–17604

Chilès JP (1988) Fractal and geostatistical methods for modeling of a fracture network. Math Geol 20:631–654

Chinnery MA (1966a) Secondary faulting I. Theoretical aspects. Can J Earth Sci 3:163–174

Chinnery MA (1966b) Secondary faulting II Geological aspects. Can J Earth Sci 3:175–190

Chossat P, Stewart CA (1992) Spherical symmetry-breaking bifurcations and thermal convection. In: Yuen DA (ed) The earth's mantle. Chaotic processes in the geological sciences. Springer, Berlin Heidelberg New York, pp 11–42

Christensen NI (1978) Ophiolites, seismic velocities and oceanic crustal structure. Tectonophysics 4:131–157

Christensen NI, Smewing JD (1981) Geology and seismic structure of the northern section of the Oman ophiolite. J Geophys Res 86:2545–2555

Christeson GL, Purdy GM, Fryer GJ (1994) Seismic constraints on shallow crustal emplacement processes at the fast spreading East Pacific Rise. J Geophys Res 99:17957–17973

Chu D, Gordon RG (1999) Evidence for motion between Nubia and Somalia along the southwest Indian ridge. Nature 398:64–67

Cladouhos TT, Allmendinger RW (1993) Finite strain and rotation from fault-slip data. J Struct Geol 15:771–784

Clague DA, Straley PF (1977) Petrologic nature of the oceanic Moho. Geology 5:133–136

Cloetingh S (1982) Evolution of passive continental margins and initiation of subduction zones. Geologica Ultraiectina, no 29. PhD Thesis, University of Utrecht, Utrecht, 111 pp

Cloetingh S, Banda E (1992)Europe's lithosphere – physical properties: mechanical structure. In: Blundell D, Freeman R, Mueller SA (eds) Continent revealed the European geotraverse. Cambridge University Press, Cambridge, pp 80–91

Cloetingh S, Wortel R (1982) Finite element models of passive continental margins with implications for the initiation of subduction zones. Geol Mijnbouw 61:281–292

Cloetingh S, Wortel R, Vlaar NJ (1989) On the initiation of subduction zones. PAGEOPH 129:7–25

Cloos H (1928) Experimenten zur Inneren Tektonik. Centralbl f Mineral Abt B:609–621

Cluff L (1978) Geologic considerations for seismic microzonation. Proc 2nd Int Conf on Microzonation, San Francisco, vol 1, pp 135–152

Cluff LS, Coppersmith KJ, Knuepper PL (1982) Assessing degrees of fault activity for seismic microzonation. 3rd Int Earthquake Microzonation Conf Proc, Seattle, vol I/II, pp 113–118

Cobbold P (1976) Mechanical effects of anisotropy during large finite deformations. Bull Soc Géol Fr 18:1497–1510

Cobbold PR (1979) Removal of finite deformation using strain trajectories. J Struct Geol 1:67–72

Cobbold PR (1982) Buckling of an embedded layer in rubberlike materials: a critical comparison of theories. Tectonophysics 89:7–15

Cobbold PR, Barbotin E (1988) The geometrical significance of strain trajectory curvature. J Struct Geol 10:211–218

Cobbold PR, Davy P (1988) Indentation tectonics in nature and experiment. 2. Central Asia. Bull Geol Inst Uppsala 14:143–162

Cobbold PR, Gapais D (1986) Slip-system domains I. Plane-strain kinematics of arrays of coherent bands with twinned fibre orientations. Tectonophysics 131:113–132

Cobbold PR, Means WD, Bayly MB (1984) Jumps in deformation gradients and particle velocities across propagating coherent boundaries. Tectonophysics 108:283–298

Cobbold PR, Brun JP, Davy P, Fiquet G, Basile C, Gapais D (1989) Some experiments on block rotation in the brittle upper crust. In: Kissel C, Laj C (eds) Paleomagnetic rotations and continental deformation. Kluwer, Dordrecht, pp 145–155

Coleman PJ (1971) Plate tectonics emplacement of upper mantle peridotites along continental edges. J Geophys Res 76:1212–1222

Coleman RG (1977) Ophiolites. Springer, Berlin Heidelberg New York, 229 pp

Condie K (1997) Plate tectonics and central evolution, 4th edn. Butterworth, Heinemann, Oxford, 282 pp

Coney PJ, Jones DL, Monger JWH (1980) Cordilleran suspect terranes. Nature 288:329–333

Connerney JEP, Acuña MH, Wasilewski PJ, Ness NF, Rème H, Mazelle C, Vignes D, Lin RP, Mitchell DL, Cloutier PA (1999) Magnetic lineations in the ancient crust of Mars. Science 284:794–798

Converse DR, Holland HD, Edmond JM (1984) Flow rates in the axial hot springs of the East Pacific Rise (210 N); implications for the heat budget and the formation of massive sulfide deposits. Earth Planet Sci Lett 2 875:159–175

Cosgrove JW (1993) The interplay between fluids, folds and thrusts during the deformation of a sedimentary succession. J Struct Geol 15:491–500

Cottrell AH (1964) The mechanical properties of matter. Wiley, New York, 430 pp

Courtillot V (1995) La Vie en catastrophes. Fayard, 279 pp

Cowan J, Cann JR (1988) Supercritical two phase separation of hydrothermal fluids in the Troodos ophiolite. Nature 333:259–261

Coward M (1994) Inversion tectonics. In: Hancock P (ed) Continental deformation. Pergamon Press, Oxford, pp 289–304

Coward M (1994) Continental collision. In: Hancock P (ed) Continental deformation. Pergamon Press, Oxford, pp 264–288

Coward MP (1983) Thrust tectonics thin skinned of thick skinned, and the continuation of thrusts to deep in the crust. J Struct Geol 5:113–123

Cowie PA, Scholz CH (1992a) Physical explanation for the displacement-length relationship of faults using a post-yield fracture mechanics model. J Struct Geol 14:1133–1148
Cowie PA, Scholz CH (1992b) Displacement-length scaling relationship for faults: data synthesis and discussion. J Struct Geol 14:1149–1156
Cowie PA, Scholz CH, Edwards M, Malinverno A (1993) Fault strain and seismic coupling on mid-ocean ridges. J Geophys Res 98:17911–17920
Cowie PA, Malinverno A, Ryan WBF, Edwards MH (1994) Quantitative fault studies on the East Pacific Rise: a comparison of sonar imaging techniques. J Geophys Res 99:15.205–15.208
Cowie PA, Knipe RJ, Main IG (1996) Introduction to the special issue. Scaling laws for fault and fracture populations-analyses and applications. J Struct Geol 18:v–xi
Cowie PA (1997) Cracks in the Earth's surface. Phys World:31–35
Cox A, Hart RB (1986) Plate tectonics: how it works. Blackwell, Palo Alto, 392 pp
Crétaux JF, Soudarin L, Cazenave A, Bouillé F (1998) Present-day tectonic plate motions and crustal deformations from the DORIS space system. J Geophys Res 108:30167–30181
Cronin VS (1987) Cycloid kinematics of relative plate motion. Geology 15:1006–1009
Cronin VS (1991) The cycloid relative-motion model and the kinematics of transform faulting. Tectonophysics 187:215–249
Cronin VS (1992) Types and kinematic stability of triple junctions. Tectonophysics 207:287–300
Das S (1992) Reactivation of an oceanic fracture by the Macquarie Ridge earthquake of 1989. Nature 357:150–153
Davaille A (1999) Simultaneous generation of hotspots and superswells by convection in a heterogeneous planetary mantle. Nature 402:754–760
Davies GF (1992) On the emergence of plate tectonics. Geology 20:465–466
Davies GF (1999) Dynamic earth: plates, plumes, and mantle convection. Cambridge University Press, Cambridge, 458 pp
Davies GF, Richards MA (1992) Mantle convection. J Geol 100:151–206
Davies R, England P, Parsons B, Billiris H, Paradissis D, Veis G (1997) Geodetic strain of Greece in the interval 1892–1992. J Geophys Res 102:24571–24588
Davis DM, Suppe J, Dahlen FA (1983) Mechanics of fold-and-thrust belts and accretionary wedges. J Geophys Res 88:1153–1172
Davis T, Nanson J (1994) A balanced cross-section of the 1994 Northridge earthquake, southern California. Nature 372:167–169
Davison L (1994) Linked fault systems; extensional, strike-slip and contractional. In: Hancock P (ed) Continental deformation. Pergamon Press, Oxford, pp 121–142
Davy P, Cobbold PR (1988) Indentation tectonics in nature and experiment. 1. Experiments scaled for gravity. Bull Geol Inst Uppsala 14:129-141
Davy P, Cobbold PR (1988) Indentation tectonics in nature and experiment. 1. Central Asia. Bull Geol Inst Uppsala 14:143-162
Davy P, Hansen A, Bonnet E, Zhang S-Z (1995) Localization and fault growth in layered brittle-ductile systems: implications for deformations of the continental lithosphere. J Geophys Res 100:6281–6294
Dawers NH, Anders MH, Scholz CH (1993) Growth of normal faults: displacement-length scaling. Geology 21:1107–1110
De Bremaecker JC (1977) Is the oceanic lithosphere elastic or viscous? J Geophys Res 82:2001–2004
Delaney JR, Kelley DS, Lilley MD, Butterfield DA, Baross JA, Wilcock WSD, Embley RW, Summit M (1998) The quantum event of oceanic crustal accretion, impacts of diking at mid-ocean ridges. Science 261:222–230
DeLong SE, Dewey JF, Fox PJ (1977) Displacement history of oceanic fracture zones. Geology 5:199–202

DeMets C, Dixon TH (1999) New kinematic models for Pacific-North America motion from 3 Ma to present. I. Evidence for steady motion and biases in the NUVEL-IA model. Geophys Res Lett 26:1921–1924

DeMets C, Gordon RG, Argus DF, Stein S (1990) Current plate motions. Geophys J Int 100:425–478

DeMets C, Gordon RG, Argus DF, Stein S (1994) Effect of recent revisions to the geomagnetic reversal time scale on estimates of current plate motions. Geophys Res Lett 21:2091–2094

DePaor DG (1983) Orthographic analysis of geological structures – I. Deformation theory. J Struct Geol 5:255–277

DePaor DG (1986) Orthographic analysis of geological structures – II. Practical applications. J Struct Geol 8:87–100

Deplus C, Diament M, Hebert H, Bertrand G, Dominguez S, Dubois J, Malod J, Patriat PH, Pomtoise B, Sibila J-J (1998) Direct evidence of active deformation in the eastern Indian oceanic plate. Geology 36:131–134

Detrick RJ, Buhl P, Vera E, Mutter J, Orcutt J, Madsen J, Brocher T, (1986) Multichannel seismic imaging of a crustal magma chamber along East Pacific Rise. Nature 326:35–41

Detrick RJ, Collins J, Steven R, Swift S (1994) In situ evidence for the nature of the seismic layer 2/3 boundary in oceanic crust. Nature 370:288–290

Dewey JF (1969) Evolution of the Appalachian Caledonian orogen. Nature 222:124–129

Dewey JF (1972) Plate tectonics. Sci Am 226:56–68

Dewey JF (1976) Opholite obduction. Tectonophysics 31:93–120

Dewey JF, Bird JM (1970) Mountain belts and the new global tectonics. J Geophys Res 75:2625–2647

Dewey JF, Burke KCA (1974) Hot spots and continental break-up implications for collisional orogeny. Geology 2:57–60

Dewey JF, Holdsworth RE, Shrachan RA (1998) Transpression and transtension zones. In: Holdsworth RE, Strachan RA, Dewey JF (eds) Continental transpressional and transtensional tectonics. Geol Soc Lond Spec Publ 135:1–14

Dias R, Ribeiro A, Cabral J, Terrinha P, Brandão Silva J, Dias RP (1995) Constriction in incipient to mature curved orogenic belts with centripetal vergence. In: Curved orogenic belts: their nature and significance. Abstracts, University Buenos Aires, Buenos Aires, pp 1–4

Dias RP, Borges JF, Cabral J, Fitas A, Costa PT, Ribeiro A, Matias L, Terrinha P (1997) Constrictive deformation in Algarve region (Portugal, SW Iberia); comparison of neotectonic data and earthquake focal mechanisms. Terra Nova 9:307

Doglioni C (1993) A proposal for the kinematic modelling of W-dipping subductions -possible applications to the Tyrrhean-Apennines system. Terra Nova 3:423-434

Duffield WA (1972) A naturally occurring model of global plate tectonics. J Geophys Res 77:2549-2555

Duin EJTH, Mesdag CS, Kok PTJ (1984) Faulting in Madeira abyssal plain sediments. Mar Geol 56:299-308

Dunbar J, Sandwell DT (1988) A boudinage model for crossgrain lineations (Abstract). EOS Trans AGU 69: p. 1429

Elder J (1976) The bowels of the Earth. Oxford University Press, Oxford, 222 pp

Ellis S, Fullsack P, Beaumont C (1995) Oblique convergence of the crust driven by basal forcing: implications for length-scales of deformation and strain partitioning in orogens. Geophys J Int 126:24–44

Elósegui P, Sari DB, Dávila JM, Cárate JL, Mendes V, Ouzar D, Pagarete J, Reilinger R, Rius A, Talaya A, Bennett R, Davis JL (1999) The AMIGO project: present-day crustal deformation of the western section of the Africa-Eurasia plate boundary zone. IUGG XXII General, Birmingham, UK Abstracts, p B80

Engelder T (1993) Stress regimes in the lithosphere. Princeton University Press, Princeton, 457 pp
Engeln J, Wiens D, Stein S (1986) Mechanisms and depths of Atlantic transform earthquakes. J Geophys Res 91:548–577
England P, Houseman GA (1989) Extension during continental convergence, with application to the Tibetan plateau. J Geophys Res 94:17561–17579
England P, Molnar P (1990) Right-lateral shear and rotation as the explanation for strike-slip faulting in eastern Tibet. Nature 344:140–142
England P, Molnar P (1991) Inferences of deviatoric stress in actively deforming belts from simple physical models. Philos Trans R Soc Lond A 337:151–164
England P (1992) Deformation of the continental crust. Understanding the Earth. In: Brown G; Hawkesworth C, Wilson C (eds) Cambridge University Press, Cambridge, pp 275–300
England P, McKenzie D (1982) A thin viscous sheet model for continental deformation. Geophys J R Astron Soc 70:295–321
England P, Molnar P (1997) Active deformation of Asia: from kinematics to dynamics. Science 278:647–650
England P, Houseman G, Sonder L (1985) Length scales for continental deformation in convergent, divergent, and strike-slip environments; analytical and approximate solutions for a thin viscous sheet model. J Geophys Res 90:3551–3557
Erickson SG (1993) Sedimentary loading, lithospheric flexure, and subduction initiation at passive margins. Geology 21:125–128
Erickson SG, Arkani-Hamed JA (1993) Subduction initiation at passive margins: the Scotian basin, eastern Canada as a potential example. Tectonics 12:678–687
Escartín J, Hirth G, Evans B (1997a) Non-dilatant brittle deformation of serpentinites: implications for Mohr-Coulomb theory and the strength of faults. J Geophys Res 102:2897–2913
Escartín J, Hirth G, Evans B (1997b) Effects of serpentinization on the lithospheric strength and the style of normal faulting at slow-spreading ridges. Earth Planet Sci Lett 151:181–189
Etchecopar A (1984) Étude des États de Contraintes en Tectonique. Cassante et Simulations de Déformations Plastiques (Approche Mathématique). Thèse Doct Etat, Univ Sci Tech Languedoc, Montpellier, France, 270 pp
Etchecopar A, Mattauer M (1988) Méthodes dynamiques d'analyse des populations de failles. Bull Soc Géol Fr 2:289–302
Etchecopar A, Vasseur G, Daignères M (1981) An inverse problem in microtectonics for the determination of stress tensors from fault striation analysis. J Struct Geol 3:51–55
Evernden JF (1970) Study of regional seismicity and associated problems. Bull Seismol Soc Am 60:393–446
Faccenna C, Giardini D, Davy P, Argentieri A (1999) Initiation of subduction at Atlantic-type margins: Insights from laboratory experiments. J Geophys Res 104:2749–2766
Fehn U, Cathles L (1979) Hydrothermal convection at slow-spreading midocean ridges. Tectonophysics 55:239–260
Fehn U, Cathles L (1986) The influence of plate movement on the evolution of hydrothermal convection cells in the oceanic crust. Tectonophysics 125:289–312
Feynman R, Leighton RB, Sands M (1963) The Feynman lectures on physics, vol 1. Addison-Wesley, Reading, MA
Fisher AT (1998) Permeability within basaltic oceanic crust. Rev Geophys 36:143–182.
Fleitout L, Moriceau C (1992) Short-wavelength geoid, bathymetry and the convective pattern beneath the Pacific Ocean. Geophys J Int 110:6629
Fleitout L, Dalloubeix C, Moriceau C (1989) Small-wavelength geoid and topography anomalies in the South Atlantic Ocean – a clue to new hot-spot tracks and lithospheric deformation. Geophys Res Lett 16:637–640

Fletcher RC, Pollard DD (1999) Can we understand structural and tectonic processes and their products without appeal to a complete mechanics? J Struct Geol 2:1071–1088
Flinn EA, Engdhal ER, Hill AR (1974) Seismic and geographical regionalization. Bull Seismol Soc Am 64:771–993
Forsyth DW (1992) Finite extension and low-angle normal faulting. Geology 20:27–30
Forsythe RD, Nelson EP, Carr MJ, Kaeding ME, Herve M, Mpodozis C, Soffia JM, Harambour S (1986) Pliocene near-trench magmatism in southern Chile: a possible manifestation of ridge collision. Geology 14:23–27
Fowler CMR (1990) The solid Earth an introduction to global geophysics. Cambridge University Press, Cambridge, 472 pp
Francis TJG (1981) Serpentinization faults and their role in tectonics of slow spreading ridges. J Geophys Res 86:11616–11622
Frank FC (1968) Curvature of island arcs. Nature 220: p. 363
Frank IA, Siggwarth JB, Craven JD (1986a) On the influx of small comets into the Earth's upper atmosphere I. Observations. Geophys Res Lett 13:303–306
Frank IA, Stigwarth JB, Craven JD (1986b) On the influx of small comets into the Earth's upper atmosphere II. Interpretation. Geophys Res Lett 13:307–310
Freund R (1974) Kinematics of transform and transcurrent faults. Tectonophysics 21:95–134
Frohlich C, Apperson KD (1992) Earthquake focal mechanisms, moment tensors and the consistency of seismic activity near plate boundaries. Tectonics 11:279–296
Frohlich C, Davis SD (1993) Teleseismic b values; or, much ado about 1.0. J Geophys Res 98:631–644
Fryer P (1992) Mud volcanoes of the Marianas. Sci Am 266:26–32
Fryrer P, Ambros EJ, Hassong DM (1985) Origin and emplacement of Mariana Foreare seamounts. Geology 13:774–777
Fullsack P (1995) An arbitrary Lagrangian-Eulerian formulation for creeping flows and is application in tectonic models. Geophys J Int 126:1–23
Fyfe WS (1976) Chemical aspects of rock deformation. Philos Trans R Soc Lond Ser A 283:221–226
Fyfe WS (1986) Fluids in deep continental crust. Am Geophys Union Geodyn Ser 14:33–39
Fyfe WS (1988) Granites and a wet convecting ultramafic planet. Trans R Soc Edinb Earth Sci 79:339–346
Fyfe WS (1992) Geosphere interactions on a convecting planet: mixing and separation. In: Hutzinger O (ed) The handbook of environmental chemistry, vol 1, part F. Springer, Berlin Heidelberg New York, pp 1–26
Fyfe WS (1994) The water inventory of the Earth: fluids and tectonics. In: Parnell J (ed) Geofluids origin migration and evolution of fluids in sedimentary basins. Geol Soc Lond Spec Publ 78:1–7
Fyfe WS (1996) The biosphere is going deep. Science 273:448
Fyfe WS (1997) Deep fluids and volatile recycling crust to mantle. Tectonophysics 275:243–251
Fyfe WS, Kerrich R (1985) Fluids and thrusting. Chem Geol 49:555–562
Fyfe WS, Leonardos OH Jr (1977) Speculations on the causes of crustal rifting and subduction. With applications to the Atlantic margin of Brazil. Tectonophysics 42:29–36
Fyfe WS, Price NJ, Thompson AB (1978) Fluids in the Earth's crust. Elsevier, Amsterdam, 376 pp
Gable CW, O'Connell J, Travies BJ (1991) Convection in three dimensions with surface plates: generation of toroidal flow. J Geophys Res 96:8391–8405
Gallahan WE, Duncan RA (1994) Spatial and temporal variability in crystallization of celadonites within Troodos ophiolite, Cyprus: implications for low-temperature alteration of the oceanic crust. J Geophys Res 99:3147–3161
Garmany J (1989) Accumulations of melt at the base of young ocean crust. Nature 340:620–632

Gass IG (1968) Is the Troodos massif of Cyprus a fragment of the Mesozoic ocean floor? Nature 220:39–42
Gennes PG, Badoz J (1996) Fragile objects Soft Matter, Hard Science and the Thrill of Discovery, Springer, Berlin Heidelberg New York, 189 pp
Genrich JF, Bock Y, McCaffrey R, Calais E, Stevens CW, Subarya C (1996) Accretion of the southern Banda arc to the Australian plate margin determined by global positioning system measurements. Tectonics 15:288–295
Gerbault M, Burov EB, Poliakov ANB, Daignières M (1999) Do faults trigger folding in the lithosphere? Geophys Res Lett 26:271–274
Ghosh SK (1993) Structural geology: fundamentals and modern developments. Pergamon Press, Oxford, 598 pp
Gilbert GK (1893) The moon's face. A study of the origin of its features. Bull Philos Soc Wash 12:241–292
Gillis KM, Robinson PT (1988) Distribution of alteration zones in the upper oceanic crust. Geology 16:262–266
Gills KM, Roberts MD (1999) Cracking at the magma-hydrothermal transition: evidence from the Troodos Ophiolite, Cyprus. Earth Planet Sci Lett 169:227–244
Girardeau J, Comen G, Beslier MO, Le Gall B, Monnier C, Agrinier P, Dubuisson G, Pinheiro L, Ribeiro A, Whitechurch H (1998) Extensional tectonics in the Gorringe Bank rocks, Eastern Atlantic ocean: evidence of an oceanic ultra-slow mantellic accreting. Terra Nova 10:330–336
Glennie KW, Clarke MWH, Boeuf MGA, Pilaar WFH, Reinhardt BM (1990) Inter-relationship of Makran-Oman Mountains belts of convergence. In: Robertson A, Slarle M, Rios A (eds) The geology and tectonics of the ocean region. Geol Soc Lond Spec Publ 49:773–786
Goff JA (1998) Finding chaos in abyssal hills. Nature 350:224–225
Goltz C (1997) Fractal and chaotic properties of earthquakes. Springer, Berlin Heidelberg New York, 178 pp
Goodwillie AM (1995) Short wavelength brevity lineations and unusual flexure results at the Puka Puka volcanic ridge system. Earth Planet Sci Lett 136:297-314
Goodwillie A, Parsons B (1992) Placing bounds on lithospheric deformation in the central Pacific Ocean. Earth Planet Sci Lett 111:123–139
Gordon RD, DeMets C, Royer JY (1998) Evidence for long-term diffuse deformation of the lithosphere of the equatorial Indian Ocean. Nature 395:370–374
Gordon RG (1995) Plate motions, crustal and lithospheric mobility, and paleomagnetism prospective viewpoint. J Geophys Res 100:24367–24392
Gordon RG (1998) The plate tectonic approximation: plate nonrigidity, diffuse plate boundaries, and global plate reconstructions. Annu Rev Earth Planet Sci 26:615–642
Gordon RG (2000a) The Antarctic connection. Nature 404:139–140
Gordon RG (2000b) Diffuse oceanic plate boundaries: strain rates, vertically averaged rheology, and comparisons with narrow plate boundaries and stable plate interiors. The history and dynamics of global plate motions geophysical monograph 121. Am Geophys Union:143–159
Gordon RG, Stein S (1992) Global tectonics and space geodesy. Science 256:333–341
Govers R, Wortel MJR, Cloetingh SAPL, Stein CA (1992) Stress magnitude estimates from earthquakes in oceanic plate interiors. J Geophys Res 97:11749–11759
Grady DE, Kipp E (1987) Dynamic rock fragmentation. In: Atkinson B (ed) Fracture mechanics. Academic Press, London, pp 429–476
Grand SP (1994) Mantle shear structure beneath the Americas and surrounding areas. J Geophys Res 99:1590–1620
Gratier J-P (1993) Le fluage des roches par dissolution-cristallisation sous contrainte, dans la croûte supérieure. Bull Soc Géol Fr 164:267–287

Gratier JP, Gamond JF (1990) Transition between seismic and aseismic deformation in the upper crust. In: Knipe RJ, Rutter EH (eds) Deformation mechanisms, rheology and tectonics. Geol Soc Lond Spec Publ 54:461–473

Gratier JP, Renard F, Labaume P (1999) How pressure solution creep and fracturing processes interact in the upper crust to make in behave in both a brittle and viscous manner. J Struct Geol 21:1189–1197

Gregory RT, Taylor HP (1981) An oxygen isotope profile in a section of Cretaceous oceanic crust, Samail ophiolite, Oman: evidence for ^{18}O buffering of the oceans by deep (>km) seawater-hydrothermal circulation at mid-oceans ridges. J Geophys Res, 86:2737–2755

Grieve RAF (1990) Impact cratering on the Earth. Sci Am 1:44–59

Grieve RAF, Pilkington M (1996) The signature of terrestrial impacts. J Aust Geol Geophys 16:399–420

Griggs DT (1939) A theory of mountain building. Am J Sci 237:611–650

Griggs DT, Blacic JD (1964) The strength of quartz in ductile regime. Trans Am Geophys Union 45:102–103 (Abstr)

Griggs D, Blacic JD (1965) Quartz-anomalous weakness of synthetic crystals. Science 247:292–295

Grimson NL, Chen WP (1986) The Azores-Gibraltar Plate Boundary: Focal mechanisms, depth of earthquakes and their tectonic implications, J Geophys Res 91:2029–2047

Gripp AE, Gordon RG (1990) Current plate velocities relative to the hotspots incorporating the nuvel-1 global plate motion model. Geophys Res Lett 17:1109–1112

Groshong RH, Teufel LW, Gasteiger C (1984) Precision and accuracy of the calcite strain-gauge technique: Geol Soc Am Bull 95:357–363

Gutenberg B, Richter CF (1954) Seismicity of the Earth and associated phenomena, 2nd edn. Princeton University Press, Princeton, 310 pp

Hacker BR (1991) The role of deformation in the formation of metamorphic gradients: ridge subduction beneath the Oman Ophiolite. Tectonics 10:455–473

Hadley D, Kanamori H (1977) Seismic structure of the Transverse Ranges, California. Geol Soc Am Bull 88:1469–1478

Haines AJ (1982) Calculating velocity fields across plate boundaries from observed shear rates. Geophys J R Astron Soc 68:203–209

Haines J (1998) Continental mechanics. Nature 391:634–635

Hancock PL (1994) (ed) Continental deformation. Pergamon Press, Oxford, 421 pp

Handschumacher DW, Sager WW, Hilde TWC, Bracey DR (1988) Pre-Cretaceous tectonic evolution of the Pacific plate and extension of the geomagnetic polarity reversal time scale with implications for the origin of the Jurassic Quiet Zone. Tectonophysics 155:365–380

Hardebeck JL, Hauksson E (1999) Role of fluids in faulting inferred from stress field signatures. Science 285:236–239

Harding AJ, Kent GM, Orcutt JA (1993) A multichannel seismic investigation for upper oceanic crust at 90 N on the East Pacific Rise: implications for crustal accretion. J Geophys Res 98:13925–13944

Harland WB (1971) Tectonic transpression in Caledonian Spitzbergen. Geol Mag 108:27–42

Harper GD (1985) Tectonics of slow-spreading mid-ocean ridges and consequences of a variable depth to the brittler-ductile transition. Tectonics 93:4625–4656

Harper TR, Szymanski JS (1991) The nature and determination of stress in the accessible lithosphere. Philos Trans R Soc Lond 337:5–24

Hatcher RD Jr, Hooper RJ (1992) Evolution of crystalline thrust sheets in the internal parts of mountain chains. In: McClay KR (ed) Thrust tectonics. Chapman and Hall, London, pp 217–233

Hayes DE, Kane KA (1991) The dependence of seafloor roughness on spreading rate. Geophys Res Lett 18:1425–1428

Hayward N, Watts AB, Westbrook GK, Collier S (1999) A seismic reflection and GLORIA study of compressional deformation in the Gorringe Bank region, eastern North Atlantic. Geophys J Int 138:831–850

Head JW, Wilson L, Smith DK (1996) Mid-ocean ridge eruptive vent morphology and substructure: evidence for dike widths, eruption rates, and evolution of eruptions and axial volcanic ridges. J Geophys Res 101:28265–28280

Heard H (1968) Experimental deformation of rocks and the problem of extrapolation to nature. In: Ricker R (ed) NSF advanced science seminar in rock mechanics. Air Force Cambridge Res Lab Bedford 14:439–508

Heki K (1996) Horizontal and vertical crustal movements from three-dimensional very long baseline interferometry kinematic reference frame: implication for the reversal timescale revision. J Geophys Res 100:3187–3198

Heki K, Miyazaki S, Tsuji H (1997) Silent fault slip following an interplate thrust earthquake at the Japan Trench. Nature 386:595–598

Henstock TJ, Levander A, Hole JA (1997) Deformation in the lower crust of the San Andreas Fault System in northern California. Science 278:650-653

Herraiz M, De Vicente G, Lindo-Ñaupari R, Giner J, Simón JL, Casado JMG, Vadillo O, Pascua MAR, Cicuéndez JI, Casas A, Cabañas L, Rincón P, Cortés AL, Ramirez M, Lucimi M (2000) The recent (upper Miocene to Quaternary) and present tectonic stress distributions in the Iberian Peninsula. Tectonics 19:762–786

Hess HH (1962) History of ocean basins. In: Engel AEJ, James HL, Leonard BF (eds) Petrologic studies; a volume in honor of A.F. Buddington. Geol Soc Am, Boulder, CO, pp 599–620

Hey RN, Menard HW, Atwater TM, Caress DW (1988) Changes in direction of seafloor spreading revisited. J Geophys Res 93:2803–2811

Hickman SH (1990) The strength of active faults in the earth's crust and the high stress/low stress controversy. EOS 71:1579–1580

Hickman SH (1991) Stress in the lithosphere and the strength of active faults: Rev Geophys Suppl:759–775

Hill DP, Reaserberg PA, Michael A, Arabaz WJ, Beroza G, Brumbaugh D, Brune JN, Castro R, Davis S, DePolo D, Ellsworth WL, Gomberg J, Hamsen S, House L, Jackson M, Johnston MJS, Jones L, Keller R, Malone S, Munguia L, Nava S, Pechmann JC, Sanford A, Simpson RW, Smith RB, Stark M, Stickney M, Vidal A, Walter S, Wong V, Zollweg J (1993) Seismicity remotely triggered by the magnitude 7.3 Landers, California, Earthquake. Science 263:1617–1623

Hindle D, Burkhard M (1999) Strain, displacement and rotation associated with the formation of curvature in fold belts; the example of the Jura arc. J Struct Geol 21:1089–1100

Hirata T (1989) Fractal dimension of fault systems in Japan: fractal structure in rock fracture geometry at various scales. PAGEOPH 131:157–170

Hirn A, Haessler H, Hoang Trong P, Wittlinger G, Mendes Victor. LA (1980) Aftershock sequence of the January 1st 1980, earthquake and present-day tectonics in the Azores. Geophys Res Lett 7:501–504

Holbrook W, Brocher TM, Brink US, Hole JA (1996) Crustal structure of a transform plate boundary: San Francisco Bay and the central California continental margin. J Geophys Res 101:22311–22334

Holmes A (1944), The Machinery of Continental Drift: the search for a mechanism. In: Cox A (ed) Plate tectonics and geomagnetic reversals. Freeman, 1973, San Francisco, pp 19–22

Horton RE (1945) Erosional development of streams and their drainage basins, hydrophysical approach to quantitative morphology. Bull Geol Soc Am 56:275–370

Hossack JR (1979) The use of balanced cross-sections in the calculation of orogenic contraction: a review. J Geol Soc Lond 136:705–711

Houseman GA, McKenzie DP, Molnar P (1981) Convective instability of a thickened boundary layer and its relevance for the thermal evolution of continental convergent belts. J Geophys Res 86:6115–6132

Hsu KJ (1994) Uniformitarianism vs. catastrophism. In: Glen W (ed) The mass-extinction debates: how science works in a crisis. Stanford University Press, Stanford, pp 215–229

Hsu TC (1966) The characteristics of coaxial and non-coaxial strain paths. J Strain Anal 2:216–222

Hubbert MK, Rubey WW (1959) Role of fluid pressure in mechanics of overthrust faulting. I. Mechanics of fluid-filled porous solids and its application to overthrust faulting. Bull Geol Soc Am 70:115–166

Huchon P, Bourgois J (1990) Subduction-induced fragmentation of the Nazca Plate of Peru: Mendana fracture zone and Trujillo trough revisited. J Geophys Res 95:8419–8436

Instituto Geografico Nacional (1992) Análisis sismotectónico de la Península Ibérica, Baleares y Canarias. Inst Geogr Nac Publ No 26, 43 pp

Isacks BL, Barazangi M (1977) Geometry of Benioff zones: lateral segmentation and downwards bending of the subducted lithosphere. In: Talwani M, Pitman WC III (eds) Island arcs, deep-sea trenches and back-arc basins. Am Geophys Union, Maurice Ewing Series 1, Washington, DC, pp 99–114

Ito K (1995) Punctuated-equilibrium model of biological evolution is also a self-organized-criticality modal of earthquakes. Phys Rev E 52:3232–3233

Ito K, Matsuzaki M (1990) Earthquakes as self-organized critical phenomena. J Geophys Res 95:6853–6860

Jackson J, McKenzie D 1983) The geometrical evolution of normal fault systems. J Struct Geol 5:471–482

Jackson J, McKenzie D (1988) The relationship between plate motions and seismic moment tensors, and the rates of active deformation in the Mediterranean and Middle East. Geophys J 93:45–73

Jacobson RS (1992) Impact of crustal evolution on changes of the seismic properties of the uppermost oceanic crust. Res Geophys 30:23–42

Jaeger JC, Cook NGW (1981) Fundamentals of rock mechanics,3 edn. Chapman and Hall, London 593 pp

Janecky DR, Seyfried WE Jr (1986) Hydrothermal serpentinization of peridotite within the oceanic crust: experimental investigations of mineralogy and major element chemistry. Geochim Cosmochim Acta 50:1357–1378

Jaume SC, Sykes LR (1992) Changes in state of stress on the southern San Andreas Fault resulting from the California earthquake sequence of April to June 1992. Science 258:1325–1328

Jin Y, McNutt MK, Zhu Y (1994) Evidence from gravity and topography data for folding of Tibet. Nature 371:669–674

Johnson AM (1970) Physical processes in geology. Freeman, Cooper and Cop, San Francisco, 577 pp

Johnson HP, Becker K, Herzen RPV (1993) Near-axis heat flow measurements on the northern Juan de Fuca Ridge: implications for fluid circulation in oceanic crust. Geophys Res Lett 20:1875–1978

Johnston AC (1989) The seismicity of stable continental interiors. In: Gregersen S, Basham PW (eds) Earthquakes at North-Atlantic passive margins: neotectonics and postglacial rebound. Kluwer, Dordrecht, pp 299–327

Johnston AC, Kanter R (1990) Earthquakes in stable continental crust. Sci Am 262:42–49

Jones BW (1999) Discovering the solar system. Wiley, Chichester, 416 pp

Juteau T, Maury R (1997) Géologie de la croûte océanique-Pétrologie et dynamique endogènes. Masson, Paris, 367 pp

Kahle HG, Staub C, Reilinger R, McClusky S, King R, Hurst K, Veis G, Kastens K, Cross P (1998) The strain rate field in the eastern Mediterranean region, estimated by repeated GPS measurements. Tectonophysics 294:237–252

Kanamori H, Anderson DL (1975) Theoretical basis of some empirical relations in seismology. Bull Seismol Soc Am 65:1073–1095

Karato S, Spetzler HA (1990) Defect microdynamics in minerals and solid-state mechanisms pf seismic wave attenuation and velocity dispersion in the mantle. Rev Geophys 28:399–421

Karner GD, Weisel JK (1990) Factors controlling the location of compressional deformation of oceanic lithosphere in the central Indian Ocean. J Geophys Res 95:19795–19800

Karson JA (1990) Seafloor spreading on the Mid-Atlantic Ridge: implications for the structure of ophiolites and oceanic lithosphere produced in slow-spreading environments. In: Malpas J, Moores EM (ed) Proceedings of the symposium on ophiolites and oceanic lithosphere. Geological Survey, Nicosia, Cyprus, pp 547–555

Karson J, Hurst S, Lawrence R, Gao D (1996) Time/space variations in the seafloor spreading in the MARK area the past era of crustal accretion along 100 km of the Mid-Atlantic ridge. FA-RA Meeting J Conf Abstr 1:805–806

Karson JA, Rona PA (1990) Block-tilting, transform faults and structural control of magmatic and hydrothermal processes in the TAG area, mid-Atlantic 260North. Geol Soc Am Bull 102:1635–1645

Karson RL, Searle RC, Kleinrock MC, Schouten H, Bird RT, Naar DF, Rusby RI, Hooft EE, Lasthiotakis H (1992) Roller-bearing tectonic evolution of the Juan Fernandez microplate. Nature 356:571–576

Kearey PH, Vine FJ (1996) Global tectonics, 2nd edn, Blackwell, Oxford, 333 pp

Keilis-Borok, V. I VIK (1990) The lithosphere of the earth as a nonlinear system with implications for earthquake prediction. Rev Geophys 28:19–34

Keith ML (1993) Geodynamics and mantle flow, an alternative earth model. Earth Sci Rev 33:153–337

Kent CC (1997) Plate tectonics and crustal evolution. Butterworth Heinemann, Oxford, 282 pp

Kidd RGW (1977) A model for the process of formation of the upper oceanic crust. Geophys. J R Astron Soc 50:149–183

King G, Brewer J (1983) Fault related folding near the Wind River thrust, Wyoming, USA. Nature 306:147–150

King G CP, Stein RS, Rundle JB (1988) The growth of geological structures by repeated earthquakes I. Conceptual framework. J Geophys Res 98:13307–13308

Kirby SH (1984) Introduction and digest to the special issue on chemical effects of water on the deformation and strengths of rocks. J Geophys Res 89:3991–3995

Kissel C, Laj C (1989) Paleomagnetic rotations and continental deformation. Kluwer, Dordrecht, 516 pp

Klemperer SL, BIRPS Group (1987) Reflectivity of the crystalline crust: hypotheses and test. Geophys. J R Astron Soc 89:217–222

Knapp RB, Knight JE (1977) Differential thermal expansion of pore fluids: fracture propagation and microearthquake production in hot pluton environments. J Geophys Res 82:2515–2522

Kohlstedt DL, Evans B, Mackwell SJ (1995) Strength of the lithosphere: constraints imposed by laboratory experiments. J Geophys Res 100:17587–17602

Kong LS, Solomon SC, Purdy GM (1992) Microearthquake characteristics of a mid-ocean ridge along-axis high. J Geophys Res 97:1659–1685

Korvin G (1992) Fractal models in the earth sciences. Elsevier Science Publishers-BW, Amsterdam, 396 p

Kranz RL (1983) Microcracks in rocks: a review. Tectonophysics 100:449–480

Kullberg MC (1996) Estudos tectónicos e fotogeológicos na Serra de Sintra e Arrábida. PhD Thesis, Dep Geol Fac Ciências Univ. Lisboa, 187 pp

Kumagai H, Fukao Y, Watanabe S, Baba Y 1999) A self-organized model of earthquakes with constant stress drops and the b-value of 1. Geophys Res Lett 26:2817–2820

Kusznir N, Park RG (1986) The extensional strength of the continental lithosphere: is dependence on geothermal gradient, and crustal composition and thickness. In: Coward MP, Dewey JH, Hancock PH (eds) Continental extensional tectonics. Geol Soc Lond Spec Publ 28:35–52

Lachenbruch AH (1973) A simple mechanical model for oceanic spreading centers. J Geophys Res 76:400–401

Lachenbruch AH, Thompson GA (1972) Oceanic ridges and transform faults: their intersection angles and resistance to plate motions. Earth Planet Sci Lett 15:116–122

Lachenbruch AH, Sass JH (1980) Heat flow and energetics of the San Andreas fault zone. J Geophys Res 85:6185–6222

Lagabrielle Y, Le Moigne J, Maury RC, Cotton J, Bourgois J (1994) Volcanic record of the subduction of an active spreading ridge, Taitao Peninsula (southern Chile). Geology 22:515–518

Lallemand S, Mozé JP, Monti S, Sibvet JC (1985) Carte bathymétrique de l' Atlantique Nord-Est. Échelle 1/2,400,000. Ifremer, Bust

Lamb S (1989) Rotations about vertical axes in part of the New Zealand plate-boundary zone, theory and observation. In: Kissel C, Laj C (eds) Paleomagnetic rotations and continental deformation. Kluwer, Dordrecht, pp 473–488

Lambeck K (1988) The slow deformations of the Earth. Geophysical Geodesy, Oxford Science Publications, Clarendon Press, Oxford, 718 pp

Landau L, Lifshitz E (1986) Theory of elasticity Course of theoretical physics, vol 7, 3rd edn. (Translated by Sykes JB, Rheid WH) Pergamon Press, Oxford, 187 pp

Langmuir CH, Bender JF, Batiza R (1986) Petrological and tectonic segmentation of the East Pacific Rise 5°30′–14°30′N. Nature 322:422–429

Langton C (1990) Computation at the edge of chaos phase transitions and emergent computation. Physica D 42:12–37

Larson KM, Freymueller JT, Philipsen S (1997) Global plate velocities from the global positioning system. J Geophys Res 102:9961–9981

Larson RL, Chase CG (1972) Late Mesozoic evolution of the western Pacific Ocean. Geol Soc Am Bull 83:3627–2644

Larson RL, Pitman WC (1972) World-wide correlation of Mesozoic. Bull 83:3645–3662

Laubscher H (1971) The large-scale kinemations of the western Alps and the northern. Apennines and its palinspastic implications. Am J Sci 271:193–226

Laughton AS, Roberts DG, Graves R (1975) Bathymetry of the northeast Atlantic: Mid-Atlantic Ridge to southwest Europe. Deep Sea Res 22:791–810

Lay T, Wallace T (1995) Modern global seismology. Academic Press, San Diego, 521 pp

Le Pichon X (1968) Sea Floor Spreading and continental drift. J Geophys Res 73(12):3661–3705

Le Pichon X, Francheteau J, Bonnin J (1973) Plate tectonics. Elsevier, Amsterdam, 300p p

Le Pichon X (1980) La lithosphère océanique: fondement de la tectonique globale. Mem H-Ser Soc Géol Fr 10:339–350

Lepvrier C, Mougenot D (1984) Déformations cassantes et champs de contrainte posthercyniens dans l'Ouest de l'Ibérie (Portugal). Rev Géol Dyn Géogr Phys 25:291–305

Letouzey J, Tremolières P (1980) Paleo-stress fields around the Mediterranean since the Mesozoic derived from microtectonics: comparisons with plate tectonic data. Mèm BRGM 115 Orléaus:261–273

Lin J, Purdy GM, Schouten H, Sempéré J-C, Zervas C (1990) Evidence from gravity data for focused magmatic accretion along the Mid-Atlantic ridge. Nature 344:627–632

Lin M, Yang Y, Stein S, Zhu Y, Engeln J (2000) Crustal shortening in the Andes: why do GPS rates differ from geological rates?. Geophys Res Lett 27:3005–3008

Lin S, van Staal CR, Dube B (1994) Promontory-promontory collision in the Canadian Appalachians. Geology 22:897–900

Lin S, Jiang D, Williams PF (1998) Transpression (or transtension zones of triclinic symmetry: natural example and theoretical modelling. In: Holdsworth RE, Strachan RA, Dewey JF (eds) Continental transpressional and transtensional tectonics. Geol Soc Lond Spec Publ 135:41–57

Lisle R (1994) Paleostrain analysis. In: Hancock P (Ed) Continental deformation. Pergamon Press, Oxford, pp 28–42

Lisowski M (1991) Recent plate motions and crustal deformation. Rev Geophys Suppl:162–171

Lister CRB (1975) Gravitational drive on oceanic plates caused by thermal contraction. Nature 257:663–665

Lister CRB (1977) Qualitative models of spreading center processes including hydrothermal penetration. Tectonophysics 37:203–218

Lister CRB (1983) The basic physics of water penetration into hot rocks. In: Rona PA, Bostrom K, Laubier L, Smith L (eds) Hydrothermal processes at seafloor spreading centers. Plenum Press, New York, pp 141–168

Lliboutry L (1992) Very slow flows of solids. Martinus Nijhoff, Dordrecht, 510 pp

Lliboutry L (1998) Géophysique et géologie. Masson, Paris, 462 pp

Lockwood JP (1972) Possible mechanisms for the emplacement of Alpine-type serpentinite. Geol Soc Am Mem 132:273–287

Lorenz EN (1963) Deterministic nonperiodic flow. J Atmos Sci 20:130–141

Lorenzo JM, O'Brien GW, Stewart J, Tandon K (1998) Inelastic yielding and forebulge shape across a modern foreland basin: North West Shelf of Australia, Timor Sea. Geophys Res Lett 25:1455–1458

Louden KE (1995) Variations in crustal structure related to intraplate deformation: evidence from seismic refraction and gravity profiles in the central Indian Ocean Basin. Geophys J Int 120:375–392

Lourenço JN (1997) Morfotectónica da Junção Tripla dos Açores: Estudo de Pormenor do Segmento Lucky Strike (37°17′N, 32°16′O), a Partir de Dados de Batimetria Multi-Feixe e Sonar Lateral de Alta Resolução. MSc Thesis, Departamento de Geologia da Faculdade de Ciências da Universidade de Lisboa, 144 pp

Lourenço N, Miranda JM, Luís JF, Ribeiro A, Victor LAM, Madeira J, Needham HD (1998) Morpho-tectonic analysis of the Azores Vulcanic Plateau from a new bathymetric compilation of the area. Mar Geophys Res 30:144–156

Lowell RP (1980) Topographically drive subcritical hydrothermal convection in the oceanic crust. Earth Planet Sci Lett 49:21–28

Lowell RP (1991) Modeling continental and submarine hydrothermal systems. Rev Geophys 29:457–476

Lowell RP Germanovich LN (1995) Dike injection and the formation of megaplumes t ocean ridges. Science 267:1804–1807

Lowell RP, Rona PA, Herzen RP (1995) Seafloor hydrothermal systems. J Geophys Res 100:327–352

Lowrie W (1997) Fundamentals of geophysics. Cambridge University Press, Cambridge, 349 pp

Luís JF (1996) Le Plateau des Açores et le point triple associé. Analyse géophysique et évolution. PhD Thesis, Paris VII University, Paris France, 203 pp

Luís JF, Miranda JM, Galdeano A, Patriat P, Rossignol JC, Mendes Victor LA (1994) The Azores triple junction evolution since 10 Ma from aeromagnetic survey of the Mid-Atlantic Ridge. Earth Planet Sci Lett 125:439–459

Luís JF, Miranda JM, Galdeano A, Patriat P (1998) Constraints on the structure of the Azores spreading center from gravity. Mar Geophys Res 20:157–170

Lyell C (1830) Principles of geology I, 1st edn. John Murray, London (Facsimile reproduction of first edition, with an introduction by MJS Rudwick, 1990, Univ of Chicago Press, Chicago), 511 pp
Lyell C (1832) Principles of geology II, 1st edn. John Murray, London (Facsimile reproduction of first edition, 1991, Univ of Chicago Press, Chicago), 330 pp
Lynch MA (1999) Linear ridge groups: evidence for tensional cracking in the Pacific Plate. J Geophys Res 104:29321–29333
Lynnes CS, Ruff LJ (1985) Source process and tectonic implications of the great 1975 North Atlantic earthquake. Geophys J R Astron Soc 62:497–500
MacDonald AH, Fyfe WS (1985) Rate of serpentinization in seafloor environments. Tectonophysics 116:123–135
Macdonald KC (1989) Tectonic and magmatic processes on the East Pacific Rise. The geology of North America, vol N. The eastern Pacific Ocean and Hawaii. The Geological Society of America, Boulder, CO, pp 93–110
MacDonald KC, Fox PJ, Perram LJ, Eisen MF, Haymon RM, Miller SP, Carbotte CM, Cormin M-H, Shor AN (1988) A new view of the mid-ocean ridge from the behavior of the ridge-axis discontinuities. Nature 335:217–225
Macdougall JO, Winterer EL, Sandwell D, Lynch A (1994) Tholeitic and alkali volcanism along extensional ridges on the western flank of the superfast. East Pacific Rise. EOS Trans 75:582
Madeira J (1998) Estudos de Neotectónica nas Ilhas do Faial, Pico e S. Jorge: Uma contribuição para o conhecimento Geodinâmico da Junção Tripla dos Açores. PhD Thesis, Faculdade de Ciências da Universidade de Lisboa, Departamento de Geologia, 428 pp
Madeira J, Ribeiro A (1990) Geodynamic models for the Azores triple junction: a contribution from tectonics. Tectonophysics 184:405–415
Main IG, Meredith PHG, Sammonds PR, Jones C (1990) Influence of fractal flaw distributions on rock deformation in the britte field. In: Knipe RJ, Rutter EH (eds) Deformation mechanisms, rheology and tectonics. Geol Soc Lond Spec Publ 54:71–79
Malavieille J (1984) Modélisation expérimentale des chevauchements imbriqués; application aux chaînes de montagnes. Bull Soc Géol Fr 7:129–138
Malpas J, Sporli KB, Black PHM, Smith IEM (1992) Northland ophiolite, New Zealand, and implications for plate-tectonic evolution of the southwest Pacific. Geology 20:149–152
Malvern LE (1969) Introduction to the mechanics of a continuous medium. Prentice Hall, Englewood Cliffs, 713 pp
Mancktelow NS (1993) Tectonic overpressure in competent mafic layers and the development of isolated eclogites. J Metamorph Geol 11:801–812
Mandelbrot BB (1982) The fractal geometry of nature. Freeman, New York, 468 pp
Mandl G (1987) Discontinuous fault zones. J Struct Geol 9:105–110
Markus M, Kusch I, Ribeiro A, Almeida P (1996) Class IV behavior in cellular automata models of physical systems. Int J Bifurcation Chaos 6:1817–1827
Markus M, Czajka A, Bõhm D, Hahn T, Schulit T, Ribeiro A (1999) Phenomenology of cellular automata simulations of natural processes. In: Goles E, Martinez S (eds) Cellular automata and complex systems. Kluwer, Dordrecht, pp 55–105
Marrett R, Allmendinger RW (1991) Estimates of strain due to brittle faulting: sampling of fault populations. J Struct Geol 13:734–738
Martinod J, Davy P (1992) Periodic instabilities during compression of extension of the lithosphere I. Deformation modes from an analytical perturbation method. J Geophys Res 97:1999–2014
Martinod J, Davy P (1994) Periodic instabilities during compression of the lithosphere 2. Analogue experiments. J Geophys Res 99:10067–10069

Martinod J, Molnar P (1995) Lithospheric folding in the Indian Ocean and the rheology of the oceanic plate. Bull Soc Géol Fr t 166 6:813–821
Martins I, Victor LAM (1990) Contribuição para o estudo da Sismicidade de Portugal Continental, vol 18. Universidade de Lisboa, Instituto Geofísico do Infante D Luís, 69 pp
Maruyama S (1999) Leaking water into the Earth. AGU Fall Meeting Search Results, pp 1–2
Massonnet D (1997) Satellite radar interferometry. Sci Am 265:32–39
Massonnet D, Rossi M, Carmona C, Adragna F, Peltzer G, Felgl K, Rabaute T (1993) The displacement field of the Landers earthquake mapped by radar interferometry. Nature 364:138–142
Massonnet D, Felgl K, Rossi M, Adragna F (1994) Radar interferometric mapping of deformation in the year after the Landers earthquake. Nature 369:227–230
Mateus A (1995) Evolução tectono-térmica e potencial metalogenético do troço transmontano da Zona de Falha Manteigas-Vilariça-Bragança. PhD Thesis, Lisboa Univ, 195 pp (v I); 994 pp (v II)
Mateus A, Gonçalves M (1993) Caracterização Geométrica e Distribuição Fractal da Fracturação Adjacente à Falha da Vilariça no Sector da Quinta da Terrimcha (NE de Portugal). Gaia 6:43–58
Mattauer M (1981) La formation des Chaînes de montagnes. Pour/la /Science 46:40–55
Mattauer M (1986) Intracontinental subduction, crust-mantle décollement and crustal-stackin wedge in the Himalayas and other collision belts. In: Coward M, Ries A (eds) Collision tectonics. Geol Sci Lond Spec Publ 19:57–60
Mattauer M, Proust F, Tapponnier P (1980) Tectonic mechanism of obduction in relation with high pressure metamorphism. Les orogènes, Colloques Internationaux du CNRS no 272, pp 197–201
Mattauer M, Matte PH, Olivet JL (1999) A3D model of the India-Asia collision at plate scale. Earth Planet Sci Lett 328:499–508
Matte PH, Ribeiro A (1975) Forme et orientation de l'el-lipsoide de déformation dans la virgation hercynienne de Galice. Relations avec le plissement et hypothèse sur la genèse de l'arc Ibérico-Armoricain. CR Acad Sci Paris 280:2825–2828
Matte PH, Maluski H, Caby R, Nicolas A, Kepezhinskas P, Sobolev S (1993) Geodynamic model and 39Ar/40Ar dating for the generation and emplacement of the high pressure metamorphic rocks in the SW Urals. CR Acad Sci Paris 317:1667–1674
Matte PH, Mattauer M, Olivet JM, Griot DA (1997) Continental subductions beneath Tibet and the Himalayan orogeny; a review. Terra Nova 9:264–270
McAdoo DC, Sandwell DT (1985) Folding of oceanic lithosphere. J Geophys Res 90:8563–8569
McClain JS, Atallah CA (1986) Thickening of the oceanic crust with age. Geology 14:574–576
McClusky S, Balassanian S, Barka A, Demir C, Ergintay S, Georgiev I, Gurkan O, Hamburger M, Hust K, Kahle H, Kastens K, Kekelidze G, King R, Kotzev V, Lenk O, Mahmoud S, Mishin A, Nadariya M, Ouzounis A, Paradissis D, Peter Y, Prilepin M, Reilinger R, Sanli I, Seeger H, Tealeb A, Toksôz N, Veis G (2000) Global positioning system constraints em plate kinematics and dynamics in the eastern Mediterranean and Caucasus. J Geophys Res 105:5695–5719
McGarr A (1991) On a possible connection between three major earthquakes in California and oil production. Bull Seismol Soc Am 81:948–970
McKenzie DP (1969) Speculations on the consequences and causes of plate motion. Geophys J R Astron Soc 18:1–32
McKenzie DP (1972a) Active tectonics of the Mediterranean region. Geophys. J R Astron Soc 30:109–185
McKenzie DP (1972b) Plate tectonics. In: Robertson EC (ed) The nature of solid Earth. McGraw-Hill, New York, pp 323–360
McKenzie DP (1978a) Some remarks on the development of sedimentary basins. Earth Planet Sci Lett 40:25–32

McKenzie DP (1978b) Active tectonics of the Alpine-Himalayan belt: the Aegean Sea and surrounding regions. Geophys J R Astron Soc 55:217–254
McKenzie DP (1983) The Earth's mantle. Sci Am 249:51–67
McKenzie D (1990) Spinning continents. Nature 344:109–110
McKenzie D, Jackson J (1983) The relationship between strain rates, crustal thickening, palaeomagnetism, finite strain and fault movements within a deforming zone. Earth Planet Sci Lett 65:182–202
McKenzie D, Jackson J (1986) A block model of distributed deformation by faulting. J Geol Soc Lond 143:349–353
McKenzie D, Jackson J (1989) The kinematics and dynamics of distributed deformation. In: Kissel C, Laj C (eds) Paleomagnetic rotations and continental deformation. Kluwer, Dordrecht, pp 17–31
McKenzie DP, Morgan WJ (1969) Evolution of triple junctions. Nature 224:125–133
McKenzie D, Parker RL (1974) Plate tectonics in ω Space. Earth Planet Sci Lett 22:285–293
McNutt MK (1998) Superswells. Rev Geophys 26:211–244
McNutt MK, Kogan MG (1987) Isostasy in the USSR, 2, interpretation of admittance data. In: Fychs K, Froideveaux C (ed) The composition structure, dynamics of the lithosphere-asthenosphere system. Geodyn Ser 16:309–327
McNutt MK, Diament M, Kogan MG (1988) Variations of elastic plate thickness at continental thrust belts. J Geophys Res 93:8825–8838
Means WD (1976) Basic concepts of continuum mechanics for geologists. Stress and strain. Springer, Berlin Heidelberg New York, 339 p
Means WD (1990) Kinematics, stress, deformation and material behavior. J Struct Geol 12:953–971
Melosh HJ (1978) Dynamic support of the outer rise. Geophys Res Lett 5:321–324
Melosh HJ (1989) Impact cratering a geologic process. Oxford University Press, Oxford, 245 pp
Mendes AF, Kullberg JC (1992) Drawing and use of auxiliary projection nets (the program STEGRAPH). Ciênc Terra (UNL) 11:275–291
Meredith PG, Man IG, Jones C (1990) Temporal variations in seismicity during quasistatic and dynamic rock failure. Tectonophysics 175:249–268
Mian ZU, Tozer DC (1990) No water, no plate tectonics: convective heat transfer and the planetary surfaces of Venus and Earth. Terra Nova 2:455–459
Michard A, Le Mer O, Coffe B, Montigny R (1989) Mechanism of the Oman mountains obduction onto the Arabian continental margin, reviewed. Bull Soc Géol Fr 8 (5):241–252
Michard A, Chopin C, Henry C (1993) Compression versus extension in the exhumation of the Dora-Maira coesite bearing unit, Western Alps, Italy. Tectonophysics 221:173–193
Middleton G (1991) (ed) Nonlinear dynamics, chaos and fractals. With applications to geological systems. Geological Association of Canada, Short Course Notes, Toronto, Ontario, vol 9, 235 pp
Middleton GV, Wilcock PR (1994) Mechanics in the earth and environmental sciences. Cambridge University Press, Cambridge, 459 pp
Ministerio de Obras Publicas y Transportes (1992) Analisis Sismotectonico de La Peninsula Iberica Baleares y Canarias, vol 26, 43 pp
Minster JBH, Jordan TH, Hager BH, Agnew DC, Royden LH (1990) Implications of precise positioning. Geodesy in the year 2000. Natl Acad Press, pp 23–52
Miranda JM, Luís JF, Abreu I, Mendes Victor LA (1991) Tectonic framework of the Azores triple junction. Geophys Res Lett 18:1421–1424
Miyamura S (1962) Magnitude frequency relation of earthquakes and its bearing on geotectonics. Proc Jpn Acad Sci 38:27–30
Miyashiro A, Aki K, Sengör AMC (1982) (eds) Orogeny. Wiley, Chichester, 242 pp

Mogi K (1962) Study of elastic shocks caused by the fracture of heterogeneous materials and their relation to earthquake phenomena. Bull Earthquake Res Int Univ Tokyo 40:125–173
Molnar P (1983) Average regional strain due to slip on numerous faults of different orientations. J Geophys Res 88:6430–6432
Molnar P (1986) The structure of mountain ranges. Sci Am 255:70–79
Molnar P (1988) Continental tectonics in the aftermath of plate tectonics. Nature 385:131–137
Molnar P (1992) Brace-Goetze strength profiles, the partitioning of strike-slip and thrust faulting at zones of oblique convergence, and the stress-heat flow paradox of the San Andreas Fault. In: Evans B, Wong T-F (eds) Fault mechanics and transport properties of rocks. Academic Press, London, pp 435–459
Molnar P, Deng Q (1984) Faulting associated with large earthquakes and the average rate of deformation in central and eastern Asia. J Geophys Res 89:6203–6227
Molnar P, Gipson J M (1994) Very long baseline interferometry and active rotations of crustal blocks in the Western Transverse Ranges, California. Geol Soc Am Bull 106:594–606
Molnar P, Gipson J M (1996) A bound on the rheology of continental lithosphere using very long baseline interferometry: the velocity of south China with respect to Eurasia. J Geophys Res 100:545–553
Molnar P, Gray D (1979) Subduction of continental lithosphere: some constraints of certainties. Geology 7:58–62
Molnar P, Tapponnier P (1975) Cenozoic tectonics of Asia effects of a continental collision. Science 189:419–426
Molnar P, England PH, Martinod J (1993) Mantle dynamics, uplift of the Tibetan Plateau, and the Indian Monsoon. Rev Geol:357–396
Monteiro JF, Munhá J, Ribeiro A (1998) Impact ejecta horizon near the Cenomanian-Turonian boundary, north of Nazaré, Portugal. Comun Int Geol Min 84:111–114
Moore DE, Lockner DA, Ma S, Summers R, Byerlee JD (1997) Strengths of serpentine gouges at elevated temperatures. J Geophys Res 102:14787
Moore JMCM, Al Shanti AM (1973) The use of stress trajectory analysis in the elucidation of part of the najd fault system, Saudi/Arabia. Proc Geol Assoc 84:383–403
Moores EM (1982) Origin and emplacement of ophiolites. Rev Geophys (Space Phys) 20:735–760
Moores EM, Vine FJ (1971) The Troodos massif, Cyprus, and other ophiolites as oceanic crust: evaluations and implications. Philos Trans R Soc Lond Ser A 268:443–466
Moores M, Twiss RJ (1995) Tectonics. Freeman, New York, 415 pp
Moresi L, Solomatov W (1998) Mantle convection with a brittle lithosphere: thoughts on the global tectonic styles of the Earth and Venus. Geophys J Int 133:669–682
Morgan J, Wagner M, Chicxulub Working Group, Brittan J, Buffer R, Camargo A, Christeson G, Denton P, Hidebrand A, Hobbst R, Macintyre H, Mackenzie G, Maguire P, Martins L, Nakamura Y, Plikington M, Sharpton V, Snyder D, Suarez G, Trejo A (1997) Size and morphology of the Chicxulub impact crater. Nature 391:472–476
Morgan WJ (1968) Rises, trenches, great faults and crustal blocks. J Geophys Res 73:1959–1982
Morrison D (1993) Exploring planetary worlds. Scientific American Library, New York, 240 pp
Mottl MJ, Wheat CG (1994) Hydrothermal circulation through mid-ocean ridge flanks: fluxes of heat and magnesium. Geochim Cosmochim Acta 58:2225–2237
Mount VS, Suppe J (1987) State of stress near the San Andreas fault implications for wrench tectonics. Geology 15:1143–1146
Mueller S, Phillips RJ (1991) On the initiation of subduction. J Geophys Res 96:651–665
Muraoka H, Kamata H (1983) Displacement distribution along minor fault traces. J Struct Geol 5:483–495
Murphy JB, Nance RD (1992) Mountain belts and the supercontinent cycle. Sci Am 260:34–41

Mutter J, Karson JA (1992) Structural processes at slow spreading ridges. Science 257:627–634

Myers JS, Kroner A (1994) Archaean tectonics. In: Hancock P (ed) Continental deformation. Pergamon Press, Oxford, pp 355–369

Nakanishi M, Winterer E L (1998) Tectonic history of the Pacific-Farallon-Phoenix triple junction from Late Jurassic to Early Cretaceous: an abandoned Mesozoic spreading system in the central Pacific Basin. J Geophys Res 105:12453–12468

Namson J, Davis T (1988) Structural transect of the western Transverse Ranges, California: implications for lithospheric kinematics and seismic risk evaluation. Geology 16:675–679

Nance RD, Morsley TR, Moody JB (1988) The supercontinent cycle. Sci Am 259:44–52

Nascimento U (1990) Um modelo de Fluência Friccional aplicável em Geotecnia e Sismotectónica. Geotecnia 60:1–24

National Academy of Sciences (1992) The great heat engine: modeling Earth's dynamics Grenwood, Addison. Ed Natl Acad Press, Washington, pp 1–25

Nehlig P (1993) Interactions between magma chambers and hydrothermal systems: oceanic and ophiolitic constraints. J Geophys Res 98:19621–19633

Nehlig P (1994) Fracture and permeability analysis in magma-hydrothermal transition zones in the Samail ophiolite (Oman). J Geophys Res 99:589–601

Nehlig P, Juteau T (1988) Flow porosities, permeabilities and preliminary data on fluid-inclusions and fossil thermal gradients in the crustal sequence of the Samail Ophiolite (Oman). Tectonophysics 151:199–221

Nehlig P, Juteau T, Bendel V, Cotton J (1994) The root zones of oceanic hydrothermal systems: constraints from the Samail ophiolite (Oman). J Geophys Res 99:4703–4713

Nelson KD (1992) Are crustal thickness variations in old mountain belts like the Appalachians a consequence of lithospheric delamination? Geology 30:498–502

Neumann GA, Forsyth DW (1993) The paradox of the axial profile: isostatic compensation along the Mid-Atlantic ridge? J Geophys Res 98:17891–17910

Nicolas A (1989) Structures of ophiolites and dynamics of oceanic lithosphere. Kluwer, Dordrecht 367 pp

Nicolas A, Poirier JP (1976) Crystalline plasticity and solid state flow in metamorphic rocks. Wiley, London, 444 pp

Nicolas A, Reuber I, Benn K (1988) A new magma chamber model based on structural studies in the Oman ophiolite. Tectonophysics 151:87–105

Nicolas A, Freydier C, Godard M, Vauchez A (1993) Magma chambers at oceanic ridges: how large? Geology 21:53–56

Nikishin AM, Cloetingh S, Lobkovsky LI, Burov EB, Lankreijer AL (1993) Continental lithosphere folding in Central Asia (I): constraints from rheological observations. Tectonophysics 226:59–72

Norabuena EO, Dixon TH, Stein S, Harrison CGA (1999) Decelerating Nazca-South America and Nazca-Pacific plate motions. Geophys Res Lett 26:3405–3408

Nur A (1982) The origin of tensile fracture lineaments. J Struct Geol 4:31–40

Nur A, Ben-Avraham Z (1982a) Displaced terranes and mountain building. Mountain building processes. In: Hsu K (ed) Academic Press, London, pp 73–84

Nur A, Ben-Avraham Z (1982b) Oceanic plateaus, the fragmentation of continents and mountain building. J Geophys Res 87:3644–3661

Ocampo A, Pope KO, Fischer AG (1996) Ejecta blanket deposits of the Chicxulub crater from Albion Island, Belize. In: Ryder G, Fastovsky D, Gartner S (eds) New developments regarding the KT event and other catastrophes in Earth history. Geol Soc Am Spec Pap 307:75–88

Ogawa I, Seno H, Akiyosh H, Tokuyama K, Fujioka, Tamiguchi H (1989) Structure and development of the Sagami Trough and the Boso triple junction. Tectonophysics 160:135–150

Okal EA, Talandier J (1980) Seismicity and tectonic stress in the south-central Pacific. J Geophys Res 85:6479–6495
Okubo PG, Aki K (1987) Fractal geometry in the San Andreas Fault System. J Geophys Res 92:345–355
Olami Z, Feder HJS, Christensen K (1992) Self-organized criticality in a continuous, nonconservative cellular automaton modeling earthquakes. Phys Rev Lett 68:1244–1247
Oldow JS, Bally AW, Avé Lallemant HG (1990) Transpression, orogenic float, and lithospheric balance. Geology 18:991–994
Oliver J (1992) The spots and stains of plate tectonics. Earth Sci Rev 32:77–106
Olson J, Pollard DD (1989) Inferring paleostresses from natural fracture patterns: a new method. Geology 17:345–348
Oppenheimer DH (1990) Aftershock slip behavior of the 1989 Loma Prieta, California earthquake. Geophys Res Lett 17:1199–1202
Ord A, Hobbs BE (1989) The strength of the continental crust, detachment zones and the development of plastic instabilities. Tectonophysics 158:269–289
Orth C, Atrep M, Quintana LR, Elder WP, Kauffman EG, Diner R, Willamil T (1993) Elemental abundance anomalies in the late. Cenomanian extinction interval: a search for the source (s). Earth Planet Sci Lett 117:189–214
Oxburgh ER (1974) The plain man's guide to plate tectonics. Proc Geol Assoc 85:299–357
Oxburgh ER (1972) Flake tectonics and continental collision. Nature 238:202–204
Pacheco JF, Schoz CH, Sykes L (1992) Changes in frequency-size relationship from small to large earthquakes. Nature 385:71–73
Pagarete J, Pinto JT, Mendes VB, Antunes C, Ribeiro A (1998) The importance of classical geodetic observations for analyzing the geodynamic behaviour of the Açores archipelago. Tectonophysics 294:281–290
Park RG (1988) Geological structures and moving plates. Blackie, Glasgow, 337 pp
Park RG (1997) Foundations of structural geology. Chapman and Hall, London, 202 pp
Park RG, Jaroszewski W Craton (1994) tectonics, stress and seismicity. In: Hancock P (ed) Continental deformation. Pergamon Press, Oxford, pp 220–272
Parsons B, Sclater IG (1977) An analysis of the variation of ocean floor bathymetry and hear flow with age. J Geophys Res 82:803–827
Parsons T, Thompson GA (1991) The role of magma overpressure in suppressing earthquakes and topography: worldwide examples. Science 253:1399–1402
Passchier CW (1997) The fabric attractor. J Struct Geol 19:115–127
Patchett PJ (1996) Scum of the Earth after all. Nature 382:758–759
Paterson MS (1987) Problems in the extrapolation of laboratory rheological data. Tectonophysics 188:38–48
Patriat PH, Courtillot V (1984) On the stability of triple junctions and its relation to episodicity in spreading. Tectonics 3:317–332
Paulssen H, Visser J (1993) The crustal structure in Iberia inferred from P-Wave Coda. Inst Geogr Nac Publ Ser Monogr 10:3–18
Pedlosky J (1987) Geophysical fluid dynamics. 2nd Edn. Springer, Berlin Heidelberg New York, 710 pp
Pierce C, Barton PJ (1991) Crustal structure of the Madeira-Tore Rise, eastern North Atlantic-results of a DOBS wide-angle and normal incidence seismic experiment in the Josephine Seamount region. Geophys J Int 106:357–370
Penrose Conference Participants (1972) Geotimes 17:24–25
Perez-Estaún A, Bastida F, Alonso JL, Marquinez J, Aller J, Alvarez MJ, Marcos A, Pulgar JA (1988) A thin-skinned tectonics model for an arcuate fold and thrust belt; the Cantabrian Zone (Variscan Ibero-Armorican Arc). Tectonics 7:517–537

Peterson MD, Wesnousky SG (1994) Fault slip rates and earthquake histories for active faults in southern California. Seismol Soc Am Bull 84:1608–1649

Pfiffner OA, Ramsay JG (1982) Constraints em geological strain rates; arguments from finite strain states of naturally deformed rocks. J Geophys Res 87:311–321

Pickering KT, Taira A (1994) Tectonosedimentation: with examples from the tertiary-recent of southeast Japan. In: Hancock P (ed) Continental deformation . Pergamon Press, Oxford, pp 320–354

Pinheiro LM, Wilson RCL, Reis RP, Whitmarsh RB, Ribeiro A (1996) I. The Western Iberia margin: a geophysical and geological overview. In: Whitmarsch RB, Sawyer DS, Klaus A, Masson DG (eds) Proceeding Ocean Drilling Program Scientific Results, vol 149, pp 3–23

Pollard DD, Saltzer SD (1993) Stress inversion methods: are they based on faulty assumptions? J Struct Geol 15:1045–1054

Poston T, Stewart I (1978) Catastrophe theory and its applications. Pitman, Boston, 491 pp

Price NJ (1968) A dynamic mechanism for the development of second order faults. Proc Conf on Research in Tectonics. Geol Surv Can:68–52

Price NJ (1975) Rates of deformation. J Geol Soc Lond 131:553–575

Price NJ, Cosgrove JW (1990) Analysis of geological structures. Cambridge University Press, Cambridge, 502 pp

Qiang ZW, Zhu W, Xiong Y (1999) Global plate motion models incorporating the velocity field of ITRF96. Geophys Res Lett 26:2813–1816

Quinlan GM, Beaumont C (1984) Appalachian thrusting lithospheric flexure and the Paleozoic stratigraphy of the Eastern Interior of North America Can J Earth Sci 21:973–996

Radjai F (1997) La Double Vie Du Sable La Recherche, Paris, pp 44–46

Radjai F, Jean M, Moreau JJ, Roux S (1996) Force distributions in dense two-dimensional granular systems. Phys Rev Lett 77:274–277

Raleigh CB, Paterson MB (1965) Experimental deformation of serpentinite and its tectonic implications. J Geophys Res 70:3965–3985

Ramberg H (1975) Particle paths, displacement, and progressive strain applicable to rocks. Tectonophysics 28:1–37

Ramberg H (1988) Thermodynamics applied to deformation structures. Bull Geol Inst Univ Uppsala 14:1–12

Rampino M, Ernston K, Anguita F, Claudin F (1997) Striations, polish, and related features from clasts in impact-ejecta deposits and the tillite-problem. Abstracts, Large Impacts Conference, Sudbury, Canada

Ramsay JG (1967) Folding and fracturing of rocks. McGraw-Hill, New York, 568 pp

Ramsay JG (1992) Some geometric problems of ramp-flat thrust models. In: McClay K (ed) Thrust tectonics. Chapman and Hall, London, pp 191–200

Ramsay JG (1969) The measurement of strain and displacement in orogenic belts. In: Kent PE, Satterthwaite GE, Spencer AM (eds): Time and place in orogeny. Geological Society, London, pp 43–79

Ramsay JG (1980) Shear zone geometry: a review. J Struct Geol 2:83–99

Ramsay JG, Graham RH (1970) Strain variation in shear belts: Can J Earth Sci 7:786–813

Ramsay JG, Huber M (1983) The techniques of modern structural geology. Analysis Strain, vol 1, pp 1–308

Ramsay JG, Huber MI (1987) The Techniques of modern structural geology, vol 2. Folds and fractures. Academic Press, London pp 309–700

Ramsay JG, Lisle R (2000) The techniques of modern structural geology, vol 3. Applications of continuum mechanics in structural geology. Academic Press, London pp 701–1061

Ranalli G (1980) A stochastic model for strike-slip faulting, Math Geol 12:399–341

Ranalli G (1995) Rheology of the Earth: deformation and flow processes in geophysics and geodynamics. Chapman and Hall, London, 413 pp

Ranalli G (1997) Rheology of the lithosphere in space and time. In: Bwy JP, Ford M (eds) Orogeny through time. Geol Soc Lond Spec Publ 121:19–37
Ranalli G, Murphy DC (1987) Rheological stratification of the lithosphere. Tectonophysics 152:281–295
Reches Z (1978) Analysis of faulting in three-dimensional strain field. Tectonophysics 47:109–129
Reches Z (1983) Faulting of rocks in three-dimensional strain fields. II. Theoretical analysis. Tectonophysics 95:133–156
Reches Z (1988) Evolution of fault patterns in clay experiments. Tectonophysics 145:141–156
Reches Z, Dietrich J (1983) Faulting of rocks in three-dimensional strain fields. I. Failure of rocks in polyaxial, servo-control experiments. Tectonophysics 95:111–132
Reis RPP, Corrochano A, Armenteros L (1997) El paleokarst de Nazaré (Cretácico Superior de la Cuenca Lusitana, Portugal). Geogaceta 22:149–152
Rey P, Burg J-P, Casey M (1997) The Scandinavian Caledonides and their relationship to the Variscan belt. In: Burg J-P, Ford M (eds) Geol Soc Spec Publ 121:179-200
Ribeiro A (1974) Contribution à l'étude tectonique de Trás-os-Momtes Oriental. Serv Geol Portugal Mem, vol 24, 168 pp
Ribeiro A (1986) A stochastic model to estimate maximum expectable magnitude of earthquakes from fault dimensions and slip-rate. Terra Cognita 6:611–615
Ribeiro A (1987) Tectónica de placas (uma visão pouco ortodoxa) Protecção Civil 1:13–16
Ribeiro A (1989) Global Tectonics with a viscoelastic oceanic lithosphere. Terra Abstr 1:240 pp
Ribeiro A (1990/1991) Order and chaos in the magnitude of earthquake faulting. Memórias da Academia das Ciências de Lisboa, Classe de Ciências, Tomo XXXI, pp 537–549
Ribeiro A (1991) Global tectonics with deformable plates. The John Ramsay Meeting. Mit Geol Inst ETH Zurich Neue Folge 239b:59–60
Ribeiro A (1993/1994) Global tectonics with deformable plates. Memórias da Academia das Ciências de Lisboa, Classe de Ciências Tomo XXXIII, Lisboa, pp 51–67
Ribeiro A (1994a) Deformable plate tectonics of the Azores-Gibraltar boundary-Where the next 1755 earthquake will strike again? Gaia 9:109–113
Ribeiro A (1994b) Towards a non-linear geodynamics. Geo Sistemas no 3, Lisboa, pp 29–36
Ribeiro A (1996) Soft plate tectonics reviewed. Gaia 12:33–36
Ribeiro A (1997) Soft plate tectonics: the example of the Pacific. Gaia 14:99–102
Ribeiro A (1998a) Uma Breve História Tectónica da Terra. A brief tectonic history of the Earth. EXPO '98, Lisboa, 53 pp
Ribeiro. A (1998b) What triggers subduction. Episodes 21:45–46
Ribeiro A (2000) Geodynamic Evolution of Iberian Variscides; unsolved questions. Variscan-Appalachian dynamics; the building of the Upper Paleozoic basement Tectonics 15 A Coruña, Spain. Program and Abstracts, vol 7, pp 21–22
Ribeiro A, Cabral J (1987) The neotectonic regime of West-Iberia continental margin: a transition from passive to active? Abstracts, EUG IV, Strasbourg, Terra Cognita, vol 7, 120 pp
Ribeiro A, Cabral J (1997) Geomorfologia tectónica e sismotectónica de Trás-os-Montes Oriental. Sismos. Inst Politécnico Bragança, pp 33–37
Ribeiro A, Silva JB (1997) Portugal Encyclopedia of European and Asian regional geology. In: Moores E, Fairbridge RH (eds) Chapmann and Hall, London, pp. 611–618
Ribeiro A, Antunes MI, Ferreira MB, Rocha RB, Soares AF, Zbyszewski G, Almeida FM, Carvalho D, Monteiro JH (1979) Introduction à la Géologie Générale du Portugal. Serviços Geológicos de Portugal, 114 pp
Ribeiro A, Kullberg MC, Kullberg JC, Manuppella G, Phipps S (1990) A review of Alpine tectonics in Portugal: Foreland detachment in basement and cover rocks. Tectonophysics 184:357–366

Ribeiro A, Mateus A, Moreira MC, Coutinho MF (1991a) The fractal geometry of an active fault (Vilariça Strike-Slip Fault, NE Portugal) and its implications on earthquake generation. In: Peitgen HO, Henriques IM, Penedo JF (eds) Fractal in the fundamental and applied sciences. Elsevier, Amsterdam, pp 367–377
Ribeiro A, Pereira E, Iglesias M (1991b) Flake tectonics in the Northwest Iberian Variscides. Tectonophysics 191:437
Ribeiro A, Kullberg MC, Kullberg JC, Rocha R, Phipps S, Manuppella G (1992) Tectonic inversion of the Lusitanian Basin. First Congress RCANS, pp 43–44
Ribeiro A, Dias R, Silva JB (1995) Genesis of the Ibero-Amorican arc. Geodinamica Acta Paris 8:175–184
Ribeiro A, Cabral J, Baptista R, Matias L(1996) Stress pattern in Portugal mainland and the adjacent Atlantic region, West Iberia. Tectonics 15:641–659
Ribeiro A, Sanderson D, SW-Iberian Colleagues (1996) SW-Iberia transpressional orogeny in the Variscides. In: Gee DG, Zeyen HJ (eds) Europrobe 1996 Lithosphere Dynamics: origin and evolution of continents. EURPROBE Secretariat, Uppsala University, 138 pp
Ribeiro A, Terrinha P, Zittelini N, Victor LM, Dañobeitia J, Carrilho F, Matias L, Pinheiro L, BIGSETS team (2000) Estrutura e sismotectónica da Margem Sudoeste Portuguesa. 30. Simpósio sobre a Margem Ibérica Atântica, Faro, pp 183–184
Ribeiro A; Matias L; Taborada R; Mender V (subm.) – A test of rigidity of the Pacific Plate.
Ricard Y Doglioni C, Sabadini R (1991) Differential rotation between lithosphere and mantle: a consequence of lateral mantle viscosity variations. J Geophys Res 96:8407–8415
Richard P, Cobbold P (1990) Experimental insights into partitioning of fault motions in continental convergent wrench zones. Ann Tectonica Spec issue IV:35–44
Richardson CJ, Cann JR, Richards HG, Cowan JG (1987) Metal depleted root zones of the Troodos ore-forming hydrothermal systems, Cyprus. Earth Planet Sci Lett 84:243–253
Riedel W (1929) Zur Mechanik geologischer Brucherscheinungen. Zentralbl F Mineral Geol Pal:354–328
Ries AC, Shackleton RM (1976) Patterns of strain variation in arcuate fold belts. Philos Trans R Soc Lond A 283:281–288
Robbins JW, Smith DE, Ma C (1993) Horizontal crustal deformation and large scale plate motions. Contribution of space geodesy to geodynamics. Crustal Dyn Geodyn 213:21–36
Roberts A, Yielding G (1994) Continental Extensional Tectonics. Hancock P (ed) Continental deformation. Pergamon Press, Oxford, pp 223–250
Rock NM S (1982) The late cretaceous alkaline igneous province in the Iberian Peninsula, and its tectonic significance. Lithos 15:111–131
Rodgers J (1987) Chains of basement uplifts within cratons marginal to orogenic belts. Am J Sci 387:661–695
Rodrigues A, Dias JA, Ribeiro A (1992) First appraisal of active faults in the north Portuguese continental shelf. Gaia 4:25–30
Roeder DH (1973) Subduction and orogeny. J Geophys Res 78:5005–5024
Roeser HA, Bargeloh HO, Eilers G, Fritsch J, Keppler H, Kewitsch P, Klein A, Pinheiro LM, Schreckenberger B (1992) SONNE cruise SO-75: geophysical investigations of the crustal structure of the North Atlantic off Portugal. BGR Cruise Report 109997, Bundesanstalt für Geowissenschaften and Rohstoffe, Hannover
Roest WR, Srivastava SP (1991) Kinematics of the plate boundaries between Eurasia, Iberia, and Africa in the North Atlantic from the Late Cretaceous to the present. Geology 19:615–616
Rompert R, Hansen U (1998) Mantle convection simulations with rheologies that generate plate-like behaviour. Nature 395:686–689
Rona PA (1988) Hydrothermal mineralisation at mid-ocean ridges. Can Mineral 25:1089–1114
Rona PA ,Widenfalk L, Boström K (1987) Serpentinized ultramafic and hydrothermal activity at the Mid-Atlantic Ridge crest near 15°N. J Geophys Res 92:1417–1427

Rona PA, Denlinger RP, Fisk MR, Howard KJ, Taghon GL, Klitgord KD, McClain JS, McMurray GR, Wiltshire JC (1990) Major off-axis hydrothermal activity on the northern Gorda Ridge. Geology 18:493–496

Rona PA, Bougault H, Charlou JL, Appriou P, Nelsen TA, Tefry JH, Eberhart GL, Barone A, Needham HD (1992) Hydrothermal circulation, serpentinization, and degassing at a rift-valley-fracture zone intersection: Mid-Atlantic Ridge near 15°N,45°W. Geology 20:783–786

Rowlett H (1981) Seismicity at intersections of spreading centers and transform faults. J Geophys Res 86:3815–3820

Royden L (1996) Coupling and decoupling of crust and mantle in convergent orogens: implications for strain partitioning in the crust. J Geophys Res 100:17679–17705

Royden LH (1993) Evolution of retreating subduction boundaries formed during continental collision. Tectonics 12:629–638

Royer JY, Gordon R G (1997) The motion and boundary between the Capricorn and Australian Plates. Science 277:1268–1274

Ruelle D, Takens F (1971) On the nature of turbulence. Commun Math Phys 20:167–192 (1971); 23:343–344 (1971)

Rundle JB (1989) Derivation of the complete Gutenberg-Richter magnitude-frequency relation using the principle of scale invariance. J Geophys Res 94:12337–12340

Rusby RI, Searle RC (1993) Intraplate thrusting near the Easter microplate. Geology 21:311–314

Rutter E (1983) Pressure solution in nature, theory and experiment. J Geol Soc Lond 140:725–740

Rutter EH (1976) The kinetics of rock deformation by pressure solution. Philos Trans R Soc 283:43–54

Rutter EH (1986) On the nomenclature of mode of failure transitions in rocks. Tectonophysics 122:381–387

Rutter EH, Brodie KH (1991) Lithosphere rheology – a note of caution. J Struct Geol 13:363–367

Rutter EM (1976) The kinetics of rock deformation by pressure solution. Philos Trans R Soc Ser A 283:203–219

Ryan PD, Dewey JF (1997) Continental eclogites and the Wilson cycle. J Geol Soc Lond 154:437–442

Sager WW, Handschumacher DW, Hilde TWC, Bracey DR (1988) Tectonic evolution of the northern Pacific plate and Pacific-Farallon-Izanagi triple junction in the Late Jurassic and Early Cretaceous (M21-M10). Tectonophysics 155:345–364

Sanderson DJ (1977) The algebraic evaluation of two-dimensional finite strain rosettes. Math Geol 9:483–496

Sanderson DJ, Marchini WRD (1984) Transpression. J Struct Geol 6:449–458

Sandwell DT, Smith WHS (1995) exploring the ocean basins with satellite altimeter data. http://julius.Ngdc.Noaa.Gov/mgg/announcements/text predictHTML

Sandwell DT, Smith WHF (1997) Marine gravity anomaly from Geosat and ERS 1 satellite altimetry. J Geophys Res 102:10039–10054

Sandwell DT, Winterer EL, Mammericx J, Duncan RA, Lynch MA, Levitt DLA, Johnson C (1995) Evidence for diffuse extension of the Pacific plate from Pukapuka ridges and cross-grain gravity lineations. J Geophys Res 100:15187–15199

Sartori R, Torelli L, Zitellini N, Peis D, Lodolo E(1994) Eastern segment of the Azores-Gibraltar line (central-eastern Atlantic): an oceanic plate boundary with diffuse compressional deformation. Geology 22:255–258

Savage JC, Svarc JL, Prescott WH (1999) Geodetic estimates of fault slip rates in the San Francisco Bay area. J Geophys Res 104:4995–5002

Schaer J-P (1987) Evolution and structure of the high atlas of Morocco. In: Schaer J-P, Rodgers J (eds) The anatomy of mountain ranges. Princeton University Press, Princeton, pp 59–54

Scheidegger AE (1982) Principles of geodynamics. Springer, Berlin Heidelberg New York, 395 pp
Schiling JG (1975) Azores mantle blob rare earth evidence. Earth Planet Sci Lett 25:103–115
Schmid SM, Pfiffner OA, Froitzheim N, Schönborn G, Kissking E (1996) Geophysical-geological transect and tectonic evolution of the Swiss-Italian Alps. Tectonics 15:1036-1064
Scholz CH (1990) The mechanics of earthquakes and faulting. Cambridge University Press, Cambridge, 439 pp
Scholz CH (1998) Earthquakes and friction laws. Nature 39:37–42
Scholz CH (2000) Evidence for a strong San Andreas fault. Geology 28:163–166
Scholz CH, Aviles CA (1986) The fractal geometry of faults and faulting. Geophys Monogr 37 AGU Maurice Ewing 6:147–145
Scholz CH, Cowie PA (1990) Determination of total strain from faulting using slip measurements. Nature 346:387–389
Scholz CH, Dawers NH, Yu JZ, Anders MH, Cowie PA (1993) Fault growth and fault scaling laws: preliminary results. J Geophys Res 98:21951–21961
Schrader F (1988) Symmetry of pebble-deformation involving solution pits and slip-lineations in the northern Alpine Molasse Basin. J Struct Geol 10:41–52
Schroeder M (1990) Fractals, chaos, power laws Minutes from an infinite paradise. Freeman, New York, 429 pp
Schumm SA (1991) To Interpret the Earth Ten ways to be wrong. Cambridge University Press, Cambridge, 133 pp
Schwartz DP, Coppersmith KJ (1984) Fault behavior and characteristic earthquakes: examples from the Wasatch and San Andreas Fault Zones. J Geophys Res 89:5681–5698
Schwartz DP, Coppersmith KJ (1986) Active tectonics. Natl Acad Press, Washington, DC, pp 215–230
Sclater JG, Jaupart C, Galson D (1980) The heat flow through oceanic and continental crust and the heat loss of the Earth. Rev Geophys 18:269–311
Searle R (1980) Tectonic pattern of the Azores spreading centre and triple junction. Earth Planet Sci Lett 51:415–434
Searle RC (1983) Multiple, closely spaced transform faults in fast-slipping fracture zones. Geology 11:507–610
Searle RC (1986) GLORIA investigations of oceanic fracture zones: comparative study of the transform fault zone. J Geol Soc 143:743–756
Searle RC, Bird RT, Rusby RI, Naar DF (1995) The development of two oceanic microplates: easter and Juan Fernandez microplates, East Pacific Rise. J Geol Soc Lond 150:965–976
Seber D, Barazangi M, Ibenbrahim A, Demnati A(1996) Geophysical evidence for lithospheric delamination beneath the Alboran Sea and Rif-Betic mountains. Nature 379:785–790
Sengör AMC (1979) Mid-Mesozoic closure of Permo-Triassic Tethys and its implications. Nature 279:590–593
Sengör AMC (1982) Classical theories of orogenesis. In: Miyashiro A, Aki K, Sengör C (eds) Orogens. Wiley, Chichester, pp 1–48
Sengör AMC (1990) Plate Tectonics and orogenic research after 25 years: a Tethyan perspective. Earth Sci Rev 27:1–201
Sengör AMC (1999) Continental interiors and cratons: any relation? Tectonophysics 305:1–42
Sengör AMC, Burke KC (1978) Relative timing of rifting and vulcanism on Earth and its tectonic implications. Geophys Res Lett 5:419–421
Sengör AMC, Dewey JF (1990) Terranology: vice or virtue? Philos Trans R Soc Lond 331:457–478
Shackleton R, Ries AC (1984) The relation between regionally consistent stretching lineations and plate motions. J Struct Geol 6:111–117

Shackleton RM (1987) Contrasting structural relations loops of Proterozoic ophiolites in northeastern and eastern Africa. In: Gaby S, Greiling R (eds) The Pan-African Belt of northeast Africa and adjacent areas. Earth Evol Sci, Vieweg, Wiesbaden, pp 183–193
Shaw HR (1994) Craters, cosmos, and chronicles-a new theory of Earth. Stanford University Press, Stanford, 688 pp
Shaw PR (1992) Ridge segmentation, faulting and crustal thickness in the Atlantic Ocean. Nature 358:490–493
Shaw PR, Lin J (1993) Causes and consequences of variations in faulting style at the Mid-Atlantic Ridge. J Geophys Res 98:21839–21851
Shaw PR, Lin J (1996) Models of ocean ridge lithospheric deformation: dependence on crustal thickness, spreading rate, and segmentation. J Geophys Res 101:17977–17993
Shemenda AI (1992) Horizontal lithosphere compression and subduction: constraints provided by physical modeling. J Geophys Res 97:11097–11116
Shemenda AI, Grocholsky AL (1994) Physical modeling of slow seafloor spreading. J Geophys Res 99:9137–9153
Shemenda AJ (1994) Subduction: insights from physical modeling. Kluwer, Dordrecht, 215 pp
Shen-Tu B, Holt WE (1995) Intraplate deformation in the Japanese Islands: a kinematic study of intraplate deformation at a convergent plate margin. J Geophys Res 100:24275–24298
Shephard LE, Auffret GA, Buckley DE, Schuttenheim RTK, Searle RC (1988) Feasibility of disposal of high-level radioactive waste into the seabed, vol 3. Organisation for Economic Co-operation and Development, Paris, 318 pp
Shimada M (1993) Lithosphere strength inferred from fracture strength of rocks at high confining pressures and temperatures. Tectonophysics 217:55–64
Shirey SB, Bender JF, Langmuir CH (1987) Three-component isotopic heterogeneity near the Oceanographer transform, Mid-Atlantic Ridge. Nature 325:217–223
Shoemaker EM (1960) Penetration mechanics of high velocity meteorites, illustrated by Meteor Crater, Arizona. Rep of the Int Geol Congress, XXI Session, Norden. Part. XVIII, pp 418–434
Sibson RH (1974) Frictional constraints on thrust, wrench and normal faults. Nature 249:542–544
Sieh K, Jahns R (1984) Holocene activity of the San Andreas fault at Wallace Creek, California. Geol Soc Am Bull 95:883–896
Sillard P, Altamimi Z, Boucher C (1998) The ITRF96 realization and its associated velocity field. Geophys Res Lett 25:3223–3226
Silva E (1995) Prolongamento das Estruturas Hercínicas para a Margem Oeste Ibérica. MSc Thesis, Fac Ciências, Univ Lisboa, 104 pp
Silver PG, Russo RM, Bertelloni CL (1998) Coupling of South American and African plate motion and plate deformation. Science 279:60–63
Simpson C, De Paor DG (1993) Strain and kinematic analysis in general shear zones. J Struct Geol 15:1–20
Simpson DW (1986) Triggered earthquakes. Am Rev Earth Planet Sci 14:21–42
Sinton JM, Detrick RS (1992) Mid-ocean ridge magma chambers. J Geophys Res 97:197–216
Slemmons DB (1977) Faults and earthquakes magnitude. US Army Congress of Engineers, Waterways Experimental Stations. Misc Pap S:73–1, Reprint, vol 6, pp 1–129
Slemmons DB (1982) Determination of design earthquake magnitudes for microzonation. 3rd Int Earthquake Microzonation Conf Proc, Seattle, vol 1, pp 119–130
Sloan H, Patriat PH (1992) Kinematics of the North American-African plate boundary between 28° and 29°N during the last 10 Ma: evolution of the axial geometry and spreading rate and direction. Earth Planet Sci Lett 115:525–541
Small C, Abbott D (1998) Subduction obstruction and the crack-up of the Pacific plate. Geology 36:795–798

Smith DE, Kolenkiewicz R, Dunn PL, Robbins JW, Torrence MH, Klosko SM, Williamson RG, Pavlis EC, Douglas NB, Fricke SK (1990) Tectonic motion and deformation from satellite laser ranging to Lageos. J Geophys Res 95:22013–22041

Smith DK, Cann JR (1992) The role of seamount volcanism in crustal construction along the Mid-Atlantic Ridge. J Geophys Res 97:1645–1658

Smith DK, Cann JR (1993) Building the crust at the Mid-Atlantic Ridge. Nature 365:707–715

Smith RB (1975) Unified theory of the onset of folding, boudinage, and mullion structure. Geol Soc Am Bull 86:1601–1609

Smith RB (1977) Formation of folds, boudinage, and mullions in non-Newtonian materials. Geol Soc Am Bull 88:312–320

Smith SW (1976) Determination of maximum earthquake magnitude. Geophys Res Lett 6:351–354

Smith WHF, Sandwell DT (1997) Global sea floor topography from satellite altimetry and ship depth soundings. Science 277:1956–1962

Snow JE (1995) Of Hess crust and layer cake. Nature 374:413–414

Snyder DB, Prasetyo H, Blundell DJ, Pigran CJ, Barber AJ, Richardson A, Tjokosaproetro S (1996) A dual double vergent orogen in the Banda Arc continent-arc collision zone as observed on deep seismic reflection profiles. Tectonics 15:34–53

Soares A (1966) Estudo das formações pós-Jurássicas da região de entre Sargento Mor e Montemor-o-Velho (Margem direita do Rio Mondego). Rev Fac Cienc Univ de Coimbra, XL, 343 pp

Soesoo A, Bons PD, Gray DR, Foster DA (1997) Divergent double subduction: tectonic and petrologic consequences. Geology 25:755–758

Sokolnikoff IS, Redheffer RM (1966) Mathematics of physics and modern engineering. McGraw-Hill, Kogakusha, 752 pp

Solomatov WS, Moresi LN (1997) Three regimes of mantle convection with non-Newtonian viscosity and stagnant lid convection on the terrestrial planets. Geophys Res Lett 24:1907–1910

Sonder LJ, England PH (1986) Vertical averages of rheology of the continental lithosphere: relation to thin sheet parameters. Earth Planet Sci Lett 77:81–98

Sonder LJ, England PHC, Houseman GA (1986) Continuum calculations of continental deformation in transcurrent environments. J Geophys Res 91:4797–4810

Srivastava SP, Schouten H, Roest WR, Klitgord KD, Kovacs LC, Verthoel J, Macnab R (1990) Iberian plate kinematics: a jumping plate boundary between Eurasia and Africa. Nature 344:756–759

Staudigel H, Hart SR, Richardson SH (1981) Alteration of oceanic crust: processes and timing. Earth Planet Sci Lett 52:311–327

Steacy SJ, McCloskey J (1999) Heterogeneity and the earthquake magnitude-frequency. Geophys Res Lett 26:899–902

Steacy SJ, McCloskey J, Bean CJ, Ren J (1996) Heterogeneity in a self-organized critical earthquake model. Geophys Res Lett 25:385–386

Stein C, Stein S (1994) Constrains on hydrothermal heat flux through the oceanic lithosphere from global heat flow. J Geophys Res 99:3081–3095

Stein M, Hofmann W (1994) Mantle plumes and episodic crustal growth. Nature 372:63–67

Stein RS (1999) The role of stress transfer in earthquake sequence. Nature 402:605–609

Stein RS, Yeats RS (1989) Hidden earthquakes. Sci Am 260:48–57

Stein RS, King GCP, Rundle JB (1988) The growth of geological by repeated earthquakes 2. Field examples of continental dip-slip faults. J Geophys Res 98:13319–13331

Stein RS, King GCF, Lin J (1992) Change in failure stress on the Southern San Andreas Fault system caused by the 1992 magnitude – 7.4 Landers earthquake. Science 258:1328–1332

Stein S (1993) Space geodesy and plate motions. In: Smith D, Turcotte D (eds) Contribution of space geology to geodynamics. Crustal Dyn Geodyn Ser 25:5–20

Stein S, Okal EA (1986) Seismological studies of the deformation of oceanic lithosphere. In: Anderson AJ, Cazenave A (eds) Space geodesy and geodynamics. Academic Press, London, pp 407–450

Stein S, Cloetingh S, Sleep N, Wortel R (1989) Passive margin earthquakes, stresses and rheology. In: Gregersen S, Basham PW (eds) Earthquakes at North-Atlantic passive margins: neotectonics and postglacial rebound. Kluwer, Dordrecht, pp 231–259

Stephen RA (1981) Seismic anisotropy observed in upper oceanic crust. Geophys Res Lett 8:865–868

Stephenson RA, Cloetingh SAPL (1991) Some examples and mechanical aspects of continental lithospheric folding. Tectonophysics 188:27–57

Stirling MW, Wesnousky STG, Shimazaki K (1996) Fault trace complexity, cumulative slip, and the shape of the magnitude-frequency distribution for strike-slip faults: a global survey. Geophys J Int 124:833–868

Su W, Mutter C, Mutter JC, Buck WR (1994) Some theoretical predictions on the relationships among spreading rate, mantle temperature, and crustal thickness. J Geophys Res 99:3215–3227

Suppe J (1983) Geometry and kinematics of fault-bend folding. Am J Sci 283:648–721

Sylvester P (2000) Continent formation, growth and recycling. Tectonophysics 322:vi–vii

Tackley PJ (2000) Mantle convection and plate tectonics. Toward an integrated physical and chemical theory. Science 288:2002–2007

Tackley PJ, Stevenson DJ, Glatzmaier GA, Schubert G (1993) Effects of an endothermic phase transition at 670 km depth in a spherical model of convection in the Earth's mantle. Nature 361:699–704

Tapponnier P, Molnar P (1976) Slip-line field theory and large-scale continental tectonics. Nature 264:319–324

Tapponnier P, Molnar P (1977) Active faulting and tectonics in China. J Geophys Res 82:2905–2930

Tapponnier P, Peltzer G, Le Dain AY, Armijo R, Cobbold P (1982) Propagating extrusion tectonics in Asia: new insights from simple experiments with plasticine. Geology 10:611–615

Taylor B, Crook K, Sinton J (1994) Extensional transform zones and oblique spreading centers. J Geophys Res 99:19707–19718

Taylor SR, McLennan SM (1996) The evolution of continental crust. Sci Am 264:60–65

Tchalenko JS, Ambraseys NN (1970) Structural analysis of the Dasht-e Bayaz (Iran) earthquake fractures. Geol Soc Am Bull 81:41–60

Terrinha P (1998) Structural geology and tectonic evolution of the Algarve Basin, South Portugal. PhD Thesis, Imperial College, 425 pp

Terrinha P, Dias RP, Ribeiro A, Cabral J (1999) The Portimão Fault, Algarve Basin, South Portugal. Comun Inst Geol Mineiro 86:107–120

Terrinha PAG, Coward MP, Ribeiro A (1990) Salt tectonics in the Algarve Basin: the Loulé diapir. Comun Serviços Geol Portugal 76:33–40

Teyssier C, Tikoff B (1998) Strike-slip partitioned transpression of the San Andreas fault system: a lithospheric-scale approach. In: Holdsworth R, Strachan R, Dewey J (eds) Continental transpressional and transtensional tectonics. Geol Soc Lond Spec Publ 135:143–158

Thatcher W (1984) The earthquake deformation cycle, recurrence, and the time-predictable model. J Geophys Res 89:5674–5680

Thatcher W (1995) Microplate versus continuum descriptions of active tectonic deformation. J Geophys Res 100:3885–3894

Thom R (1982) Plate tectonics as a catastrophe theoretic model. Predictive geology. In: De Marsily GG, Merriam D (eds) Computers and geology, vol 4. Pergamon Press, Oxford

Thomas PG, Allemand P, Mangold N (1997) Rheology of planetary lithospheres: a review from impact cratering mechanics. In: Burg J-P, Ford M (eds) Orogeny through time. Geol Soc Lond Spec Publ 121:39–62

Thompson J, Stewart H (1986) Non-linear dynamics and chaos - geometrical methods for engineers and scientists. Wiley, Chichester, 376 pp

Tikoff B, Fossen H (1995) Simultaneous pure and simple shear the unifying deformation matrix. Tectonophysics 227:267–283

Tikoff B, Teyssier C (1994) Strain modeling of displacement-field partitioning in transpressional orogens. J Struct Geol 16:1575–1588

Tolstoy M, Harding AJ, Orcutt JA (1993) Crustal thickness on the Mid-Atlantic ridge: Bull's eye gravity anomalies and focused accretion. Science 262:726–729

Tortella D, Torne M, Pérez-Estaún A (1997) Geodynamic evolution of the eastern segment of the Azores-Gibraltar Zone: the Gorringe Bank and the Gulf of Cadiz Region. Mar Geophys Res 19:211–230

Travis BJ, Janecky DR, Rosenberg ND (1991) Three-dimensional simulation of hydrothermal circulation at mid-ocean ridges. Geophysi Res Lett 18:1441–1444

Treagus SH, Lisle RJ (1997) Do principal surfaces of stress and strain always exist? J Struct Geol 19:997–1000

Tritton DJ (1988) Physical Fluid Dynamics, 2nd edn. Clarendon Press, Oxford, 519 pp

Tucholke BE, Lin J (1994) A geological model for the structure of ridge segments in slow-spreading ocean crust. J Geophys Res 99:11937–11958

Tucholke BE, Ludwig WJ (1982) Structure and origin of the J Anomaly Ridge, western North Atlantic Ocean. JGR 87 B11:9389–9407

Turcotte DL (1982) Driving mechanisms of mountain building. In: Hsu K (ed) Mountain building processes. Academics Press, London, pp 141–146

Turcotte DL (1997) Fractals and chaos in geology and geophysics. University Press, Cambridge,398 pp

Turcotte DL, Schubert G (1982) Geodynamics applications of continuum physics to geological problems. Wiley, New York, 450 pp

Turner FJ, Weiss IE (1963) Structural analysis of metamorphic tectonics. McGraw-Hill, New York, 545 pp

Twiss RJ, Gefell MJ (1990) Curved slickenfibers: a new brittle shear sense indicator with application to a sheared serpentinite J Struct Geol 12:471–481

Twiss RJ, Moores EM (1992) Structural geology. Freeman, New York, 532 pp

Twiss RJ, Unruh JR (1998) Analysis of fault slip inversions: do they constrain stress of strain rate? J Geophys Res 105:12205–12222

Twiss RJ, Protzman GM, Hurst SD (1991) Theory of slickenline patterns based on the velocity gradient tensor and microrotation. Tectonophysics 186:215–239

Twiss RJ, Souter BJ, Unruh JR (1993) The effect of block rotations on the global seismic moment tensor and the patterns of seismic P and T axes. J Geophys Res 98:645–674

US Geological Survey Staff (1990) The Loma Prieta, California, earthquake; an anticipated event: Science 24:285–293

Udías A, Buforn E (1994) Seismotectonics of the Mediterranean region. Adv Geophys 36:121–209

Udías A, Lopez Arroyo A, Mezcua J (1976) Seismotectonic of the Azores-Alboran region. Tectonophysics 31:259–289

Unruh JR, Twiss RJ, Hauksson E (1997) Kinematics of postseismic relaxation from aftershock focal mechanisms of the 1994 Northridge, California, earthquake. J Geophys Res 102:24589–24605

Uyeda S (1978) The new view of the Earth. Freeman, San Francisco, 217 pp

Uyeda S (1982) Subduction zones: an introduction to comparative subductology. Tectonophysics 81:133–159

Van Andel TH (1985) New views on an old planet. Continental drift and the history of earth. vol 1 Cambridge University Press, Cambridge, 324 pp

Van Andel TH (1992) Sea floor spreading and plate tectonics. Understanding the earth – a new synthesis. In: Brown GC; Hawkesworth CJ, Wilson RCL (eds) Cambridge University Press, Cambridge, pp 167–186

Van Bemmellen RW (1972) Geodynamics models – an evaluation and a synthesis. Developments in geotectonics 2. Elsevier, Amsterdam, 267 pp

Van der Hilst RD, Widiyantoro S, Engdahl ER (1991) Evidence for deep mantle circulation from global topography. Nature 86:578–584

Van der Pluijm B, Marsha K (1997) Earth structure: an introduction to structural geology and tectonics. WCB/McGraw-Hill, New York, 495 pp

Van der Pluijm BA, Graddock JP, Graham BR, Harris JH (1997) Paleostress in cratonic North America: implications for deformation of continental interiors. Science 277:794–796

Van der Voo R, Spakman W, Bjwaard H (1999) Mesozoic subducted slabs under Siberia. Nature 397:246–249

Vanko D (1988) Temperature, pressure, and composition of hydrothermal fluids with their bearing on the magnitude of tectonic uplift at mid-ocean ridges, inferred from fluid inclusions in oceanic layer 3 rocks. J Geophys Res 93:4595–4611

Varga RJ, Moores EM (1985) Spreading structure of the Troodos ophiolite, Cyprus. Geology 13:846–850

Walcott RI (1984) The kinematics of the plate-boundary zone through New Zealand: a comparison of short and long-tem deformation. Geophys J R Astron Soc 79:613–633

Walcott RI (1987) Geodetic strain and the deformational history of New Zealand during the late Cainozoic. Philos Trans R Soc Lond A 321:153–181

Walcott RI (1989) Paleomagnetically observed rotations along the Hikurangi Margin of New Zealand. In: Kissel C, Laj C (eds) Paleomagnetic rotations and continental deformation. Kluwer, Dordrecht, pp 459–471

Walsh JJ, Watterson J (1988) Analysis of the relationship between displacements and dimensions of faults. J Struct Geol 10:239–247

Walsh JJ, Watterson J (1991) Geometric and kinematic coherence and scale effects in normal fault systems. In: Roberts A, Yielding G, Freeman B (eds) The geometry of normal faults. Spec Publ Geol Soc Lond 56:193–206

Watts AB (1982) Seamounts and flexure of the lithosphere. Nature 297:182–183

Wdowinski S (1998) A theory of intraplate tectonics. J Geophys Res 105:5057–5059

Wees JD van, Cloetingh S (1994) A finite-difference technique to incorporate spatial variations in rigidity and planar faults into 3-D models for lithospheric flexure. Geophys J Int 117:179–195

Wegener A (1915) Die Entstehung der Kontinente und Ozeane. Friedrich Vieweg, Braunschweig (viewy), 94 pp

Wegener A (1966) The origin of continents and oceans. Translation from the 4th revised German edition, 1929. Dover, New York, 246 pp

Wells DL, Coppersmith KJ (1994) New empirical relationships among magnitude, rupture length, rupture width, rupture area, and surface displacement. Bull Seismol Soc Am 84:974–1002

Wernicke B (1985) Uniform-sense normal simple shear of the continental lithosphere. Can J Earth Sci 22:108–125

White H, Knipe RJ (1978) Transformation and reaction enhanced ductility in rocks. J Geol Soc Lond 135:513–516

Whitmarsh RB, Pinheiro LM, Miles PR, Recq M, Sibuet JC (1993) Thin crust at the western Iberia ocean-continent transition and ophiolites. Tectonics 12:1230–1239

Wiemer S, Mc Nutt R (1997) Variations in the frequency magnitude distribution with depth in two volcanic areas: Mount St. Helens, Washington, and Mt. Spurr, Alaska. Geophys Res Lett 24:189–192

Wiemer S, Wyss M (1997) Mapping the frequency-magnitude distribution in asperities: an improved technique to calculate recurrence times? J Geophys Res 102:15115–15128

Wiens DA, Stein S (1985) Implications of oceanic intraplate seismicity for plate stresses, driving forces and rheology. Tectonophysics 116:143–162

Wilcock WSD, Solomon SC, Purdy GM, Toomey DR (1992) The seismic attenuation structure for a fast-spreading mid-ocean ridge. Science 258:1470–1474

Willet SD, Beaumont C (1994) Subduction of Asian lithospheric mantle beneath Tibet inferred from models of continental collision. Nature 369:642–645

Williams SRJ (1987) Faulting in abyssal-plain sediments, Great Meteor East, Madeira Abyssal Plain. In: Weaver PPE, Thomson J (eds) Geology and geochemistry of abyssal plain. Geol Soc Lond Spec Publ 31:87–104

Wilson JI (1965) A new class of faults and their bearing on continental drift. Nature 207:363–367

Wilson JT (1966) Did the Atlantic close and then reopen? Nature 211:675–681

Wilson RCL, Hiscott RN, Willis MG, Gradstein FM (1989) The lusitanian basin of west-central Portugal: Mesozoic tertiary tectonic, stratigraphic and subsidence history. In: Tankard AJ, Balk-Will HR (eds) Extensional tectonics and stratigraphy of the North Atlantic margins. AAPG Mem 46: 341–361

Windley BF (1993) Uniformitarianism today: plate tectonics is the key to the past. J Geol Soc Lond 150:7–19

Windley BF (1995) The evolving continents. 3rd edn. Wiley, Chichester, 526 pp

Wolery TJ, Sleep NH (1976) Hydrothermal circulation and geochemical flux at mid-ocean ridges. J Geol 84:249–275

Wolfe C, Purdy GM, Toomey DR, Solomon SC (1995) Microearthquake characteristics and crustal velocity structure at 29°N of the Mid-Atlantic Ridge: the architecture of a slow-spreading segment. J Geophys Res 100:24449–24472

Wolfram S (1984) Universality and complexity in cellular automata. Physica 10D:91–125

Wyss M (1979) Estimating maximum expectable magnitude of earthquakes from fault dimensions. Geology 7:336–340

Wyss M, Shimazaki K, Wiemer S (1997) Mapping active magma chambers by b values beneath the off-Ito volcano, Japan. J Geophys Res 102:20413–20422

Wyss M, Schorlemmer D, Wiemer S (2000) Mapping asperities by minima of local recurrence time: San Jacinto-Elsinore fault zones. J Geophys Res 105:7829–7844

Yeats RS (1981) Quaternary flake tectonics of the California transverse ranges. Geology 9:16–20

Yeats RS, Huftile GJ (1995) The Oak Ridge fault system and the 1994 Northridge earthquake. Nature 373:418–420

Yeats RS, Sieh K, Allen CR (1997) Geology of earthquakes. Oxford University Press, Oxford, 568 pp

Ziegler PA (1983) Inverted basins in the Alpine Foreland. In: Bally AW (ed) Seismic expression of structural styles. Am Assoc Petrol Geol Stud Geol 15:3.3.3–3.3.12

Ziegler PA, Cloetingh S, Van Wees JD (1995) Dynamics of intra-plate compressional deformation: the Alpine foreland and other examples. Tectonophysics 252:7–59

Ziegler PA, van Wees JD, Cloetingh S (1998) Mechanical controls on collision-related compressional intraplate deformation. Tectonophysics 300:105–109

Zitelini N, Terrinha P, Danobeitia J, Gracia E, Sartori R, Torelli L, Rovere M, Chierici F, Ribeiro A, Matias L, Victor LM (2000) Mapa estrutural da margemSW Iberia. 2ª Assembleia Luso-Espanhola de Geodesia e Geofísica, Lagos, pp 145–146

Zitelini N, Chierici F, Sartori R, Torelli L (1999) The tectonic source of the 1755 Lisbon earthquake and tsunami. Ann Geofisica 42:49–55

Zoback M (2000) Strength of the San Andreas. Nature 405:31–32

Zoback MD, Zoback ML, Mount VS (1987) New evidence on the state pf stress of the San Andreas fault system. Science 238:1105–1111

Zoback MD, Emmermann R (1994) (eds) Scientific rationale for establishment of an international program of continental scientific drilling. Int Lithosph Program Potsdam, 194 pp

Zoback ML, Zoback MD, Adams J, Assumpção M, Bell S, Bergman EA, Brereton NR, Denham D, Ding J, Fuchs K, Gay N, Gregersen S, Gupta HK, Gvishiani A, Jacob K, Klein R, Knoll P, Magee M, Mercier JL Paquin C, Rajendran K, Stephansson O, Suarez G, Suter M, Udias A, Xu ZH, Zhizhin M (1989) Global patterns of tectonic stress. Nature 341:291–298

Zoback ML (1992) First and second-order patterns of stress in the lithosphere: the world stress map project. J Geophys Res 97:11703–11728

Zoback ML, Zoback MD (1980) State of stress in the conterminous United States. J Geophys Res 85:6113–6156

Zoback ML, Jachens RC, Olson JA (1999) Abrupt along-strike change in tectonic style: San Andreas fault zone, San Francisco Peninsula. J Geophys Res 104:10709–10742

Zuber MT (1987) Compression of the oceanic lithosphere: an analysis of intraplate deformation in the Central Indian Basin. J Geophys Res 92:4817–4825

Subject Index

C

F

R

U

MIX
Papier aus verantwortungsvollen Quellen
Paper from responsible sources
FSC® C105338

If you have any concerns about our products,
you can contact us on
ProductSafety@springernature.com

In case Publisher is established outside the EU,
the EU authorized representative is:
Springer Nature Customer Service Center GmbH
Europaplatz 3, 69115 Heidelberg, Germany

Printed by Libri Plureos GmbH
in Hamburg, Germany